Molecular Similarity and Reactivity

Understanding Chemical Reactivity

Volume 14

The titles published in this series are listed at the end of this volume.

Molecular Similarity and Reactivity

From Quantum Chemical to Phenomenological Approaches

Edited by

RAMON CARBÓ

Institute of Computational Chemistry
University of Girona
Girona
Spain

Kluwer Academic Publishers
Dordrecht / Boston / London

Library of Congress Cataloging-in-Publication Data

Molecular similarity and reactivity : from quantum chemical to phenomenological approaches / edited by Ramon Carbó.
p. cm. -- (Understanding chemical reactivity ; 14)
Includes index.
ISBN 0-7923-3309-8
1. Molecular structure. 2. Similarity (Physics) 3. Reactivity (Chemistry) I. Carbó, Ramon. II. Series.
QD461.M6274 1995
541.2'2--dc20 94-46540

ISBN 0-7923-3309-8

Published by Kluwer Academic Publishers,
P.O. Box 17, 3300 AA Dordrecht, The Netherlands.

Kluwer Academic Publishers incorporates
the publishing programmes of
D. Reidel, Martinus Nijhoff, Dr W. Junk and MTP Press.

Sold and distributed in the U.S.A. and Canada
by Kluwer Academic Publishers,
101 Philip Drive, Norwell, MA 02061, U.S.A.

In all other countries, sold and distributed
by Kluwer Academic Publishers Group,
P.O. Box 322, 3300 AH Dordrecht, The Netherlands.

Printed on acid-free paper

Printed in the Netherlands

Contents

Preface

In the study of various phenomena in nature, the concept of similarity plays a fundamental role. Chemistry is no exception; the similarity of molecules, both in their physical properties and in their chemical reactions, provides a basis for their classification, characterization, and scientific description. Ultimately, the recognition and analysis of molecular similarities serve as the basis of an understanding of molecular structures and properties, and represent the first steps in the development of theoretical models explaining chemical behavior. In this role, molecular similarity is the foundation of predictive models in chemistry.

Molecular similarity and molecular reactivity are strongly related. Studying the reactivities of molecules is an important tool for detecting molecular similarities and differences; alternatively, similar molecular properties often imply similar reactivities. This latter aspect is of special value, allowing chemists to make predictions concerning the outcomes of chemical reactions based on molecular similarities.

In this book, the central theme, molecular similarity, is discussed from a variety of viewpoints covering the range from rigorous quantum chemical approaches to phenomenological observations expressed within appropriate physical and mathematical frameworks. The authors of the various chapters represent some of the most current fields of research on molecular similarity. It is the hope of the editor that by bringing these subjects together under the cover of this book will provide the readers with a broad perspective and a handy reference of the contemporary approaches to similarity analysis of molecules and reactions.

Ramon Carbó
Girona

R. Carbó (ed.), Molecular Similarity and Reactivity: From Quantum Chemical to Phenomenological Approaches, vii.

Section A

Quantum Chemical Foundation

Theoretical Foundation of Quantum Molecular Similarity[1]

RAMON CARBÓ* and EMILI BESALÚ
Institut de Química Coputacional, Universitat de Girona, 17071 Girona, Spain
*(*author for correspondence)*

Key words: Mendeleev Postulates, Density Matrix, Density Transform, Expectation functions, Quantum Similarity Measures and Indices, Molecular Point Clouds, Nested Summation Symbols

Abstract. Starting from the description of the Mendeleev Postulates, a comprehensive theoretical framework involving the complete as far as possible aspects of Quantum Similarity is developed. The notation of implicit formulae in terms of Nested Summation Symbols is proposed as a means to achieve complete generality and easy high level programming language translation. General definitions of Quantum Similarity Measures and Indices are given. Description of the procedures giving particular Quantum Object Set ordering is shown.

1. Introduction

1.1. Historical background

In our Laboratory a progressive development has been made in successive papers [1] during the past twelve years in order to construct a rigorous definition of Quantum Similarity Measures (QSM) and provide the basis of their quantum mechanical meaning. The present study scheme and contents are related to the interest the subject has aroused, trying to complete the development of various new computational techniques produced by other groups [2], and attempt to complement other methodological issues on Molecular Similarity studies, as those exposed on reference [2k], for example.

1.2. General outline

A comprehensive definition of Similarity Measures in the realm of quantum mechanical objects, needs to be unambiguously interpreted on both the mathematical or philosophical sides. In order to achieve this goal, the point of view of the present work is essentially focused on the following lines:

a) Definition and interpretation of QSM according to quantum mechanical principles.
b) Construction of a QSM theoretical framework as general as possible.
c) Manipulation and visualization of QSM results in terms of a pure geometrical basis.

[1] A contribution of the Grup de Química Quàntica de l'Institut d'Estudis Catalans

R. Carbó (ed.), Molecular Similarity and Reactivity: From Quantum Chemical to Phenomenological Approaches, 3–30.

In this manner this paper intends to go one step further in the directions signaled by the mentioned threefold path. Accordingly the present work is divided into three main parts. A general scheme is built up in first instance, afterwards the various possible computational issues are discussed, and finally the interpretation consequences and utility of the previous theoretical body is presented.

1.3. Mendeleev Postulates

Starting from the previous considerations, the authors intention is to solidify the foundations of Quantum Similarity (QS) describing a set of basic rules we will refer to as *Mendeleev Postulates*. The meaning of these postulates will be clarified along the text but we will list them here as a short scheme of the present work threefold general purposes.

P0: *Every quantum system in a given state can be described by their Density Matrix Elements*. This postulate is obviously borrowed from the one related to Quantum Mechanics. In Section 3 the concept of Density Transform (DT) will be studied: It is simply an extension of the density matrix element definition. This DT can be also used as a tool to describe quantum systems. Once it has been recognized that the density matrix elements are the container of any kind of information attached to a given quantum system, it is easy to accept the idea consisting in the fact that DT must be at the beginning of any QS development.

P1: *Quantum systems can be compared by means of a QSM*. A QSM transformation into a Quantum Similarity Index (QSI) may be also used. The definition of QSI will be found in Section 5. But it must be always kept in mind that a QSI is no more than a manipulation of the QSM, which must be taken as the origin of all the theoretical framework to be developed in this paper. The meaning of QSM or QSI can be connected with Fuzzy Set Theory [1c, 2k]. As described in Section 6, Quantum Systems can be collected as Quantum Sets, whose elements become Quantum Objects afterwards, when a DT is defined and a QSM or QSI computed.

P2: *Projection of a Quantum Object Set into some* n-dimensional space is always feasible. By means of the computation of QSM Quantum Object Sets can be manipulated somehow as to be represented in an euclidean *n*-dimensional space transforming in this way into Quantum Object Point Clouds. *n*-dimensional Point Clouds constructed in such a manner can be rotated and projected into 2- and 3-dimensional spaces in order to be easily visualized. In Section 6 all these concepts will be clarified and several Optimal Visualization techniques will be described.

P3: *A Quantum Object Set Ordering exists.* Zermelo's Theorem [3], see also Appendix A1, can be invoked over any Quantum Object Set. Quantum Object ordering can be achieved by means of particular QSM or QSI computed over the Quantum Object Set elements. In Sections 5 and 6 the existence of order inside Quantum Sets will be studied and some practical ordering methods described. The *Mendeleev Conjecture* will be introduced at this stage: The object ordering induces an order over the implicit relationships which exists between the set elements and their properties.

2. The structure of density matrices

2.1. General theory: Löwdin-McWeeny scheme

Density functions of any order can be constructed by means of Löwdin or McWeeny descriptions [4]. Using a normalized n-electron wavefunction $|\Phi(\boldsymbol{j})\rangle$, usually taken as a Slater Determinant (SD) or a linear combination of them, the n-th order density matrix element $\rho^{(n)}(\boldsymbol{j}, \boldsymbol{j}')$ can be written as:

$$\rho^{(n)}(\boldsymbol{j}, \boldsymbol{j}') = |\Phi(\boldsymbol{j})\rangle\langle\Phi(\boldsymbol{j}')|, \tag{1}$$

where $\boldsymbol{j}$ and $\boldsymbol{j}'$ are n-dimensional vectors collecting the coordinates of the n particles.

A recurrent procedure can be defined to obtain the remaining lesser order density matrix elements. The $(n-1)$-th order density matrix element is obtained from the n-th order one, integrating over the coordinates of one electron. The result is:

$$\rho^{(n-1)}(\boldsymbol{j}, \boldsymbol{j}') = n \iint \delta(j'_n - j_n)\Phi(\boldsymbol{j} \oplus j_n)\Phi^*(\boldsymbol{j}' \oplus j'_n)dj'_n\, dj_n, \tag{2}$$

where δ is the Dirac Delta Function. The n-dimensional vector argument of the system wavefunction has been expressed in terms of the direct sum of two subvectors, this requires a n-dimensional vector partition in a $(n-1)$-dimensional vector plus a monodimensional one. The last one has been taken arbitrarily as the last index of the original n-dimensional vector.

Using this recurrence formula, it is possible to generate all the possible density matrix elements of order $n - m$ up to the first. The explicit expression for the $(n - m)$-th order matrix element is:

$$\rho^{(n-m)}(\boldsymbol{j}, \boldsymbol{j}') = \binom{n}{m} \iint \left(\prod_{k=1}^{m} \delta(j'_{n-k+1} - j_{n-k+1})\right) \Phi(\boldsymbol{j} \oplus \boldsymbol{i})\Phi^*(\boldsymbol{j}' \oplus \boldsymbol{i}')d\boldsymbol{i}'\, d\boldsymbol{i}, \tag{3}$$

where the vector $\boldsymbol{i}$ is constructed as $\boldsymbol{i} = (j_{n-m+1}, j_{n-m+2}, \ldots, j_n)$ and the differential d$\boldsymbol{i}$ is defined as d$\boldsymbol{i} = dj_{n-m+1}\, dj_{n-m+2} \ldots dj_n$. Similar expressions hold for the primed terms.

Diagonal terms of the density matrix are obtained when the primed arguments become equal to the unprimed ones. The n-th order diagonal element of the density matrix is then expressed as:

$$\rho^{(n)}(\boldsymbol{j}, \boldsymbol{j}) = \rho^{(n)}(\boldsymbol{j}) = |\Phi(\boldsymbol{j})\rangle\langle\Phi(\boldsymbol{j})|, \tag{4}$$

and can be named the n-th order density function. Lesser order density functions can be obtained in a similar way as described in Equation (3):

$$\rho^{(n-m)}(\boldsymbol{j}) = \binom{n}{m}\int\int \Phi(\boldsymbol{j} \oplus \boldsymbol{i})\Phi^*(\boldsymbol{j} \oplus \boldsymbol{i})d\boldsymbol{i}. \tag{5}$$

The integration of the first order density function is equal to n, the number of involved particles. The p-th order density function has the following meaning: the term $\rho^{(p)}(\boldsymbol{j})d\boldsymbol{j}$ gives, multiplied by n, the probability to find, within the volume element $d\boldsymbol{j}$, p particles having the space-spin coordinates $\{j_1, j_2, \ldots, j_p\}$ respectively, averaged over all the space-spin positions of the remaining n–p particles.

Here we will give alternative explicit expressions for non-diagonal terms of the density matrix, which will be general in the sense that diagonal elements appear as particular cases. The present expressions are written in such a way that they:

a) Serve to construct compact forms of density matrix elements.
b) Are immediately programmable, according to the methodological recipes we have built up [5].

2.2. Density matrix elements and N-particle operators

Density matrices are well suited in order to express expectation values of operators as it is shown in reference [4a]. Here, a reformulation in terms of Nested Summation Symbols (NSS), see Appendix A2, will be given.

Let be an n-particle operator expressible in terms of other zero-, one-, . . . , many-particle operators. This general n-particle operator can be written in terms of NSS formalism as:

$$\Omega(1, 2, \ldots, n) = \Omega(\boldsymbol{n}) = \sum_{i=0}^{m} \Sigma_i(\boldsymbol{j})\Omega^{(i)}(\boldsymbol{j}). \tag{6}$$

If the operators are symmetrical with respect to the particle indices, then, an equivalent expression for the operator (6) can be written:

$$\Omega(\boldsymbol{n}) = \sum_{i=0}^{m} \Sigma_i(\boldsymbol{j}, \Delta(\boldsymbol{L}))\Omega^{(i)}(\boldsymbol{j}), \tag{7}$$

where the Generalized Logical Kronecker Delta, $\Delta(\boldsymbol{L})$, see Appendix A2, included in the NSS's is used to avoid redundant terms in Equation (7). This requires the Logical Vector, $\boldsymbol{L}$, see Appendix A2, to be defined as:

$$\boldsymbol{L} = \{j_p < j_q; \forall_p = 1, n-1; q = p+1, n\} \tag{8}$$

In the general case, shown in Equation (6), the expectation value of the operator can be obtained by means of the following integral:

$$\begin{aligned}\langle\Omega(\boldsymbol{n})\rangle &= \langle\Omega(\boldsymbol{j})\,|\sum_{i=0}^{m}\Sigma_i(\boldsymbol{j})\Omega^{(i)}(\boldsymbol{j})|\,(\boldsymbol{j})\rangle \\ &= \sum_{i=0}^{m}\Sigma_i(\boldsymbol{j})\int\Phi^*(\boldsymbol{j})\Omega^{(i)}(\boldsymbol{j})\Phi(\boldsymbol{j})d\boldsymbol{j} \\ &= \sum_{i=0}^{m}\Sigma_i(\boldsymbol{j})\int\delta(\boldsymbol{j}'-\boldsymbol{j})\Omega^{(i)}(\boldsymbol{j})\rho^{(i)}(\boldsymbol{j}',\boldsymbol{j})d\boldsymbol{j}'d\boldsymbol{j},\end{aligned} \tag{9}$$

where $|\Phi(\boldsymbol{j})\rangle$ is the system wavefunction and it is assumed that the primed variables are just a copy of the unprimed ones, but the many-particle operator only acts over the later ones.

With respect to the above comments, it can be said that density matrix elements can be considered as n-particle operators. This feature will be used when developing the QS theoretical framework in Section 4 below.

2.3. Slater determinants

Usually the basis functions needed to construct polyelectronic wavefunctions are made of SD. Thus, an analysis of the structure of SD in terms of NSS formalism is studied here.

As it is shown in Appendix A2, we introduce the SD computation in terms of previous work [5a, 5b, 5e]. Using NSS terminology a general compact expression for the determinant structure is obtained. In this manner, this formulation can be transferred to the study of SD [6], constructed by n spinorbitals associated to n electrons. Normalized SD of this kind adopt the following expression when using NSS formalism:

$$\begin{aligned}|D(\boldsymbol{n})\rangle &= (n!)^{-1/2}\,Det|\psi_1\psi_2\ldots\psi_n| \\ &= (n!)^{-1/2}\,\Sigma_n(\boldsymbol{j}=\mathbf{1},\,n\mathbf{1},\,\Delta(\boldsymbol{L}))\;S(\boldsymbol{j})\;|\Psi(\boldsymbol{j})\rangle,\end{aligned} \tag{10}$$

where the $|\Psi(\boldsymbol{j})\rangle$ terms are built up with the spinorbital products:

$$|\Psi(\boldsymbol{j})\rangle = \prod_{k=1}^{n}\psi_{j_n}(k). \tag{11}$$

Next, we will use formula (10) in order to obtain explicit expressions for the density matrix elements.

2.4. NSS explicit expression for density matrix elements

Using a normalized n-electron SD $|D(\boldsymbol{j})\rangle$ as a system wavefunction, constructed as discussed in Section 2.3, one can write the n-th order density matrix element

$\rho^{(n)}(\boldsymbol{j},\boldsymbol{j}')$ as:

$$\begin{aligned}\rho^{(n)}(\boldsymbol{j},\boldsymbol{j}') &= |D(\boldsymbol{j})\rangle\langle D(\boldsymbol{j}')| \\ &= (n!)^{-1}\,\Sigma_n(\boldsymbol{p}=\mathbf{1},n\mathbf{1},\Delta(\boldsymbol{L}(\boldsymbol{p})))\,\Sigma_n(\boldsymbol{q}=\mathbf{1},n\mathbf{1},\Delta(\boldsymbol{L}(\boldsymbol{q}))) \\ &\qquad S(\boldsymbol{p})\,S(\boldsymbol{q})\,|\Psi(\boldsymbol{j}[\boldsymbol{p}])\rangle\langle\Psi(\boldsymbol{j}'[\boldsymbol{q}])|,\end{aligned} \tag{12}$$

where, the $|\Psi(\boldsymbol{j}[\boldsymbol{p}])\rangle$ terms are spinorbital products as they were defined in Equation (11); the symbol $\boldsymbol{j}[\boldsymbol{p}]$ means a permutation $\boldsymbol{p}$ is performed over the parameter vector $\boldsymbol{j}$ subindices.

Generalization of this one term function result is straightforward. If the system wavefunction is a linear combination of SD $\{|D_p(\boldsymbol{j})\rangle\}$ as in the expression:

$$|\Phi(\boldsymbol{j})\rangle = \sum_{p=1}^{m} C_p|D_p(\boldsymbol{j})\rangle, \tag{13}$$

then, the n-th order density matrix element takes the following form:

$$\rho^{(n)}(\boldsymbol{j},\boldsymbol{j}') = |\Phi(\boldsymbol{j})\rangle\langle\Phi(\boldsymbol{j}')| = \sum_{p=1}^{m}\sum_{q=1}^{m} C_pC_q^*\,|D_p(\boldsymbol{j})\rangle\langle D_q(\boldsymbol{j}')|. \tag{14}$$

The determinant products appearing in the rightmost side of Equation (14) bear the same structure of those appearing in the expression (12), they differ in that the two involved determinants are constructed with two non-equivalent sets of spinorbitals.

Here we will only deal with monodeterminantal functions to describe a simplified theoretical development. The interesting practical result concerning n-th order density matrix elements, constructed using SD as basis sets, appears once spinorbitals are described by means of the LCAO approach. This possibility will be described now.

2.5. LCAO FORM OF DENSITY MATRIX ELEMENTS

Let us write the LCAO approach for a given spinorbital set $\{\psi_k\}$ as:

$$\psi_k = \sum_{a=1}^{M_k} C_{ak}\,\chi_a, \tag{15}$$

where each spinorbital has been expressed as a linear combination of atomic spinorbitals using a M_k-dimensional basis set $X = (\chi_1, \chi_2, \ldots, \chi_{M_k})$. Then, within the NSS formalism, a product of spinorbitals like (11) can be structured by means of the linear combination (15) as:

$$|\Psi(\boldsymbol{j}[\boldsymbol{p}])\rangle = \prod_{i=1}^{n}\left(\prod_{a_i=1}^{M_{j[p_i]}} c_{a_i\,j[p_i]}\chi_{a_i}\right) = \Sigma_n(\boldsymbol{a}=\mathbf{1},\boldsymbol{M})\boldsymbol{C}(\boldsymbol{a},\boldsymbol{j}[\boldsymbol{p}])\boldsymbol{X}(\boldsymbol{a}), \tag{16}$$

where a simplified NSS notation has been used, see Appendix A2. $\boldsymbol{M}$ is the vector containing the elements $\{M_{j[p_i]}\}$, The symbol:

$$\boldsymbol{C}(\boldsymbol{a},\boldsymbol{j}[\boldsymbol{p}]) = \prod_{i=1}^{n} c_{a_i\,j[p_i]}, \tag{17}$$

is a product of LCAO coefficients, and

$$X(\boldsymbol{a}) = \prod_{i=1}^{n} \chi_{a_i}, \tag{18}$$

is a basis function product respectively. When using a NSS formalism, it can be seen how the product of LCAO spinorbital expressions can be written in a similar form like the unique function of Equation (15). Using the Equation (16) in the spinorbital product appearing in the rightmost side of Equation (12), one obtains:

$$|\Psi(\boldsymbol{j}[\boldsymbol{p}])\rangle\langle\Psi(\boldsymbol{j}'[\boldsymbol{q}])| = \Sigma(\boldsymbol{a} \oplus \boldsymbol{b})\ C(\boldsymbol{a}, \boldsymbol{j}[\boldsymbol{p}])\ C^*(\boldsymbol{b}, \boldsymbol{j}[\boldsymbol{q}]\ X(\boldsymbol{a})\ X^*(\boldsymbol{b}'), \tag{19}$$

where a simplified NSS notation has been used again. Primed indices note the spinorbital dependence with respect to the bra coordinates.

Finally, the n-th order density matrix elements can be expressed in terms of the atomic spinorbital products as:

$$\rho^{(n)}(\boldsymbol{j}, \boldsymbol{j}') = \Sigma(\boldsymbol{a} \oplus \boldsymbol{b})\ \boldsymbol{Q}^{(n)}(\boldsymbol{a}, \boldsymbol{b})\ X(\boldsymbol{a})\ X^*(\boldsymbol{b}'), \tag{20}$$

being the n-th order charge and bond order hypermatrices, $\boldsymbol{Q}^{(n)}(\boldsymbol{a}, \boldsymbol{b})$, defined in turn as:

$$\begin{aligned} \boldsymbol{Q}^{(n)}(\boldsymbol{a}, \boldsymbol{b}) = \Sigma_n(\boldsymbol{p} = \mathbf{1}, n\mathbf{1}, \Delta(L(\boldsymbol{p})))\ \Sigma_n(\boldsymbol{q} = \mathbf{1}, n\mathbf{1}, \Delta(L(\boldsymbol{q}))) \\ \omega^{(n)}(\boldsymbol{p}, \boldsymbol{q})\ C(\boldsymbol{a}, \boldsymbol{j}[\boldsymbol{p}])\ C^*(\boldsymbol{b}, \boldsymbol{j}[\boldsymbol{q}]), \end{aligned} \tag{21}$$

using the *occupation* hypermatrix elements:

$$\omega^{(n)}(\boldsymbol{p}, \boldsymbol{q}) = (n!)^{-1}\ S(\boldsymbol{p})\ S(\boldsymbol{q}). \tag{22}$$

Equation (20) has the same structure as the well known LCAO form of the first order density function [6]. This is not surprising, because a characteristic of NSS formalism consists of reproducing the structure of monodimensional formulae expressions within a complete general n-dimensional framework [5].

3. Beyond the expectation value concept

The role played by the expectation value concept is well known in Quantum Mechanics [7]. Here it is shown how this concept also applies to function structures providing the concept of *Expectation functions*. This idea leads towards one of the innermost definitions of the QSM theory: General Density Transformations, which we will discuss next.

3.1. General density transformations

Let us define the *transform* of a function $f(\boldsymbol{r}, \boldsymbol{u}, \boldsymbol{p})$ as the following integral [8]:

$$T(\boldsymbol{r}, s, \boldsymbol{p}) = \int \Omega(\boldsymbol{r}, s, \boldsymbol{u}, \boldsymbol{p})\, f(\boldsymbol{r}, \boldsymbol{u}, \boldsymbol{p})\, d\boldsymbol{u}, \tag{23}$$

where the operator $\Omega(\boldsymbol{r}, s, \boldsymbol{u}, \boldsymbol{p})$ is the *kernel* of the transform. The optional vector $\boldsymbol{p}$ provides the dependence of the function, the operator or both with respect of a parameter set.

Starting from Equation (23), A *n-th order Density Integral Transform* can be defined as the transformation of a density matrix element of the same order:

$$P^{(n)}(\boldsymbol{r}, s, \boldsymbol{p}) = \int \Omega(\boldsymbol{r}, s, \boldsymbol{u}, \boldsymbol{p})\, \rho^{(n)}(\boldsymbol{r}, \boldsymbol{u}, \boldsymbol{p})\, d\boldsymbol{u}, \tag{24}$$

where $\boldsymbol{p}$ is the optional parameter vector.

This last integral in Equation (24) will be widely used in the following General Quantum Similarity Theory development.

3.2. Expectation functions and expectation values

It can be said that the DT in Equation (24) constructs *n-th order Expectation Functions* of the operator Ω. From an expectation function of order n it can be obtained some *Expectation Value* by means of the evaluation of the expectation function at a fixed point $(\boldsymbol{r}_o, s_o, \boldsymbol{p}_o)$, using the integral definition:

$$\langle P^{(n)}(\boldsymbol{r}_o, s_o, \boldsymbol{p}_o)\rangle = \iiint \delta(\boldsymbol{r} - \boldsymbol{r}_o)\, \delta(s - s_o)\, \delta(\boldsymbol{p} - \boldsymbol{p}_o)\, P^{(n)}(\boldsymbol{r}, s, \boldsymbol{p})\, d\boldsymbol{r}\, ds\, d\boldsymbol{p}, \tag{25}$$

where integration over all the coordinates is obviously not compulsive, allowing in this manner to obtain expectation values which in fact are functions of some of the parameters $\boldsymbol{r}$, s or $\boldsymbol{p}$.

Some cases worthy of mention arise from the analysis of Equation (24) when the kernel takes precise particular forms. Some examples of this will be treated now.

3.3. Density as an expectation function

When the operator Ω appearing in Equation (24) is defined as $\Omega(\boldsymbol{r}, s, \boldsymbol{u}, \boldsymbol{p}) = \Omega(s, \boldsymbol{u}) = \delta(\boldsymbol{u} - s)$, then the DT becomes $P^{(n)}(\boldsymbol{r}, s, \boldsymbol{p}) = \rho^{(n)}(\boldsymbol{r}, s, \boldsymbol{p})$ and the transformation leaves invariant the density matrix element.

Moreover, if Ω is defined as $\Omega(\boldsymbol{r}, s, \boldsymbol{u}, \boldsymbol{p}) = \delta(\boldsymbol{u} - \boldsymbol{r})$, then $P^{(n)}(\boldsymbol{r}, s, \boldsymbol{p}) = P^{(n)}(\boldsymbol{r}, \boldsymbol{p}) = \rho^{(n)}(\boldsymbol{r}, \boldsymbol{p})$ and the transformation returns a diagonal element of the density matrix: the n-th order density function.

3.4. Electrostatic potentials

When within the DT definition the operator form $\Omega(\boldsymbol{r}, s, \boldsymbol{u}, \boldsymbol{p}) = |\boldsymbol{u} - s|^{-1}$ is used, then $P^{(n)}(\boldsymbol{r}, s, \boldsymbol{p}) = V^{(n)}(\boldsymbol{r}, s, \boldsymbol{p})$. That is: a generalized form of the electrostatic potential is obtained. One can call this general formulation a n-th order Electrostatic Potential. A particular case appears when using the density function $\rho^{(1)}(s, \boldsymbol{p})$ in Equation (24), then the obtained transform coincides with the usual electrostatic potential function.

More general is to consider the case where the operator Ω is defined as $\Omega(\boldsymbol{r}, s, \boldsymbol{u}, \boldsymbol{p}) = |\boldsymbol{u} - s|^{k}$, where k is a non zero integer. This choice generates some particular kinds of DT which will be studied later on.

3.5. System energy

When the operator Ω in Equation (24) is taken as the hamiltonian, that is: $\Omega(\boldsymbol{r}, s, \boldsymbol{u}, \boldsymbol{p}) = \delta(\boldsymbol{u} - \boldsymbol{r})H(\boldsymbol{u}, s, \boldsymbol{p})$, then $P^{(n)}(\boldsymbol{r}, s, \boldsymbol{p}) = E(\boldsymbol{r}, s, \boldsymbol{p})$ and the transformation returns the energy of the system if n equals the particle number. In this case, the vector $\boldsymbol{p}$ can be chosen as the nuclear coordinates treated as a parameter of the integral (24), according to the usual Born-Oppenheimer approximation. It is assumed that the differential operators present in $H(\boldsymbol{u}, s, \boldsymbol{p})$ only act over the s coordinate, in a similar manner as in Equation (9) with respect to the operator $\Omega(\boldsymbol{n})$.

4. Quantum similarity measures

In this Section, the General QSM integral will be defined. Particular forms will be listed and analyzed. Finally, a practical computer implementation of the integral will be given within the LCAO MO framework.

4.1. General structure of quantum similarity measures

The *n-th order Quantum Similarity Measure* between a set of m systems collected in the vector $\boldsymbol{A} = (A_1, A_2, \ldots, A_m)$, with respect to an operator $\boldsymbol{W}$, can be defined as an integral of the following kind:

$$Z_A^{(n)}(\boldsymbol{W}, \boldsymbol{R}^{(m+1,q)}, \boldsymbol{S}^{(m+1,q)}, \boldsymbol{p}) = \iint \boldsymbol{W}(\boldsymbol{R}^{(1,q)}, \boldsymbol{S}^{(1,q)}, \boldsymbol{p}) \left(\prod_{i=1}^{m} P_{A_i}^{(n_i)}(\boldsymbol{r}_i, s_i, \boldsymbol{p})\right) d\boldsymbol{R}^{(1,m)} d\boldsymbol{S}^{(1,m)}. \tag{26}$$

Considering the previous Equation (26), some particular aspects of the theory and comments on the nature of the QSM involved elements have to be taken into account:

a) The vector $\boldsymbol{n} = (\boldsymbol{n}_1, n_2, \ldots, n_m)$ collects the order of the DT of the set of systems which are active in the vector $\boldsymbol{A}$.
b) Vector $\boldsymbol{R}$ is defined as containing a set of vector coordinates: $\boldsymbol{R}^{(i,j)} = (\boldsymbol{r}_i, \boldsymbol{r}_{i+1}, \ldots, \boldsymbol{r}_j)$. The corresponding differential vector has a related structure: $d\boldsymbol{R}^{(i,j)} = d\boldsymbol{r}_i \, d\boldsymbol{r}_{i+1} \ldots d\boldsymbol{r}_j$. The same definitions can be considered for the vector $\boldsymbol{S}$.
c) If $m \geq q$ the measure in Equation (26) does not depend on $\boldsymbol{r}$ or $\boldsymbol{s}$ coordinates.
d) It is possible to define the operator $\boldsymbol{W}$ in Equation (26) with an embedded Dirac delta function, appearing as a multiplicative factor. The interest of this construction appears when using the Dirac functional properties: upon integration with respect one of the variables appearing in the delta function, two coordinate vectors of the whole integrand become identical. This convention will be used below to obtain expressions of QSM containing cross terms between the DT. However, this general situation restricts the order of the involved DT to be the same.
e) A word of caution must be said concerning the operator or DT definition. The introduction of pure atomic density terms as Dirac delta functions of nuclear coordinates can produce divergent integral elements when using the Equation (26) and, thus, it will be the same for any previous definition. This drawback can be avoided in some cases by introducing within the operator $\boldsymbol{W}$ some kind of weight function [9]. This warning must be taken carefully into account when using electrostatic molecular potentials as DT appearing in the similarity measure. Some recent work on quantum similarity [21, 2m, 10] seems to ignore this peculiar fact.
f) The operator $\boldsymbol{W}$ is usually taken as a positive definite operator, but in a recent work [1i] the possibility of using any kind of operators is slightly explored. This aspect will be discussed later on.

4.2. Particular cases and simplified forms

A particular case of Equation (26) is obtained when dealing with two quantum systems A and B and their respective DT of order n and m:

$$Z_{AB}^{(n,m)}(\boldsymbol{W}, \boldsymbol{R}^{(3,q)}, \boldsymbol{S}^{(3,q)}, \boldsymbol{p}) = \iint \boldsymbol{W}(\boldsymbol{R}^{(1,q)}, \boldsymbol{S}^{(1,q)}, \boldsymbol{p}) \, P_A^{(n)}(\boldsymbol{r}_1, \boldsymbol{s}_1, \boldsymbol{p}) \, P_B^{(m)}(\boldsymbol{r}_2, \boldsymbol{s}_2, \boldsymbol{p}) \, d\boldsymbol{R}^{(1,2)} \, d\boldsymbol{S}^{(1,2)}. \tag{27}$$

Usually, $q = 2$ and $n = m$, then the integral above appears in a more explicit form:

$$Z_{AB}^{(n)}(\boldsymbol{W}, \boldsymbol{p}) = \iiiint \boldsymbol{W}(\boldsymbol{r}_1, \boldsymbol{r}_2, \boldsymbol{s}_1, \boldsymbol{s}_2, \boldsymbol{p}) \, P_A^{(n)}(\boldsymbol{r}_1, \boldsymbol{s}_1, \boldsymbol{p}) \, P_B^{(n)}(\boldsymbol{r}_2, \boldsymbol{s}_2, \boldsymbol{p}) \, d\boldsymbol{r}_1 \, d\boldsymbol{r}_2 \, d\boldsymbol{s}_1 \, d\boldsymbol{s}_2. \tag{28}$$

From Equations (26) or (27) a large number of possibilities arise. Some of these particular cases are examined next.

4.3. Triple and multiple density transform similarity measures

In a recent discussion on QSM [1k] we have presented a practical computational framework derived from a previous idea [1i], concerning the positive definite nature of the n-th order density functions. Due to this property, which can be attached to every density function, one can consider any other system's appropriate DT as the positive definite operator $\mathbf{W}$ in Equation (26). A *Triple Density Transform Similarity Measure* (TDTSM) [1k] is then constructed:

$$T^{(n)}_{C;\,AB}(\mathbf{p}) = \int P^{(n)}_{A}(\mathbf{r}, \mathbf{p})\, P^{(n)}_{C}(\mathbf{r}, \mathbf{p})\, P^{(n)}_{B}(\mathbf{r}, \mathbf{p})\, d\mathbf{r}, \tag{29}$$

In an even more general context one can study two-particle system operators which will be written as $\mathbf{W}(\mathbf{r}_1, \mathbf{r}_2, \mathbf{p})$ so it is possible to define the QSM:

$$Z^{(n,\,m)}_{AB}(\mathbf{W}, \mathbf{p}) = \int\int P^{(n)}_{A}(\mathbf{r}_1, \mathbf{p})\, \mathbf{W}(\mathbf{r}_1, \mathbf{r}_2, \mathbf{p})\, P^{(m)}_{B}(\mathbf{r}_2, \mathbf{p})\, d\mathbf{r}_1,\, d\mathbf{r}_2. \tag{30}$$

This definition seems useful in the TDTSM case where one can write:

$$T^{(n)}_{C;\,AB}(\mathbf{p}) = \int\int P^{(n)}_{A}(\mathbf{r}, \mathbf{p})\, P^{(n)}_{C}(\mathbf{r}, \mathbf{s}, \mathbf{p})\, P^{(n)}_{B}(\mathbf{s}, \mathbf{p})\, d\mathbf{r}\, d\mathbf{s}, \tag{31}$$

where $P^{(n)}_{C}(\mathbf{r}, \mathbf{s}, \mathbf{p})$ corresponds to a density matrix non-diagonal element transform [4].

Generalizing the idea of Cioslowski [2d] concerning QSM involving first order density matrix elements, one can define an alternative form as the off-diagonal DT function matrix measure:

$$Z^{(n)}_{AB}(\mathbf{p}) = \int\int P^{(n)}_{A}(\mathbf{r}, \mathbf{s}, \mathbf{p})\, P^{(n)}_{B}(\mathbf{r}, \mathbf{s}, \mathbf{p})\, d\mathbf{r}\, d\mathbf{s}, \tag{32}$$

which can be easily extended inside our scheme, defining the counterpart of equation (30) as:

$$Z^{(n)}_{AB}(\mathbf{W}, \mathbf{p}) = \int\int\int P^{(n)}_{A}(\mathbf{r}_1, \mathbf{r}_2, \mathbf{p})\, \mathbf{W}(\mathbf{r}_2, \mathbf{r}_3, \mathbf{p})\, P^{(n)}_{B}(\mathbf{r}_3, \mathbf{r}_1, \mathbf{p})\, d\mathbf{r}_1\, d\mathbf{r}_2\, d\mathbf{r}_3. \tag{33}$$

This approach may be particularly useful in thc TDTSM case:

$$T^{(n)}_{C;\,AB}(\mathbf{p}) = \int\int\int P^{(n)}_{A}(\mathbf{r}_1, \mathbf{r}_2, \mathbf{p})\, P^{(n)}_{C}(\mathbf{r}_2, \mathbf{r}_3, \mathbf{p})\, P^{(n)}_{B}(\mathbf{r}_3, \mathbf{r}_1, \mathbf{p})\, d\mathbf{r}_1\, d\mathbf{r}_2\, d\mathbf{r}_3. \tag{34}$$

Although this off-diagonal density matrix approach does not differ in essence from the initial QS integral structure, it may become interesting from a computational point of view, due to simpler final forms obtained from the involved integrals. Therefore, the off-diagonal density matrix formalism must be considered as a worth-while alternative.

It is more general to consider the whole integrand of Equation (26) as a product of m DT belonging to m quantum chemical systems in a one-to-one correspondence. In this manner one can construct a *m-th order Multiple Density Transform Similarity Measure* as an integral derived from Equation (26), having the form:

$$Z_A^{(n)}(\{P_i\}, \boldsymbol{p}) = \int\int \left(\prod_{i=1}^{m} P_{A_i}^{(n_i)}(\boldsymbol{r}_i, \boldsymbol{s}_i, \boldsymbol{p}) \right) d\boldsymbol{R}^{(1,\,m)} d\boldsymbol{S}^{(1,\,m)}. \tag{35}$$

4.4. Gravitational and derived QSM

Being so general, the definition in Equation (26) allows us to construct other kinds of QSM. As another example, if the density matrix elements are transformed to n-th order electrostatic potentials and one uses a similar Equation as (28) with the operator $\boldsymbol{W}(\boldsymbol{r}_1, \boldsymbol{r}_2, \boldsymbol{r}_3)$ form:

$$\boldsymbol{G}(\boldsymbol{r}_1, \boldsymbol{r}_2, \boldsymbol{r}_3) = 1/4\ \delta(\boldsymbol{r}_3 - (\boldsymbol{r}_1 + \boldsymbol{r}_2)/2), \tag{36}$$

then a similarity measure is obtained bearing a *gravitational* operator structure:

$$Z_{AB}^{(n)}(\boldsymbol{G}, \boldsymbol{p}) = \int\int \rho_A^{(n)}(\boldsymbol{r}_1, \boldsymbol{p})\ \rho_B^{(n)}(\boldsymbol{r}_2, \boldsymbol{p})\ |\boldsymbol{r}_1 - \boldsymbol{r}_2|^{-2}\ d\boldsymbol{r}_1\, d\boldsymbol{r}_2. \tag{37}$$

This kind of QSM provides values related to the electronic interactions between two molecular systems, and can substitute the usual ones, based on the integration of the product of two bulk electrostatic molecular potentials.

In general, it is also possible to define a DT as described in Section 3.3, using the operator:

$$\boldsymbol{G}^{(k_1+k_2)}(\boldsymbol{r}_1, \boldsymbol{r}_2, \boldsymbol{r}_3) = 2^{-(k_1+k_2)}\ \delta(\boldsymbol{r}_3 - (\boldsymbol{r}_1 + \boldsymbol{r}_2)/2), \tag{38}$$

where k_1 and k_2 are the powers of the term $|\boldsymbol{u} - \boldsymbol{s}|$ appearing in the DT operator. In this case the obtained QSM has the generic form:

$$Z_{AB}^{(n)}(\boldsymbol{G}, \boldsymbol{p}) = \int\int \rho_A^{(n)}(\boldsymbol{r}_1, \boldsymbol{p})\ \rho_B^{(n)}(\boldsymbol{r}_2, \boldsymbol{p})\ |\boldsymbol{r}_1 - \boldsymbol{r}_2|^{k_1+k_2}\ d\boldsymbol{r}_1\, d\boldsymbol{r}_2, \tag{39}$$

which has several interpretative issues corresponding to the power value. Usually $k_1 = k_2$ and the power in Equation (39) is written $k = k_1 + k_2$. This leads in some particular cases to the following QSM types according to the most usual values of the exponent k: a) $k = -1$: Coulomb repulsion-like; b) $k = -2$: Gravitational-like; c) $k = 1$: Dipole-like or Distance-like, which has been described in another context [11], involving first order Electrostatic Molecular Potentials; d) $k = 2$: Quadrupole-like.

4.5. Gradient QSM

A new class of possible similarity integrals can be defined: $Z_{AB}(\Delta)$ where Δ is a non-definite operator. This possibility has been previously described in early work [1i], but not further explored. In this case one can consider the Laplacian operator as a good candidate to fulfill this role, so one obtains:

$$\begin{aligned} Z_{AB}^{(n)}(\nabla^2, \boldsymbol{p}) &= \int P_A^{(n)}(\boldsymbol{r}, s, \boldsymbol{p})\, \nabla^2\, P_B^{(n)}(\boldsymbol{r}, s, \boldsymbol{p})\, d\boldsymbol{r}\, ds \\ &= -\int \nabla P_A^{(n)}(\boldsymbol{r}, s, \boldsymbol{p})\, \nabla P_B^{(n)}(\boldsymbol{r}, s, \boldsymbol{p})\, d\boldsymbol{r}\, ds. \end{aligned} \tag{40}$$

Similarity integrals arising from Equation (40) can be related to the density function analysis of Bader [12].

4.6. Time dependent density transforms

All the formulation presented here includes in an implicit manner the possibility to deal with time dependent density functions.

This feature also suggests the interesting possibility of working in a momentum space environment as was done in references [2a, 2b, 2c]. A reformulation of the theory in this sense can lead to a framework which overrides some of the problems we have described in Section 4.1.e, concerning the nuclear coordinates treatment.

4.7. LCAO expressions of QSM

Usually, the DT resulting form Equation (24) are the density functions themselves. In this case one can write, for example, the LCAO form of the measure (35) using the NSS formalism relative to the density functions, outlined in Section 2.5 by means of Equation (20):

$$\begin{aligned} Z_A^{(n)}(\{\rho_k\}) = &\left\{ \prod_{i=1}^{m} \Sigma_{n_i}(\boldsymbol{a}_i \oplus \boldsymbol{b}_i)\, Q_{A_i}^{(n_i)}(\boldsymbol{a}_i, \boldsymbol{b}_i) \right\} \\ &\langle \{\textstyle\prod_i X_i(\boldsymbol{b}_i')\} \,|\, \{\textstyle\prod_i X_i(\boldsymbol{a}_i)\} \rangle, \end{aligned} \tag{41}$$

where the parameter dependence in $\boldsymbol{p}$ has been omitted for simplicity's sake.

Under this formulation, a $2m$ atomic orbital product integral appears in the rightmost side of Equation (41). It can be computed as a product of overlap-like integrals, involving $2m$ basis functions and m centers. The practical implementation of this kind of integrals is studied in Appendix A3 and more details can be found in reference [5e]. In this manner the use of NSS's can be easily connected to the most general up to date definition of a QSM.

The simplificative potential of this approach may be studied in the TDTSM case when dealing with first order density matrices. In the molecular electronic case within the LCAO MO framework, a first order density matrix element can be written as:

$$\rho^{(1)}(\boldsymbol{r}, \boldsymbol{s}) = \Sigma_\mu \Sigma_\nu \, D_{\mu\nu} \, \chi_\mu(\boldsymbol{r}) \, \chi_\nu(\boldsymbol{s}), \tag{42}$$

where $\{D_{\mu\nu}\}$ is the charge-bond order matrix and $\{\chi_\mu\}$ the AO basis set functions.

Then, a TDTSM like those of Equation (34) can be expressed as:

$$\begin{aligned} T^{(1)}_{C;\,AB} &= \iiint \rho^{(1)}_A(\boldsymbol{r}, \boldsymbol{s}) \, \rho^{(1)}_C(\boldsymbol{s}, \boldsymbol{t}) \, \rho^{(1)}_B(\boldsymbol{t}, \boldsymbol{r}) \, d\boldsymbol{r} \, d\boldsymbol{s} \, d\boldsymbol{t} \\ &= \Sigma_{\mu\nu} \, \Sigma_{\alpha\beta} \, \Sigma_{\lambda\sigma} \, D^A_{\mu\nu} \, D^C_{\alpha\beta} \, D^B_{\lambda\sigma} \, S^{AB}_{\mu\sigma} \, S^{AC}_{\nu\alpha} \, S^{CB}_{\beta\lambda}, \end{aligned} \tag{43}$$

where, for instance,

$$S^{AB}_{\mu\sigma} = \int \chi^A_\mu(\boldsymbol{r}) \, \chi^B_\sigma(\boldsymbol{r}) \, d\boldsymbol{r}. \tag{44}$$

Equation (43) above shows that when a QSM contains cross terms between non-diagonal elements of the density matrix, the integral computation becomes a product of overlap integrals. This case is apparently simpler than the evaluation of QSM integrals where density functions are used. These two options can be combined with the role of the operator $\boldsymbol{W}$ coming from the general definition. The problem is always reduced to compute integrals of the type:

$$\langle \{\Pi_i \, X_i(\boldsymbol{b}_i')\} | \, \boldsymbol{W} \, | \{\Pi_i \, X_i(\boldsymbol{a}_i)\} \rangle, \tag{45}$$

where $\boldsymbol{W}$ is some operator and $\{X_i\}$ are spinorbital products. See Appendix A3 for more details.

5. Molecular quantum similarity measures

All the ideas and definitions exposed until now are applied in general to Quantum Object Sets. Usually, the practical application of the QS method is performed over Molecular Sets. Then one is dealing with Molecular Quantum Similarity. The authors will refer from now on to the QS theory of Molecular Sets. This is done with no generality loss because the extension to arbitrary Quantum Object Sets is always immediate.

5.1. Similarity matrices

As it can be deduced from the discussion of Section 4 by means of Equation (26), every system A_i can be represented by a hypermatrix, whose elements are taken as:

$$\{Z_{ij\ldots k}(\boldsymbol{W})\} = \{Z^{(n_i,\, n_j,\, \ldots,\, n_k)}_{(A_i,\, A_j,\, \ldots,\, A_k)} \, (\boldsymbol{W}, \boldsymbol{R}^{(m+1,\, q)}, \boldsymbol{S}^{(m+1,\, q)}, \boldsymbol{p})\}, \tag{46}$$

In order to obtain a computationally manageable form, it is possible to reduce the m-dimensional hypermatrix information into a matrix involving all the available pairs of molecules:

$$A(W) = \{A_{IJ}(W)\}. \tag{47}$$

This can be obtained by means of the following hypermatrix reduced product definition:

$$A_{IJ}(W) = \Sigma_{m-1}(i)\; Z_I(i)(W)\; Z_J(i)(W), \tag{48}$$

which possess a similar structure to a scalar product. There a NSS formulation has been used and the terms $Z_I(i)$ are defined in a short way as the hypermatrix elements:

$$Z_I(i)(W) = Z_{I,\, i_1,\, i_2,\, \ldots\, ,\, i_{m-1}}(W), \tag{49}$$

5.2. Approximate measures and fitting of density functions

Despite all the explicit expressions which are outlined above related to the QSM computation, there is a bottleneck when practical computations are considered. For example, when one is dealing with molecular QSM, there is implicit a relative position optimization of the involved molecules. This requires evaluating the QSM integrals many times. Recently, the authors have been performing studies in order to establish cheaper density matrix elements expansions and, in this manner, speed the whole molecular QSM computation.

A choice is to expand the n-th order density matrix element in terms of a bilinear combination of m determinants acting as basis functions $\{D_i\}$:

$$\rho^{(n)}(r, r') \approx \sum_{i=1}^{m} \sum_{j=1}^{m} c_{ij}\, |D_i(r)\rangle\langle D_j(r')| = D(r)^t\, C\, D(r'). \tag{50}$$

where the vector D collects the SD $D = (D_1, D_2, \ldots , D_m)^t$ and matrix C contains all the bilinear coefficients $C = \{c_{ij}\}$.

The optimal coefficients can be obtained by supposing both sides of Equation (50) are equal and multiplying on the left and on the right by a pair of basis functions and integrating. The following equivalent matrix notation is then obtained:

$$A = S\, C\, S, \tag{51}$$

where the matrix A is the matrix representation of the density matrix element taken as an operator in the space of the basis functions: $A = \{\langle D_i(r)|\rho^{(n)}(r, r')|D_j(r')\rangle\}$. S is the metric matrix related to the basis set: $S = \{\langle D_i(r)|D_j(r)\rangle\}$.

As S is a definite positive matrix, then an inverse will always exist: S^{-1}, which can be used to solve the Equation (51). For example, the Cholesky decomposition algorithm [13] express S as a product of a triangular matrix and its transpose:

$$S = T^t\, T, \tag{52}$$

then the inverse of S reads

$$S^{-1} = T^{-1}\,(T^{-1})^t, \tag{53}$$

and the matrix Equation (51) is solved as:

$$C = S^{-1}\, A\, S^{-1}, \tag{54}$$

Moreover, this computational scheme is the same as applying a least squares fit to the expression (50). That is, solution of the system (51) minimizes the quadratic error ε with respect to the parameters $\{c_{ij}\}$:

$$\varepsilon = |\rho^{(n)}(r, r') - D(r)^t\, C\, D(r')|^2. \tag{55}$$

In this manner, the procedure is equivalent to fit the expression (50) with the criteria to obtain the maximal overlap between the density matrix element and the bilinear expansion. Thus, the procedure is equivalent to maximize the value of the integral:

$$\xi = \int\!\!\int \rho^{(n)}(r, r')\, D(r)^t\, C\, D(r')\; dr\, dr'. \tag{56}$$

Also, one can think of adjusting the linear combination (50) to the matrix element $\rho^{(n)}(r, r')$ using as a weight some positive definite operator $\Theta(r, r')$. In this case, one has to optimize the integral:

$$\xi = \int\!\!\int \rho^{(n)}(r, r')\, \Theta(r, r')\, D(r)^t\, C\, D(r')\; dr\, dr'. \tag{57}$$

In our Laboratory, we are investigating the possibility that the choice of the appropriate operator $\Theta(r, r')$ in Equation (57) is related to the kind of QSM the density function expansion (50) will be involved with.

Density function expansion is a particular case of the previously presented algorithm. The same kind of equations hold with the difference that the involved integrals are slightly different. For example, the expansion (50) reads:

$$\rho^{(n)}(r) \approx \sum_{i=1}^{m} \sum_{j=1}^{m} c_{ij}\, |D_i(r)\rangle\langle D_j(r)| = D(r)^t\, C\, D(r). \tag{58}$$

and matrix A becomes $A = \{\langle D_i(r)|\rho^{(n)}(r)|D_j(r)\rangle\}$. The same considerations hold when interpreting the solutions of the Equation (51).

It is possible to solve the expansion by adding successively new determinants acting as basis functions in Equation (50). The iterative Cholesky decomposition allows us to do this in a cheap way. In reference [5a] this procedure is outlined. There it was applied to expand two center charge distributions. It is also shown how to optimize expansion (50) with respect to non linear coefficients as, for example, those appearing into the definition of the Slater basis functions.

5.3. Manipulation of similarity measures: Similarity indices

Once the set of quantum objects to study is chosen and the operator related to the QSM definition in Equation (26) is defined, the QSM related to the set is unique. But they can be transformed or combined in order to obtain a new kind of auxiliary terms which can be named *Similarity Indices*. A vast quantity of possible QSM manipulations leading to a variety of QSI definitions exists. The most common, which can be considered as standard ones, arise from the manipulation of QSM between two molecules as those defined in Equation (28) or, more general, from formula (26) by means of the use of the reduction (48). In this case the QSM can be referred to as the elements of the matrix $\mathbf{Z} = \{Z_{IJ}(\mathbf{W})\}$ and a Cosine-like Similarity Index between two molecules I and J is constructed as:

$$C_{IJ}(\mathbf{W}) = Z_{IJ}(\boldsymbol{i})(\mathbf{W})\ [Z_{II}(\mathbf{W})\ Z_{JJ}(\mathbf{W})]^{-1/2}. \tag{59}$$

A Distance Similarity Index, as is shown in reference [1a], can also be defined, then a Dissimilarity Index:

$$D_{IJ}(\mathbf{W}) = [Z_{II}(\mathbf{W}) + Z_{JJ}(\mathbf{W}) - (Z_{IJ}(\mathbf{W}) + Z_{JI}(\mathbf{W}))]^{1/2}, \tag{60}$$

with an euclidean distance-like form can be easily computed.

Both previous definitions as they appear in Equations (59) and (60), correspond to the oldest QSI described in this context: our originally described cosine and euclidean distance indices respectively [1a]. But other alternative choices can be considered. Once one of the entities has been computed, the other can be obtained by means of a new transformation over the first values. In general, a distance-like index A_{IJ} can be defined from a cosine-like of Equation (59) in the following way [1i]:

$$A_{IJ}(\mathbf{W}) = K\cos^{-1}(C_{IJ}(\mathbf{W})), \tag{61}$$

where K is a scale factor. Also, a cosine-like index M_{IJ} can be defined from the set of distance-like indices (60) [1i]:

$$M_{IJ}(\mathbf{W}) = 1 - D_{IJ}(\mathbf{W})\ [\max(\{D_{IJ}(\mathbf{W})\})]^{-1}. \tag{62}$$

More ways are open. It is possible to transmute a distance-like index into a correlation-like one, N_{IJ}, using the hyperbolic transformation:

$$N_{IJ}(\mathbf{W}) = 1 - \tanh(D_{IJ}(\mathbf{W})), \tag{63}$$

or to change a correlation-like index into a distance-like one, B_{IJ}, as:

$$B_{IJ}(\mathbf{W}) = K\coth(C_{IJ}(\mathbf{W})), \tag{64}$$

where K is a scale factor.

Other QSI have been defined in the literature [21] but they are not so easy to interpret as those defined here. For example, the Hodgkin-Richards [21] or Tanimoto [14] indices can be cast into a unique formula:

$$V_{IJ}(\mathbf{W}) = (2 - k)Z_{IJ}(\mathbf{W})\ [Z_{II}(\mathbf{W}) + Z_{JJ}(\mathbf{W}) - kZ_{IJ}(\mathbf{W})]^{-1}, \quad k \in [0, 2), \tag{65}$$

where with $k = 0$ the Hodgkin-Richards index is reproduced and when $k = 1$ the Tanimoto index is obtained. In Equation (65) the Z terms can be taken as QSM or reduced hypermatrix elements as those obtained by means of Equation (48).

5.4. Ordering molecular sets: Mendeleev's conjecture

Steps below justify the use of QSM or QSI as a tool to order a quantum molecular set. The QSM or QSI values can be ordered and this order may be transferred into the compared quantum molecular systems using the following algorithm [1e]:

a) Given a set of molecular structures $\boldsymbol{S} = \{S_i\}$,

b) One can compute a set of DT $\boldsymbol{D} = \{D_i\}$, associated to $\boldsymbol{S}$ in a one-to-one correspondence:

$$\boldsymbol{S} \leftrightarrow \boldsymbol{D} \Rightarrow S_i \Leftrightarrow D_i;\ \forall_i, \tag{66}$$

c) One can construct a QSM hypermatrix or a QSI matrix upon the set $\boldsymbol{D}$ as those defined in Sections 4 and 5.3 respectively.

d) Similarity hypermatrix or index matrix elements can be used to order $\boldsymbol{D}$ by means of a chosen rule R:

$$R[\boldsymbol{D}] = \{D_1 \preceq D_2 \preceq \ldots \preceq D_n\}, \tag{67}$$

where the symbol $\preceq$ indicates some ordering convention, for example, $D_i \preceq D_j$ signifies: D_j is the nearest neighbor of D_i.

e) This ordering in $\boldsymbol{D}$ can be used to induce an order in $\boldsymbol{S}$, using b):

$$R[\boldsymbol{D}] \Rightarrow R[\boldsymbol{S}] = \{S_1 \preceq S_2 \preceq \ldots \preceq S_n\}, \tag{68}$$

f) Suppose known a new element s of $\boldsymbol{S}$ whose density function transform d is also known, all QSM $\{Z_d(\boldsymbol{i})\}$ or QSI involving d can be computed. The rule R can act on the set $\boldsymbol{D} \cup \{\mathrm{d}\}$ giving:

$$R[\boldsymbol{D} \cup \{d\}] = \{D_1 \preceq D_2 \preceq \ldots \preceq D_k \preceq d \preceq D_{k+1} \preceq D_{k+2} \preceq \ldots \preceq D_n\} \tag{69}$$

thus, a place in the expanded set can be obtained for d.

g) Taking into account b), this will induce an order in the extended system set $\boldsymbol{S} \cup \{s\}$ such as:

$$R[\boldsymbol{S} \cup \{s\}] = \{S_1 \preceq S_2 \preceq \ldots \preceq S_k \preceq s \preceq S_{k+1} \preceq S_{k+2} \preceq \ldots \preceq S_n\} \tag{70}$$

h) Suppose now a new set $\boldsymbol{P} = \{P_i\}$ is known, built up with some values of some property attached in a one-to-one correspondence to the elements of set $\boldsymbol{S}$:

$$\boldsymbol{S} \leftrightarrow \boldsymbol{P} \Rightarrow S_i \Leftrightarrow P_i;\ \forall_i. \tag{71}$$

As a consequence of steps a)–g) above one can state the *Mendeleev*

Conjecture as follows: "The unknown value p of the property attached to the new structure s may be found somewhere within the order of the property set. For using the rule R one may expect:

$$R[\mathbf{P}] = \{P_1 \preceq P_2 \preceq \ldots \preceq P_n\}, \tag{72}$$

Then applying the rule R over the augmented property set:

$$R[\mathbf{P} \cup \{p\}] = \{P_1 \preceq P_2 \preceq \ldots \preceq P_k \preceq p \preceq P_{k+1} \preceq P_{k+2} \preceq \ldots \preceq P_n\} \tag{73}$$

In this way, an estimation of the property value p, can be made."

6. Visualization techniques

Once a set of Quantum Systems, usually made by molecules, is described by means of QSM or QSI matrices, the problem arises of how to extract the information contained in the set. Throughout the calculation of principal components of these matrices, the Quantum Systems can be studied as n-dimensional vectors and can be referred to as Quantum Objects.

It is a common practice to visualize the object set [1g, 1j] in order to achieve this goal. An optimal visualization is searched in order to construct clusters allowing us to classify the whole object set into subsets sharing a common range of a given property value. When dealing with a molecular set, the information which can be extracted form the object set and is related to the structure-activity or structure-property relationships, may become particularly interesting and useful to solve molecular engineering problems.

6.1. Point molecules and molecular point clouds

The matrix representation of a molecular set can be associated with a set of finite dimensional vectors representing the molecules. This point of view leads to the concept of *Point-Molecules* collected as a *Molecular Point Cloud*. If the set contains m molecules, the problem can be projected into an m-dimensional space. Of course, when one is dealing with hypermatrices, the m-dimensionality can be achieved by using the reduction (48). However, this m-dimensional projection is not compulsive.

Several representations can be obtained since the matrix of the object coordinates may come from the eigenvectors of QSM or QSI matrices of any kind.

6.2. Projection techniques. Molecular cloud spread

The following considerations clarify the steps to be done in order to extract the information of an Object Set or, in particular, of a Molecular Point Cloud.

a) *Mendeleev Second Postulate* can be redefined by saying that there exists an *optimal* finite dimensional *projection* of a given Object Set.
b) The term *optimal projection* can have several meanings. For example: 1) maximizes the distances between the points representing the elements of the set; 2) the order of the objects fits in the best way with respect to the order of some property; 3) some object clusters are obtained following some given rules; 4) a particular tree or graph is reproduced. Under this context, several classification procedures can be applied as clustering techniques or, in general, using algorithms related to pattern recognition [14, 15].
c) Principal components of the similarity matrix $\boldsymbol{Z}$, collecting QSM or QSI, are a convenient choice to represent the Objects or the Molecular Point Cloud. Usually, the operator $\boldsymbol{W}$ appearing in the general definition (26), is a n-particle *positive definite operator*. This means that the matrix $\boldsymbol{Z}$ is definite positive if it is obtained directly from QSM. The eigenvectors of $\boldsymbol{Z}$:

$$\boldsymbol{Z}\,\boldsymbol{u}_i = z_i\,\boldsymbol{u}_i \tag{74}$$

form a set $\boldsymbol{U} = \{\boldsymbol{u}_i\}$ which can be used as a finite dimensional representation of the DT set $\boldsymbol{D}$, hence of the system of molecular structures set $\boldsymbol{S}$ as outlined in Section 5.4. According to a well known variational principle, the eigenvectors of the definite positive matrix representing the set are the optimal space directions where the projections of the objects can be made. This kind of projection ensures the minimal loss of information from the original ∞-dimensional set representation where the $\boldsymbol{D}$ set belongs. Choosing representations in m eigenvectors, a m-dimensional space projection is obtained if it is necessary. Successive dimension reductions can be made by various means, see [1e] for example. Using spatial multidimensional rotations, the final visualization can take care of the multiple aspects present in a m-dimensional representation. The description of the set order can be performed by means of the existence of its corresponding optimal space projection.
d) Any set of objects can be made *homogeneous* through the computation of the similarity matrix obtained from the QSM or QSI and their principal components.
e) Once all the previous steps have been done, one can consider that the set has been properly described. Our definition of set description lean upon the idea of *set homogenization*. This can be performed by means of the set representation in the eigenvector space. In this context, we can call the eigenvectors the *homogeneous object vector set* and the object set coordinates the *homogeneous object coordinates*.

In fact, despite the existence of the *Third Mendeleev Postulate* which warrants the existence of the set ordering and the molecular representation in a finite dimensional space, there is no unique way of extracting information contained in the Molecular Point Cloud. There exists a great deal of possibilities, not only concerned with the personal preferences of the chemist, but also with the kind of information one wants to obtain in order to dilucidate the visualization of a concrete chemical problem. This is not only related to the final chosen point of view, but also to the initial DT and operator used into the definition of the QSM, this fact gives many degrees of freedom to the method we are describing here. Moreover, the freedom in the definition of a QSI is another factor to be taken into account. With all these considerations in mind, another way of study related to the Quantum Similarity and their applications pops up: search for the best QSM and QSI definitions as well as the optimal visualization procedure used to analyze a concrete chemical problem, possibly related to the study of a given molecular set, in order to obtain a concrete physicochemical or biological relationship.

6.3. Graphs and trees. Automatic representation

Internal relationships between elements of an Object Set or, in particular, the Molecular Point Clouds can also be obtained by means of the construction of Graphs and Trees. Besides using classical techniques [16] in our Laboratory we are interested in developing general drawing procedures. For example, from a set of m objects, it is possible to construct a d-dimensional polyhedron, usually a hypercube or a hyperpolyhedron, a complex. All or some of the d-dimensional polyhedron vertices represents one of the object set elements. The application of some algorithm [1c, 1d] leading to the genesis of several connections between vertices, which manifests the internal relationship connections between objects, permits us to automatically construct a graph or a tree.

The projection of the resultant figures into 2- or 3-dimensional spaces is a complementary task which enables us to catch the information in a human-like manner.

7. Conclusions

An overview on Quantum Similarity Theory and the possible application framework has been done justifying *Mendeleev Postulates* outlined at the beginning of this paper. Using an actualized form of density functions, the Density Transform concept, involving Nested Summation Symbols, a Generalized Quantum Similarity Measure and Quantum Similarity Indices have been defined. Its manipulation has been constructed following the idea of achieving the ordering of Quantum Object Sets describing known or predicted property values of its elements. Also, practical computational implementation clues and procedures have been built up.

Acknowledgments

This work has been financed by the CICYT-CIRIT, Fine Chemicals Programme of the "Generalitat de Catalunya" through a grant: #QFN91-4206. One of us (E.B.) benefits from a grant from the "Departament d'Ensenyament de la Generalitat de Catalunya". Some aspects of this work have been benefitted by many discussions and analysis performed with Profs. L. Vera and B. Calabuig. The authors deeply acknowledge them for their help.

A1. Mathematical tools

A1.1. Set theory

The interpretation of the QSM may lean upon *Zermelo's theorem* which states that every set $\boldsymbol{A}$ can be well-ordered [3]. A set $\boldsymbol{A}$ is well ordered when a binary relation $\preceq$ is defined in the set satisfying the following axioms:

a) $\forall x, y \in \boldsymbol{A}$, either $x \preceq y \vee y \preceq x$
b) if $x \preceq y \wedge y \preceq x \Rightarrow x = y$
c) if $x \preceq y \wedge y \preceq z \Rightarrow x \preceq z$
d) $\forall \boldsymbol{B} \subset \boldsymbol{A} \wedge \boldsymbol{B} \neq \varnothing, \exists x \in \boldsymbol{A} \mid x \preceq y \; \exists y \in \boldsymbol{B}$

A1.2. Measure and ordered set definitions

The notion of a Measure defined for a set arises in the theory of functions and it is related to the improvement of the notion of Integral. For a complete overview related to Measure and Ordered set definitions, the authors refer the readers to references [3, 19].

A2. Nested sums and nested do loops

Here will be briefly defined the NSS notation. More information about NSS and its applications can be found in references [5].

A2.1. Preliminary definitions

a) *Logical Vector*: Let us define a n-dimensional vector whose elements are logical expressions $\boldsymbol{L} = \{L_i\}_{i=1,n}$. A vector of this kind will be called a Logical Vector.
b) *Logical Kronecker Delta* (LKD) symbol: A LKD is a boolean function which has as argument a logical expression. It returns a value of 1 if the argument is a true expression, otherwise it returns a value of 0.
c) *Generalized Logical Kronecker Delta* (GLKD): it corresponds to a boolean symbol whose argument is a logical vector and it is defined as:

$$\Delta(\boldsymbol{L}) = \begin{cases} 1 & \text{if } L_i \text{ is true } \forall_i = 1, n. \\ 0 & \text{if } \exists_i | L_i \text{ is false.} \end{cases} \tag{75}$$

It must be noted that a GLKD symbol can be substituted by a unique LKD symbol, using a logical *and* ($\wedge$) operator n times over the elements of the logical vector $\boldsymbol{L}$:

$$\Delta(\boldsymbol{L}) = \prod_{i=1}^{n} \delta(L_i) = \delta\left(\bigwedge_{i=1}^{n} L_i \right). \tag{76}$$

The GLKD symbol returns the unit value iff all the elements of the logical vector are true. A zero value is generated in any other circumstance.

A2.2. Nested summation symbols: Definition and properties

Let us define an operator like:

$$\Sigma_n(\boldsymbol{j} = \boldsymbol{i}, \boldsymbol{f}, \boldsymbol{s}, \mathcal{L}(\boldsymbol{L})), \tag{77}$$

and let us call it a *Nested Summation Symbol*. The arguments in Equation (77) are vectors, except the last, which is a boolean function. Index n is an integer called the *Dimension of the NSS* and it signals the dimension of the involved index vectors $\boldsymbol{j}$, $\boldsymbol{i}$, $\boldsymbol{f}$ and $\boldsymbol{s}$. All these vectors can be named *Parameters of the NSS*.

The meaning of such a convention about NSS corresponds to accomplish all the sums involved in the generation of all the possible forms of *Index Vector* $\boldsymbol{j}$. The allowed forms of this vector arise from the fulfillment of the two following rules:

a) The elements of the vector $\boldsymbol{j}$ are defined by the limits:

$$\{i_k \leq j_k \leq f_k, \text{ if } s_k \geq 0\} \quad \text{or} \quad \{i_k \geq j_k \geq f_k, \text{ if } s_k \leq 0\}; \quad \forall k = 1, n, \tag{78}$$

where the j_k indices can be incremented or decremented respectively in steps of length s_k.

b) All the possible forms of vector $\boldsymbol{j}$ are attached to the boolean function $\mathcal{L}$, in such a way that, when developing the NSS, only those vectors j associated to a true value of $\mathcal{L}$ are taken into account.

So, the NSS stands for a set of summation symbols numbered from 1 to n and where the k-th summation symbol is attached to the corresponding k-th index of vectors $\boldsymbol{j}$, $\boldsymbol{i}$, $\boldsymbol{f}$ and $\boldsymbol{s}$.

NSS, beside permitting concise and compact formulae writing has the important advantage of being an operator which is easily implemented in a computational environment, see references [5] for more details and Section 3 in this Appendix for a brief description of the algorithm which implements the NSS.

Despite the form adopted here to write a NSS, sometimes it is superfluous to explicit all the involved parameters. When this circumstance does occur, some parts of the general form can be dropped in an arbitrary manner. This convention is used several times along the text.

Also, in this article, the following NSS property and notation has been used: the product of two NSS's of dimension n and m is another NSS of dimension $n + m$, or:

$$\begin{aligned} &\Sigma_n(\boldsymbol{j} = \boldsymbol{i}, \boldsymbol{f}, \boldsymbol{s}, \mathcal{L}(\boldsymbol{L}))\, \Sigma_n(\boldsymbol{j}' = \boldsymbol{i}', \boldsymbol{f}', \boldsymbol{s}', \mathcal{L}(\boldsymbol{L}')) = \\ &\qquad \Sigma_{n+m}(\boldsymbol{j} \oplus \boldsymbol{j}' = \boldsymbol{i} \oplus \boldsymbol{i}', \boldsymbol{f} \oplus \boldsymbol{f}', \boldsymbol{s} \oplus \boldsymbol{s}', \mathcal{L}(\boldsymbol{L} \oplus \boldsymbol{L}')), \end{aligned} \tag{79}$$

where the new index vectors are constructed by the direct sum of the original vectors appearing in the product (79).

A2.3. Computational algorithm. Parallelism

NSS constitutes a link between mathematical formalism and program implementation techniques, because successive generation of $\boldsymbol{j}$ index vector elements can be programmed in a general way under any high level language. This can be achieved using a unique *do* or *for* loop statement construct, which is *general and independent of the dimension of the involved nested sums*. This kind of programming structure can be called a *General Nested Do Loop* (GNDL).

In order to prove the previous claims, *Program 1* represents a simple Fortran source code, generating all the $\boldsymbol{j}$ index vectors appearing in a general NSS. This algorithm corresponds to the implementation of a GNDL structure which represents a NSS of the type $\Sigma_n(\boldsymbol{j} = \boldsymbol{i}, \boldsymbol{f}, \boldsymbol{s}, \mathcal{L}(\boldsymbol{L}))$. The step vector $\boldsymbol{s}$ can have positive or negative indices. The values of the parameters depend on the particular application one wants when running the algorithm. So, they are not explicitly specified and a question mark (?) stands for its values. The code takes into account the use of Logical Functions as, for example, a GLKD: it is assumed the presence of a Logical Function

```
      Parameter (n=?)      !   Dimension of the NSS
      Integer j(n),i(n),f(n),s(n)

*  < Initial parameter values >

      do k=1,n
         i(k)=?
         f(k)=?
         s(k)=?
         j(k)=i(k)
      end do

*  < GNDL procedure >

      k=n
      do while (k.gt.0)
         if ((j(k)-f(k))*s(k).gt.0) then
            j(k)=i(k)
            k=k-1
         else
            if (LFUN(n,j,i,f,s)) call Application(n,j)
            k=n
         end if
         if (k.gt.0) j(k)=j(k)+s(k)    ! Step
      end do
      END
```

Program 1. GNDL Fortran codification implementing the NSS: $\Sigma_n(\boldsymbol{j} = \boldsymbol{i}, \boldsymbol{f}, \boldsymbol{s}, \mathcal{L}(\boldsymbol{L}))$. See text for meaning.

called *LFUN* which has several arguments (the parameters of the NSS or others) and returns a true or false value depending on the definition of the logical expressions constituting the vector $\boldsymbol{L}$. A call to the *Application* routine is only performed when a true result is returned by the *LFUN* function. *Application* is a routine which implements the codification of the operand upon which the NSS operator acts.

A2.4. A USEFUL MATHEMATICAL EXPRESSION: EXPLICIT FORM OF THE DETERMINANT OF AN ARBITRARY SQUARE MATRIX

Using the NSS notation, one can rewrite the expression of an arbitrary $(n \times n)$ square matrix $\boldsymbol{A} = \{a_{ij}\}$ determinant, $\mathrm{Det}|\boldsymbol{A}|$, [17]. A compact formula of the determinant can be written in this way as:

$$\mathrm{Det}|\boldsymbol{A}| = \Sigma_n(\boldsymbol{j} = \mathbf{1}, n\mathbf{1}, \Delta(\boldsymbol{L}))\ S(\boldsymbol{j})\ \Pi(\boldsymbol{j}, \boldsymbol{A}), \tag{80}$$

where $\boldsymbol{L}$ is a $\binom{n}{2}$-dimensional logical vector defined as

$$\boldsymbol{L} = \{L_{pq} = [j_p \neq j_q];\quad p = 1, n-1;\quad q = p+1, n\}, \tag{81}$$

The $S(\boldsymbol{j})$ factor is a sign, which can be expressed by:

$$S(\boldsymbol{j}) = (-1)^{P(\boldsymbol{j})}, \tag{82}$$

where $P(\boldsymbol{j})$ is the parity connected to the ordinal structure of the vector $\boldsymbol{j}$ index elements. It can be computed using the following LKD expression:

$$P(\boldsymbol{j}) = \delta\left(2 \neq \sum_{p=1}^{n-1} \sum_{q=p+1}^{n} \delta(j_p > j_q)\right). \tag{83}$$

Finally, the $\Pi(j, A)$ terms are defined by means of the product:

$$\Pi(j, A) = \prod_{i=1}^{n} a_{i, j_i}. \tag{84}$$

In Section 2.3 this determinant form is used to deal with SD.

A3. Molecular quantum similarity associated integrals

A3.1. INTEGRALS OVER *n*S-STO FUNCTIONS

The authors have experience in computing QSM within the CNDO-Mulliken-like approach [1k]. In this case the integrals of the type (45) involve products between *n*S-STO functions. The following algorithm is proposed in order to compute the integrals:

a) When in Equation (45) there appears two or more *n*S STO functions sharing the same center, they can be multiplied and condensed into a unique S STO function of higher order.

b) Two center product functions as the distributions of the type:

$$\Omega_{AB} = \chi_A \chi_B, \tag{85}$$

where χ_A and χ_B are S STO functions, can be expanded following the procedures outlined in references [1k, 5a, 5b]. That is, a two center charge distribution can be expressed as a sum of two bilinear forms:

$$\Omega_{AB} = \Sigma_{\alpha \in A} \Sigma_{\alpha' \in A}\, a_{\alpha\alpha'}\, \Omega_{\alpha\alpha'} + \Sigma_{\beta \in B} \Sigma_{\beta' \in B}\, b_{\beta\beta'}\, \Omega_{\beta\beta'}. \tag{85}$$

where $\{a_{\alpha\alpha'}\}$ and $\{b_{\beta\beta'}\}$ are expansion coefficients. The terms $\{\Omega_{\alpha\alpha'}\}$ and $\{\Omega_{\beta\beta'}\}$ are one-center charge distributions.

In this manner, the multicenter integral problem reduces to a linear combinations of bicentric terms. As bicentric distributions are easily integrated, the entire problem can be considered solved.

A3.2. GENERAL INTEGRALS OVER GTO FUNCTIONS

As it was shown in previous Sections, the QSM integrals treated in the LCAO framework lead to the computation of linear combinations of multicenter integrals. Here it will be described how to explicit this kind of integrals when dealing with spinorbital functions, whose space part is made of GTO functions or linear combinations of a set of them. Also it will be seen how the use of the NSS notation provides us with a singular formulation which can be easily implemented.

In reference [5e] a General Gaussian Product Theorem is described, which allows us to obtain a convenient algorithm, in order to contract the exponential part of each of the GTO functions involved in the product. Also we provide an explicit form for the resulting angular cartesian part products. The final result gives the explicit form which adopts the product of GTO functions [5e]. Only the final results will be given here: let be known N GTO functions with scaling factors collected in the vector $\boldsymbol{a} = (\alpha_1, \alpha_2, \ldots, \alpha_N)$, angular quantum numbers $\{\boldsymbol{n}_i\}_{i=1,N}$ and centered at the points $\{\boldsymbol{A}_i\}_{i=1,N}$. The product of the whole gaussian function set can be expressed in terms of NSS notation as follows:

$$\prod_{i=1}^{N} \gamma(\boldsymbol{A}_i, \boldsymbol{n}_i, \alpha_i) = \kappa \Sigma_3 \left(\boldsymbol{j} = 0, \sum_{k=1}^{N} \boldsymbol{n}_k \right) f(\boldsymbol{j}, \boldsymbol{n}, \boldsymbol{A}, \boldsymbol{P})\, \gamma(\boldsymbol{P}, \boldsymbol{j}, \langle \boldsymbol{a} \rangle), \tag{87}$$

where $\boldsymbol{A}$ and $\boldsymbol{n}$ are $(N \times 3)$ rectangular matrices, whose rows, $\{\boldsymbol{A}_i\}_{i=1,N}$ and $\{\boldsymbol{n}_i\}_{i=1,N}$, are the coordinates of the centers and the quantum numbers of each GTO, respectively. The sum of the elements of a n-dimensional vector $\boldsymbol{a}$ is symbolized by $\langle \boldsymbol{a} \rangle$, that is:

$$\langle \boldsymbol{a} \rangle = \sum_{i=1}^{n} \alpha_i. \tag{88}$$

Equation (87) is the expression of the GGPT. The theorem generalizes the well known Two Gaussian Product Theorem (TGPT) [18]. In Equation (87), $\kappa = \exp(\langle \boldsymbol{K} \rangle)$ and the elements of vector $\boldsymbol{K}$ are defined as:

$$K_q = \langle \boldsymbol{a} \rangle^{-1} \sum_{i=1}^{N} \left\{ A_{iq}^2 \alpha_i (\langle \boldsymbol{a} \rangle - \alpha_i) - 2 \sum_{i<j=2}^{N} A_{iq} A_{jq} \alpha_i \alpha_j \right\}; \quad \forall q = 1, 3, \tag{89}$$

being the row vectors $\boldsymbol{A}_i = \{A_{iq}\}_{q=1,3}$. The point $\boldsymbol{P}$, as a row vector, is defined through the weighted mean:

$$\boldsymbol{P} = \langle \boldsymbol{a} \rangle^{-1} \sum_{i=1}^{N} \alpha_i \boldsymbol{A}_i. \tag{90}$$

The f coefficients can be defined as:

$$f(\boldsymbol{j}, \boldsymbol{n}, \boldsymbol{A}, \boldsymbol{P}) = \prod_{q=1}^{3} g(\boldsymbol{j}_q, [\boldsymbol{n}]_q, [\boldsymbol{d}]_q), \tag{91}$$

where $\boldsymbol{d}$ is a $(N \times 3)$ matrix build up by means of the difference:

$$\boldsymbol{d} = \boldsymbol{A} - \boldsymbol{Q}, \tag{92}$$

with the $(N \times 3)$ matrix $\boldsymbol{Q}$ constructed with all the rows equal to vector $\boldsymbol{P}$. The symbols $[\boldsymbol{n}]_q$ and $[\boldsymbol{d}]_q$ describe the q-th columns of the matrices $\boldsymbol{n}$ and $\boldsymbol{d}$.

The g coefficients can be computed, in general, using a NSS formalism as:

$$g(\boldsymbol{j}, \boldsymbol{u}, \boldsymbol{v}) = \Sigma_N(\boldsymbol{j} = \boldsymbol{0}, \boldsymbol{u}, \delta(j = \langle \boldsymbol{j} \rangle)) \prod_{i=1}^{N} v_i^{u_i - j_i} \binom{u_i}{j_i}, \tag{93}$$

where $\boldsymbol{u}$ and $\boldsymbol{v}$ are column vectors of dimension N, and $\langle \boldsymbol{j} \rangle$ stands for the sum of all the $\boldsymbol{j}$ vector elements.

With all these considerations in mind, one can write Equation (87) in a simplified form as:

$$\prod_{i=1}^{N} \gamma_i = \kappa \Sigma_3(\boldsymbol{j})\, f(\boldsymbol{j})\, \gamma(\boldsymbol{j}), \tag{94}$$

where the symbols γ are primitive gaussian functions and the $f(\boldsymbol{j})$ terms are factors.

So, an integral such as the one appearing in Equation (45), if each spin orbital is attached to a space GTO function, may be evaluated according to the following expression:

$$\begin{aligned} &\left\langle \prod_{p=1}^{n} \prod_{i=1}^{m_p} \gamma_i(p) \,|\, \boldsymbol{W}(1, 2, \ldots, n) \,|\, \prod_{q=1}^{n} \prod_{j=1}^{m_q} \gamma_j'(q) \right\rangle = \\ &K\, \Sigma_{3n}(\boldsymbol{k}_1 \oplus \boldsymbol{k}_2 \oplus \ldots \oplus \boldsymbol{k}_n)\, \Sigma_{3n}(\boldsymbol{l}_1 \oplus \boldsymbol{l}_2 \oplus \ldots \oplus \boldsymbol{l}_n) \left(\prod_{r=1}^{n} f(\boldsymbol{k}_r)\, f'(\boldsymbol{l}_r) \right) \\ &\left\langle \prod_{p=1}^{n} \gamma_p(\boldsymbol{k}_p)(p) \,|\, \boldsymbol{W}(1, 2, \ldots, n) \,|\, \prod_{q=1}^{n} \gamma_q'(\boldsymbol{l}_q)(q) \right\rangle, \end{aligned} \tag{95}$$

where K collects the product of the κ constants arising from Equation (87), $\boldsymbol{W}(1, 2, \ldots, n)$ is a n-electron operator and the products over p and q run over all the electrons. Each electron has associated a set of GTO which are considered by means of the products over the i and j indices. It has been shown, in this way which form takes the generalization of integrals involving an arbitrary number of electrons when dealing with GTO functions.

The extension of this formulation when the space part of each spinorbital is a linear combination of GTO functions is straightforward.

References

1. a) Carbó, R., Arnau, M., and Leyda, L., *Int. J. Quantum Chem.* **17**, 1185 (1980). b) Carbó, R. and Arnau, C., "Molecular Engineering: a general approach to QSAR", in *Medicinal Chemistry Advances*, eds. F. G. de las Heras and S. Vega (Pergamon Press, Oxford 1981), **85**. c) Carbó, R. and Domingo, Ll., *Int. J. Quantum. Chem.* **23**, 517 (1987). d) Carbó, R. and Calabuig, B., *Comp. Phys. Commun.* **55**, 117 (1989). e) Carbó, R. and Calabuig, B., "Molecular Similarity and Quantum Chemistry", chapter 6, p. 147, in reference [2k]. f) Carbó, R. and Calabuig, B., Proceedings del XIX Congresso Internazionale dei Chimici Teorici dei Paesi di Espressione Latina, Roma, Italy, September 10–14, 1990. *J. Mol. Struct. (Teochem)*, **254**, 517 (1992). g) Carbó, R. and Calabuig, B., *J. Chem. Inf. Comput. Sci.* **32**, 600 (1992). h) Carbó, R. and Calabuig, B., "Quantum Similarity", in *Structure, Interactions and Reactivity*, ed. S. Fraga (Elsevier Pub., Amsterdam 1992). i) Carbó, R. and Calabuig, B., *Int. J. Quantum Chem.* **42**, 1681 (1992). j) Carbó, R. and Calabuig, B., *Int. J. Quantum Chem.* **42**, 1695 (1992). k) Carbó, R., Calabuig, B., Besalú, E., and Martínez, A., *Molecular Engineering*, **2**, 43 (1992).
2. a) Cooper, D. L. and Allan, N. L., *J. Chem. Soc., Faraday Trans.* **83**, 449 (1987). b) Cooper, D. L. and Allan, N. L., *J. Computer-Aided Mol. Design.* **3**, 253 (1989). c) Cooper, D. L. and Allan, N. L., *J. Am. Chem. Soc.* **114**, 4773 (1992). d) Cioslowski, J. and Fleischmann, E. D., *J. Am. Chem. Soc.* **113**, 64 (1991). e) Cioslowski, J. and Challacombe, M., *Int. J. Quant. Chem.* **25**, 81 (1991). f) Ortiz, J. V. and Cioslowski, J., *Chem. Phys. Lett.* **185**, 270 (1991). g) Cioslowski, J. and Surján, P. R., *Jour. Mol. Struct. (Theochem)* **255**, 9 (1992). h) Ponec, R. and Strnad, M., *Collect. Czech. Chem. Commun.* **55**, 2583 (1990). i) Ponec, R. and Strnad, M., *J. Phys. Org. Chem.* **4**, 701 (1991). j) Ponec, R. and Strnad, M., *Int. J. Quantum Chem.* **42**, 501 (1992). k) Johnson, M. A. and Maggiora, G. (eds.), *Concepts and Applications of Molecular Similarity* (John Wiley & Sons Inc, New York 1990). l) Hodgkin, E. E. and Richards, W. G., *Int. J. Quant. Chem.* **14**, 105 (1987). m) Good, A. C., Hodgkin, E. E., and Richards, W. G., *J. Chem. Inf. Comput. Sci.* **32**, 188 (1991). n) Martín, M., Sanz, F., Campillo, M., Pardo, L., Pérez, J., and Turmo, J., *Int. J. Quant. Chem.* **23**, 1627 (1983). o) Martín, M., Sanz, F., Campillo, M., Pardo, L., Pérez, J., Turmo, J., and Aulló, J. M., *Int. J. Quantum Chem.* **23**, 1643 (1983). p) Sanz, F., Martín, M., Pérez, J., Turmo, J., Mitjana, A., and Moreno, V., in *Quantitative Approaches to Drug Design*, eds. J. C. Dearden (Elsevier, Amsterdam 1983). q) Sanz, F., Martín, M., Lapeña, F., and Manaut, F., *Quant. Struct. -Act. Relat.* **5**, 54 (1986). r) Sanz, F., Manaut, F., José, J., Segura, J., Carbó, M., and de la Torre, R., *Jour. Mol. Struct. (Theochem)* **170**, 171 (1988). s) Luque, F. J., Sanz, F., Illas, F., Pouplana, R., and Smeyers, Y. G., *Eur. J. Med. Chem.* **23**, 7 (1988).
3. *Encyclopaedia of Mathematics*, Kluwer Acad. Pub. Dordrecht (1990).
4. a) Löwdin, P. O., *Phys. Rev.* **97**, 1474 (1955). b) McWeeny, R., *Proc. Roy. Soc. A* **232**, 114 (1955). c) McWeeny, R., *Proc. Roy. Soc. A* **235**, 496 (1956). d) McWeeny, R., *Proc. Roy. Soc. A* **253**, 242 (1959).
5. a) Carbó, R. and Besalú, E., *Advances in Quantum Chemistry* **24**, 115 (1992). b) Carbó, R. and Besalú, E., *Can. J. Chem.* **70**(2), 353 (1992). c) Carbó, R. and Besalú, E., *J. Math. Chem.* (in press). d) Besalú, E. and Carbó, R., *J. Math. Chem.* (submitted). e) Carbó, R. and Besalú, E., *Int. J. Quant. Chem.* (submitted).
6. Parr, R. G., *The Quantum Theory of Molecular Electronic Structure* (W. A. Benjamin Inc., New York 1963).
7. Messiah, A., *Mechanique Quantique*, Vol I (E. Dunod., Paris 1959).
8. Zemanian, A. H., *Generalized Integral Transformations* (Dover Pub. Inc., New York 1987).
9. a) Carbó, R., Suñé, E., Lapeña, F., and Pérez, J., *Journal of Biological Physics* **14**, 21 (1986). b) Carbó, R., Lapeña, F., and Suñé, E., *Afinidad* **43**, 483 (1986).
10. a) Nakayama, A. and Richards, W. G., *Quant. Struct. Act. Relat.* **6**, 153 (1987). b) Burt, C., Richards, W. G., and Huxley, Ph., *Jour. Comput. Chem.* **11**, 1139 (1990). c) Good, A. C., So, S., and Richards, W. G., *J. Med. Chem.* **36**, 433 (1993).

11. Gill, P. M. W., Johnson, B. G., Pople, J. A., and Taylor, S. W., *J. Chem. Phys.* **96**, 7178 (1992).
12. Bader, R. F. W., *Atoms in Molecules*, International series of monographs on chemistry 22 (Oxford Science Pub., Clarendon Press, Oxford 1990).
13. a) Wilkinson, J. H. and Reinsch, C., *Linear Algebra* (Springer-Verlag, Berlin 1971). b) Carbó, R. and Domingo, Ll., *Algebra Matricial y Lineal* (McGraw-Hill, Madrid 1987).
14. Tou, J. T. and González, R. C., *Pattern Recognition Principles* (Addison-Wesley Reading, MA 1974).
15. Hardigan, J. A., *Clustering Algorithms* (Wiley, New York 1975).
16. a) Sneath, P. H. A. and Sokal, R. R., *Numerical Taxonomy* (W. H. Freeman, San Francisco 1973). b) Jardine, N. and Sibson, R., *Mathematical Taxonomy* (Wiley, New York 1977).
17. Ayres, F., *Theory and Problems of Matrices* (Schaum Pub., New York 1962).
18. a) Shavitt, I., *Methods in Computational Physics*, **2**, 1 (1963). b) Taketa, H., Huzinaga, S., and O-ohata, K., *J. Phys. Soc.* **11**, 2313 Japan (1966). c) Saunders, V. R., "An introduction to molecular integral evaluation", in *Computational Techniques in Quantum Chemistry and Molecular Physics* **15**, 347. Series C, Mathematical and Physical Sciences, eds. G. H. F. Diercksen, B. T. Sutcliffe, and A. Veillard (Reidel Pub., Dordrecht 1975).
19. a) Kolmogorov, A. N. and Fomin, S. V., *Elementos de la Teoria de Funciones y del Análisis Funcional*, Spanish version, Ed. (Mir, Moscow 1975). b) Phillips, E. R., *An Introduction to Analysis and Integration Theory* (Dover Pub. Inc., New York 1984). c) Taylor, A. E., *General Theory of Functions and Integration* (Dover. Pub. Inc., New York 1985).

Molecular Similarity and Momentum Space

DAVID L. COOPER[1] and NEIL L. ALLAN[2]
[1] *Department of Chemistry, University of Liverpool, P.O. Box 147, Liverpool L69 3BX, U.K.;*
[2] *School of Chemistry, University of Bristol, Cantocks Close, Bristol BS8 1TS, U.K.*

Abstract. This paper describes a recent development in the field of quantum similarity, in which quantitative estimates of molecular similarity are obtained from numerical comparisons of *momentum-space* electron densities. This approach places particular emphasis on the long-range slowly-varying valence electron density in position space. In addition, many of the problems associated with the more usual position-space procedures do not arise. The momentum-space approach is particularly suited to situations in which features of the long-range valence electron density are more important than the bonding topology.

We discuss the calculation of momentum-space electron densities, $\rho(\boldsymbol{p})$, from SCF wavefunctions and illustrate the form of $\rho(\boldsymbol{p})$ for different molecules taking the ground state of H_2 and the π-orbitals in large conjugated polyenes as examples. The construction and numerical evaluation of similarity and dissimilarity indices, and the relationships between them, are described in some detail.

It is possible in our approach to compare the total or valence electron densities of two molecules, or the electron densities associated with particular molecular fragments or individual (localised) molecular orbitals. Examples of all of these cases are presented. Results are reported for two model series: (i) $CH_3CH_2CH_3$, CH_3OCH_3 and CH_3SCH_3 (ii) C—F and C—H bonds in fluoromethanes. Two recent applications to larger systems are outlined. In the first, momentum-space similarity indices are used to rationalise anti-HIV virology data for a group of phospholipids. It is found that the technique has predictive value for such systems. The second application relates to a structure-activity relationship involving the hyperpolarisabilities of a range of 1,4-benzene derivatives of interest in non-linear optics.

1. Introduction

There are innumerable ways of addressing the question "How similar are two molecules?". A wide range of methodologies have been presented at this meeting, relating to an equally diverse selection of applications. In addition to graph theoretical methods and database searching [1], quantitative approaches include the comparison of position-space electron densities [2, 3], electrostatic potentials [4], and the topological analysis of the three-dimensional shapes of charge densities [5].

It has been well known since the early days of quantum mechanics that the electron density can be expressed as a function of the *momenta* of the electrons rather than as a function of their positions. This idea is frequently encountered in condensed-matter physics, and it is also very convenient for the analysis of the results of the (e, 2e) scattering experiment [6]. Despite some early work mainly by Coulson and Duncanson [7], momentum-space concepts are used much less frequently in chemistry [8]. We have suggested that momentum-space concepts provide a novel route to the quantitative estima-

R. Carbó (ed.), Molecular Similarity and Reactivity: From Quantum Chemical to Phenomenological Approaches, 31–55.

tion of molecular similarity [9–11]. Our approach involves a comparison of the *momentum-space* (p-space) electron densities, $\rho(\boldsymbol{p})$, of the molecules, orbitals, or molecular fragments of interest.

The structure of this article is as follows. The next section is concerned with the forms of the p-space wavefunction and p-space electron density for some representative systems, and includes a discussion of chemical bonding in momentum space. Technical aspects which relate to the calculation of $\rho(\boldsymbol{p})$ are also given. Section 3 deals with the construction and computation of families of similarity and dissimilarity indices. Results for the model series CH_3OCH_3, CH_3SCH_3 and $CH_3CH_2CH_3$, using *ab initio* and semi-empirical wavefunctions, are presented. Three contrasting applications are then summarised. The first of these relates to C—F and C—H bonds in fluoromethanes, the second to the similarity of anti-HIV phospholipids, and the third to the hyperpolarisabilities of a series of benzene derivatives. These involve the comparison, in turn, of localised orbitals in *different* molecules, of *total* densities, and of orbitals in the *same* molecule. The last section contains some brief remarks concerning the general applicability of our method and the prospects for future work.

2. Momentum-space electron densities

2a. Theoretical aspects

It is no harder to calculate the momentum-space electron density $\rho(\boldsymbol{p})$ than its position space counterpart $\rho(\mathbf{r})$. The wavefunction in momentum space, $\Psi(\boldsymbol{p})$, is the Fourier transform of the position-space wavefunction $\psi(\boldsymbol{r})$,

$$\Psi(\boldsymbol{p}) = (2\pi)^{-3/2} \int \psi(\boldsymbol{r}) \exp(-i\boldsymbol{p}\cdot\boldsymbol{r})\, d\boldsymbol{r}. \tag{2.1}$$

Our starting point is generally the molecular orbitals (MOs) $\psi_k(\boldsymbol{r})$ which are formed in the usual way by the overlap of atomic basis functions $\phi_\alpha(\mathbf{r})$ centred on the nuclei α which are at positions $\boldsymbol{R}_\alpha$:

$$\psi_k(\boldsymbol{r}) = \sum_\alpha c_{\alpha k}\phi_\alpha(\boldsymbol{r} - \boldsymbol{R}_\alpha). \tag{2.2}$$

The coefficients $c_{\alpha k}$ are obtained from self-consistent field (SCF) wavefunctions for the systems of interest. In the work described here, *ab initio* and semi-empirical wavefunctions are generated using the GAMESS [12] and MOPAC [13] programs respectively.

The corresponding momentum-space MO wavefunction $\Psi_k(\boldsymbol{p})$ is, from Equation (2.1),

$$\Psi_k(\boldsymbol{p}) = \sum_\alpha c_{\alpha k}\Phi_\alpha(\boldsymbol{p}) \exp(-i\boldsymbol{p}\cdot\boldsymbol{R}_\alpha), \tag{2.3}$$

where the $\Phi_\alpha(\boldsymbol{p})$ are the Fourier transforms of the $\phi_\alpha(\boldsymbol{r})$,

$$\Phi_\alpha(\boldsymbol{p}) = (2\pi)^{-3/2} \int \phi_\alpha(\boldsymbol{r}) \exp(-i\boldsymbol{p}\cdot\boldsymbol{r})\, d\boldsymbol{r}. \tag{2.4}$$

Analytic forms for many of the $\Phi_\alpha(\boldsymbol{p})$ have been presented by Kaijser and Smith [14]. Table I collects together those implemented for the applications discussed in this paper, listing the $\Phi_\alpha(\boldsymbol{p})$ corresponding to r-space Slater orbitals, and Gaussian orbitals in both spherical harmonic and cartesian form.

The momentum-space electron density $\rho(\boldsymbol{p})$ is defined analogously to its position-space counterpart $\rho(\boldsymbol{r})$,

$$\rho(\boldsymbol{p}) = \sum_{i,j} D(i|j)\Psi_i(\boldsymbol{p})\ \Psi_j(\boldsymbol{p}). \tag{2.5}$$

$D(i|j)$ is the normalised one-particle density matrix. In the special case of SCF molecular orbitals, $D(i|j) = n_i\delta_{ij}$ where n_i is the orbital occupancy (1 or 2).

There are two simple tests to show that $\rho(\boldsymbol{p})$ is evaluated correctly. The first is the normalisation condition that $\rho(\boldsymbol{p})$, like $\rho(\boldsymbol{r})$, must satisfy:

$$\int \rho(\boldsymbol{p})\, d\boldsymbol{p} = \mathrm{N}, \tag{2.6}$$

where N is the total number of electrons. The second is that the expectation value of the kinetic energy, $\langle \mathrm{T} \rangle$, is given by

$$\langle \mathrm{T} \rangle = 0.5 \int p^2 \rho(\boldsymbol{p})\, d\boldsymbol{p}, \tag{2.7}$$

where $p = |\boldsymbol{p}|$, and, of course, for any molecule at its equilibrium geometry, the virial theorem requires that

$$\langle \mathrm{T} \rangle = -E, \tag{2.8}$$

where E is the total energy of the molecule.

It is important to note that the Fourier transform preserves direction so that it is possible to refer to components of the total momentum in any particular direction – for example, one can distinguish components along a Cartesian axis, or parallel/perpendicular to a bond or plane.

The momentum density $\rho(\boldsymbol{p})$ possesses the same symmetry elements as its r-space counterpart. In addition, in the absence of net translational motion, $\psi(\mathbf{r})$ must be real and $\Psi(\boldsymbol{p})$ and $\Psi(-\boldsymbol{p})$ have the same modulus so that

$$\rho(\boldsymbol{p}) = \rho(-\boldsymbol{p}). \tag{2.9}$$

Consequently the momentum density of *any* molecule will have inversion symmetry even if $\rho(\boldsymbol{r})$ does not. It is possible to take advantage of this inversion symmetry when integrating functions of $\rho(\boldsymbol{p})$, such as the generalised overlaps to be described in Section 3.

It is clear from Table I that the momentum density $\rho(\boldsymbol{p})$ falls off very rapidly with p. It is dominated by low values of p which correspond to the slowly-varying outer valence r-space density. In r-space, on the other hand, the form of the electron density is determined largely by the core electrons and so by the positions of the nuclei, particularly the heavier ones (cf. the problem of determining H atom positions in X-ray diffraction). In marked contrast to the position-space density, the momentum density highlights some of the most chemically interesting parts of the electron distribution, without dwelling on the bonding topology.

TABLE I

Analytic forms of normalized p-space atomic basis functions. In each case α is the r-space orbital exponent. Only unique functions are given for the cartesian gaussians

		Spherical Slaters		
Common factor	n	$l = 0$		$l = 1$
$(2\alpha/\pi)^{1/2}(\alpha^2 + p^2)^{-2}$	1	$4\alpha^2 Y_{00}$		–
$(2\alpha/3\pi)^{1/2}(\alpha^2 + p^2)^{-3}$	2	$4\alpha^2(3\alpha^2 - p^2)Y_{00}$		$-i16p\alpha^3 Y_{1m}$
$(\alpha/5\pi)^{1/2}(\alpha^2 + p^2)^{-4}$	3	$32\alpha^4(\alpha^2 - p^2)Y_{00}$		$-i(32/3)p\alpha^3(5\alpha^2 - p^2)Y_{1m}$
$(2\alpha/35\pi)^{1/2}(\alpha^2 + p^2)^{-5}$	4	$16\alpha^4(p^4 - 10p^2\alpha^2 + 5\alpha^4)Y_{00}$		$-i32p\alpha^5(5\alpha^2 - 3p^2)Y_{1m}$
		Cartesian gaussians		
Common factor	1s	$2p_x$	$3d_{xx}$	$3d_{xy}$
$(2\alpha\pi)^{3/4}e^{-p^2/4\alpha}$	1	$-ip_x/\sqrt{\alpha}$	$-\frac{1}{\sqrt{3}}\left(\frac{p_x^2}{\alpha} - 2\right)$	$\frac{-p_x p_y}{\alpha}$
		Spherical gaussians		
Common factor	1s	2p	3d	4f
$(2\alpha/\pi)^{1/4}e^{-p^2/4\alpha}$	$\frac{1}{\alpha} Y_{00}$	$\frac{-ip}{\alpha\sqrt{3\alpha}} Y_{1m}$	$\frac{-p^2}{\sqrt{15\alpha^2}} Y_{2m}$	$\frac{ip^3}{\alpha^2\sqrt{105\alpha}} Y_{3m}$

2b. Examples

We illustrate momentum-space concepts by considering two relatively familiar systems – the hydrogen molecule and long, conjugated polyenes.

(I) *Hydrogen*

The $1\sigma_g$ molecular orbital is formed from the overlap of atomic 1s functions centred on the two nuclei A and B. If the position vectors of the two nuclei are $\boldsymbol{R}_A$ and $\boldsymbol{R}_B$ respectively, then, in r-space,

$$\psi(\boldsymbol{r}) = \frac{1}{\sqrt{2(1 + S)}} (\phi_{1s}(\boldsymbol{r} - \boldsymbol{R}_A) + \phi_{1s}(\boldsymbol{r} - \boldsymbol{R}_B)), \tag{2.10}$$

where S is the usual overlap integral. Using Equation (2.1),

$$\Psi(\boldsymbol{p}) = \frac{1}{\sqrt{2(1 + S)}} \Phi_{1s}(\boldsymbol{p}) (\exp(-i\boldsymbol{p} \cdot \boldsymbol{R}_A) + \exp(-i\boldsymbol{p} \cdot \boldsymbol{R}_B)), \tag{2.11}$$

where $\Phi_{1s}(\boldsymbol{p})$, a real function, is the Fourier transform of $\phi_{1s}(\boldsymbol{r})$.

The momentum density $\rho(\boldsymbol{p})$ follows immediately from Equation (2.5),

$$\begin{aligned}\rho(\boldsymbol{p}) &= \Psi^*(\boldsymbol{p})\Psi(\boldsymbol{p}) \\ &= \frac{\Phi_{1s}^2(\boldsymbol{p})}{1+S}\,(1+\cos(\boldsymbol{p}\cdot\boldsymbol{R}_{AB})),\end{aligned} \tag{2.12}$$

and $\boldsymbol{R}_{AB}(=\boldsymbol{R}_A—\boldsymbol{R}_B)$ is origin-independent.

Substituting for $\Phi_{1s}(\boldsymbol{p})$ from Table I, we obtain

$$\rho(\boldsymbol{p}) = \frac{32\alpha^5}{(1+S)\pi(\alpha^5+p^2)^4}\,(1+\cos(\boldsymbol{p}\cdot\boldsymbol{R}_{AB})). \tag{2.13}$$

It is clear from this last equation that the momentum density falls off rapidly with p (here as p^8 at large p). For this orbital there is a maximum in $\rho(\boldsymbol{p})$ at $\boldsymbol{p} = 0$. This is also evident in Figure 1b) which shows an isometric view of $\rho(\boldsymbol{p})$ in the plane $p_x = 0$. Figure 1a) is the analogous r-space plot in the plane $x = 0$. In these figures, the bond direction is the z-axis. In fact, Figure 1 was generated using a much more elaborate basis set (TZVP) than that assumed in our simple treatment, but all the features we discuss are common to the densities derived from both calculations.

The cosine term in Equation (2.13) introduces oscillatory structure into the momentum density. Using the orientation of the axes defined in Figure 1,

$$\rho(\boldsymbol{p}) = \frac{2}{(1+S)}\,\Phi_{1s}^2(\boldsymbol{p})\cos^2\left(\frac{p_z R_{AB}}{2}\right), \tag{2.14}$$

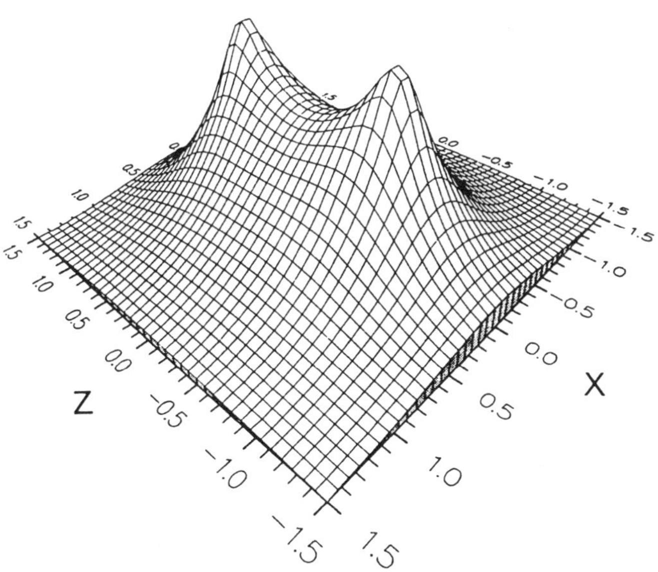

Figure 1a. Isometric view of the position-space electron density in the plane $y = 0$ for H_2. The internuclear axis is the z-axis. The view extends to ±1.5 atomic units in both x- and z-directions. The coordinates of the nuclei are (0, 0, ±0.7).

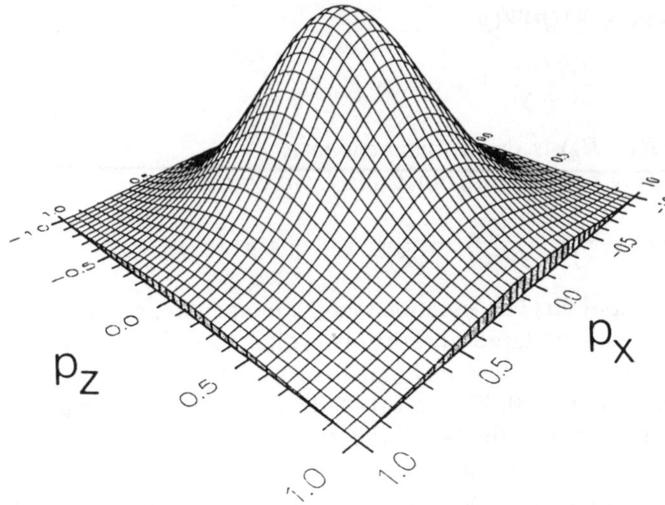

Figure 1b. Isometric view of the momentum-space electron density for the hydrogen molecule in the plane $p_y = 0$. The origin ($\boldsymbol{p} = 0$) is at the centre of the plot. The view extends to ±1.0 atomic units in both the p_x and p_z directions.

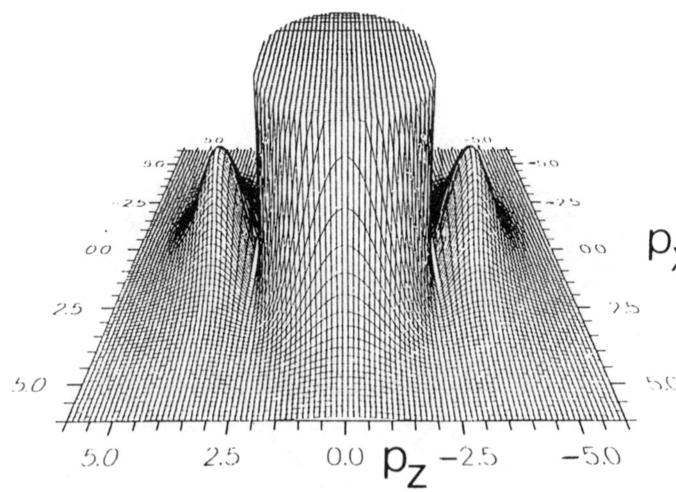

Figure 1c. Isometric view of the momentum-space electron density for H_2 in the plane $p_y = 0$. The origin ($\boldsymbol{p} = 0$) is at the centre of the plot. The view extends to ±6.0 atomic units in the p_x and p_z directions, but the peak at the origin has been omitted in order to highlight the nodal planes in $\rho(\boldsymbol{p})$ (Equation (2.14)) at $p_z = \pm\pi/R_{AB}$ (≈ ±2.2 atomic units).

and so $\rho(\boldsymbol{p}) = 0$ whenever $p_z R_{AB} = (2n + 1)\pi$ where n can take any integral value. There is a series of *nodal planes* perpendicular to the p_z axis. Those corresponding to $n = 0$ can be seen at $p_z = \pm\pi/R_{AB}$ ($\approx \pm 2.2$ atomic units) in Figure 1c) where the momentum density is plotted in the same plane as in Figure 1b) but out to larger values of $\boldsymbol{p}$. These "diffraction" effects, which have period $2\pi/R_{AB}$, contain the information about the molecular geometry. The effect of the cosine term is to change the momentum density distribution from spherical to ovaloid on bond formation.

(II) *Conjugated Polyenes* $H_2C{=}CH{-}(CH = CH)_M{-}CH{=}CH_2$

Turning now to larger molecules, we have also investigated the form of the momentum-space density for large polyenes [15]. Figure 2a) shows contour plots of the position-space, and Figure 2b) isometric views of the p-space electron densities for the ten occupied π-orbitals of the regular *trans*-polyene containing 20 carbon atoms ($C_{20}H_{22}$). The orientation of the axes is such that the x-axis points along the direction of the chain, and the z-axis is perpendicular to the plane of the molecule as shown in Figure 2a). These were calculated from SCF wavefunctions (3–21G basis set) generated using the GAMESS program [12] (r(C—C) = 1.39Å, r(C—H) = 1.09Å, all bond angles 120°). The r-space densities, plotted in the plane $z = 1$ atomic unit, show the characteristic nodal structure as expected. With decreasing binding energy, each successive π-orbital possesses one further nodal plane. The π-orbital with the largest binding energy has no nodes and, in this case, that with the lowest binding energy has nine.

The p-space plots in Figure 2b) are very different in appearance from the r-space plots and from the p-space plot for hydrogen given in Figure 1b). The momentum-space densities are plotted in the plane $p_z = 1$ atomic unit, with the *centre* of each view corresponding to $p_x = p_y = 0$. These plots show sharp wedges which, for each orbital, peak sharply at one particular value of $|p_x|$ i.e., at one value of the component of momentum *along* the chain direction. Only the π-orbital with the highest binding energy has non-zero momentum density at $p_x = 0$, and the wedges are furthest from the origin for the π-orbital with the lowest binding energy.

It is straightforward to rationalise the appearance of these plots with Hückel theory. In this simple approach, the molecular orbitals, $\psi_k(\boldsymbol{r})$, for a chain containing n-carbon atoms are

$$\psi_k(\boldsymbol{r}) = \frac{1}{\sqrt{n}} \sum_{j=1}^{n} \exp\left(\frac{2\pi ijk}{n}\right) \phi_j(\boldsymbol{r}), \tag{2.15}$$

where $\phi_j(\boldsymbol{r})$ is a normalised $2p_z$ atomic orbital on carbon atom j. To find the corresponding p-space density, $\rho_k(\boldsymbol{p})$, we use Equations (2.1) and (2.5):

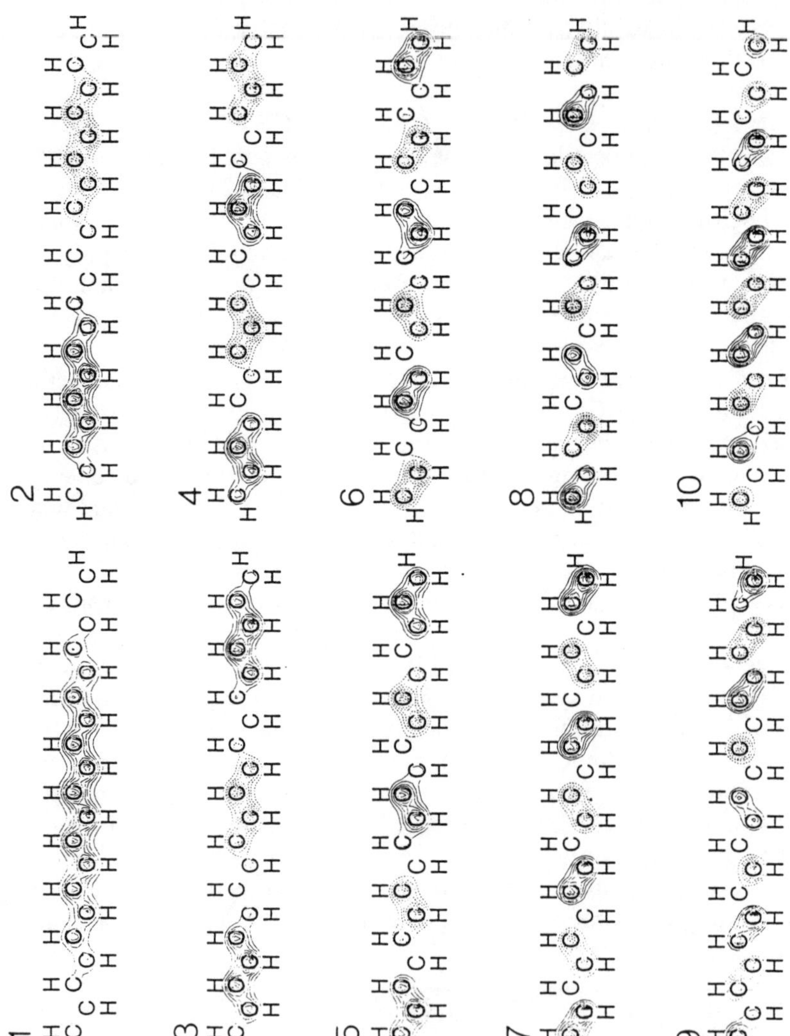

Figure 2a. Contour plots (in the plane $z = 1$ bohr) of the position-space electron density for the occupied π-orbitals for trans-$C_{20}H_{22}$. Full and broken contours denote regions where the wavefunction has opposite phase. The labels mark the positions of the nuclei (projected onto the plane $z = 1$). Orbital 1 has the highest binding energy and orbital 10 is the least strongly bound. Note also the axis orientation (the x-axis along the chain direction and the z-axis perpendicular to the molecular plane) and the marked distances d and h which are referred to in the text.

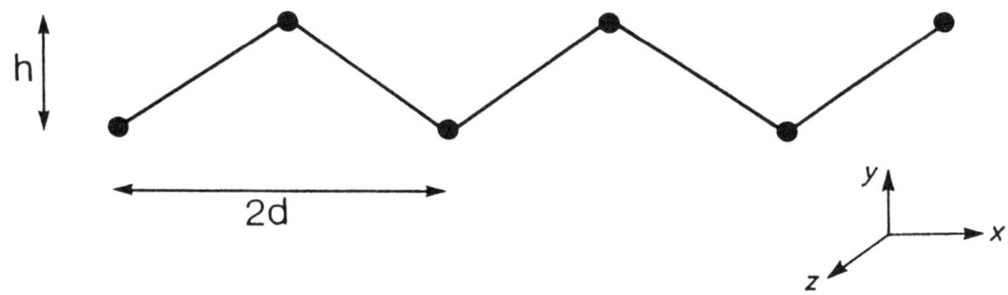

Figure 2a. (Continued).

$$\rho_k(\boldsymbol{p}) = 4\Phi(\boldsymbol{p})\Phi^*(\boldsymbol{p}) \left(\frac{\sin^2 (np_x d/2)}{n \sin^2 \left(\frac{2\pi k}{n} - p_x d \right)} \right) \cos^2 \left[\left(\frac{2\pi k}{n} - p_x d - p_y h \right) /2 \right], \qquad (2.16)$$

where the internuclear separations d and h are marked in Figure 2a) and $\Phi(\boldsymbol{p})$ is the Fourier transform of any of the $\phi_j(\boldsymbol{r})$. As in the case of H_2 (Equation (2.12)) the expression for the momentum density consists of the product of two terms – the first an "atomic term" $\Phi(\boldsymbol{p})\Phi^*(\boldsymbol{p})$, and the second a "diffraction factor" term which contains the information about the molecular geometry. Equation (2.16) shows that sharp peaks in the momentum density occur when

$$p_z = \frac{2\pi}{d} \left(\mu + \frac{k}{n} \right), \qquad (\mu \text{ integral}). \qquad (2.17)$$

The orbital binding energy decreases as k increases, and consequently the peaks for less strongly bound electrons (larger $|k|$) occur at successively larger p_x. For any non-zero value of p_z, only the π-orbital with the largest binding energy ($k = 0$) has a maximum at $p_x = p_y = 0$. This is all clearly shown in Figure 2b), where for each orbital we see the peaks corresponding to $\mu = 0$. The highest occupied molecular orbital (HOMO), for example, has a sharp peak, as expected, at $p_x = 0.9\pi/d \approx 0.7$ atomic units. Further peaks for $\mu \neq 0$ will occur at larger values of p_x (repeat distance $2\pi/d$) but, since the atomic term falls off so rapidly with p (as p^{12}), these peaks will be very much smaller.

With increasing n, the peaks in the momentum density become more marked. This is not surprising for delocalised bonding since r and p are conjugate variables. In the limit $n \to \infty$, $\rho_k(\boldsymbol{p})$ is *only* non-zero when Equation (2.17) is satisfied. The effect of the diffraction term is to restrict the range of allowed momenta in the molecular orbitals much more than in those of smaller molecules such as H_2.

The momentum-space representation also proves particularly suitable for comparing the electron distributions of systems with different nuclear frameworks. In r-space, "difference density" plots are complicated by the different nuclear positions, a complication absent in p-space. In the case of the polyenes,

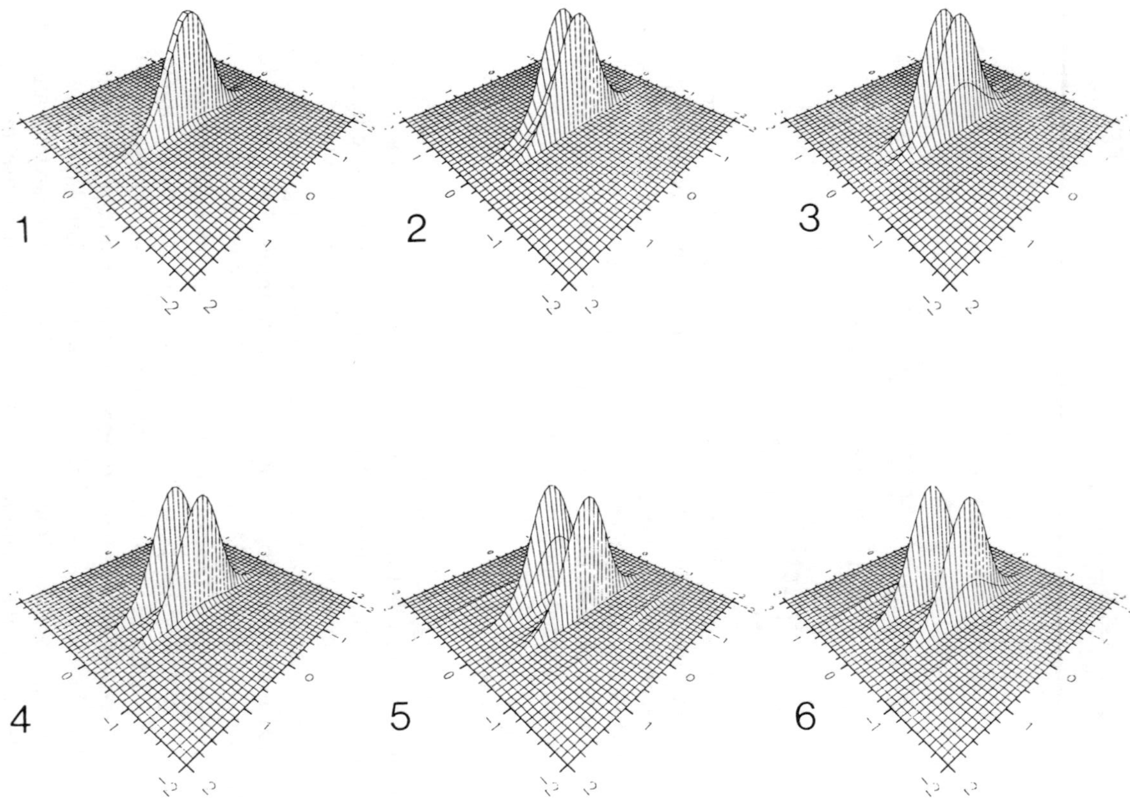

Figure 2b. *p*-space electron density in the plane $p_z = 1$ atomic unit for the orbitals shown in Figure 2a). The centre of each plot corresponds to the point (0, 0, 1) in momentum space. Each view extends to ±2 atomic units in the p_x and p_y directions. The *x*, *y* and *z*-labels refer to the same orientation of the axes as in Figure 2a).

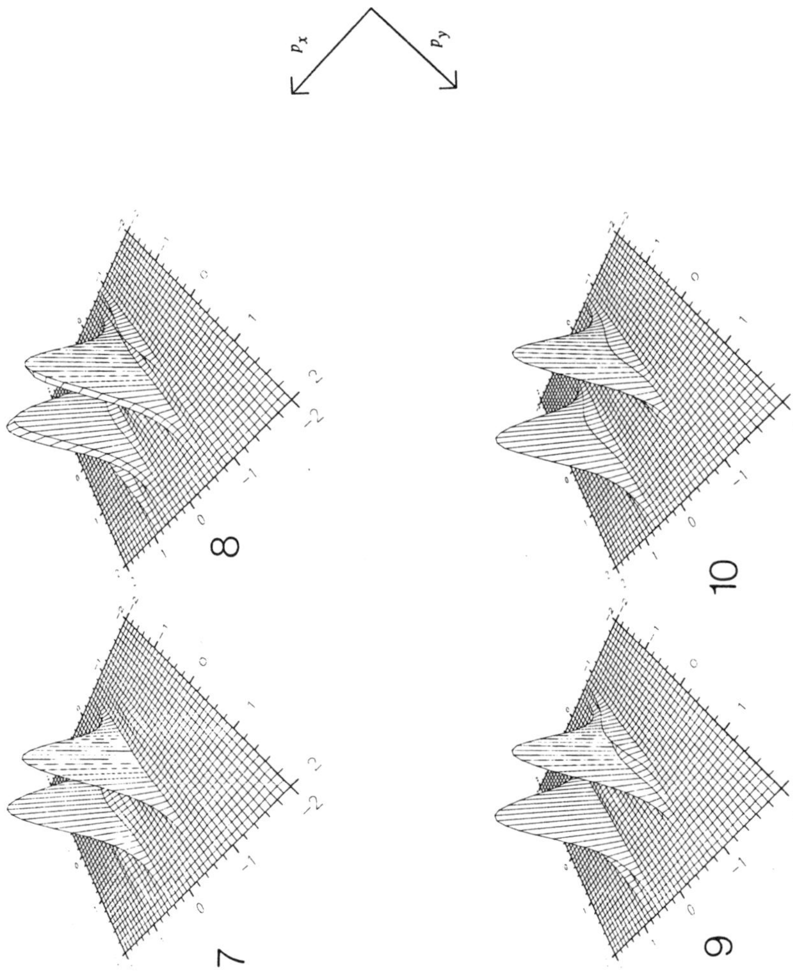

Figure 2b. (Continued).

for example, we have used p-space to examine the effects of bond alternation – an important characteristic of such systems – on the electron density.

Further discussion of the form of the momentum density, and more examples, are to be found in refs. 7, 8, 15 and 16. In particular the breaking and making of bonds is shown in a striking way in the p-space representation.

3. Molecular similarity indices

3a. Definitions

The central quantity in our approach is the generalised overlap

$$I_{AB}(n) = \int p^n \rho_A(\boldsymbol{p}) \rho_B(\boldsymbol{p})\, d\boldsymbol{p}, \tag{3.1}$$

in which $\rho_A(\boldsymbol{p})$ and $\rho_B(\boldsymbol{p})$ are the momentum-space electron densities of molecules A and B. These momentum densities can be *total* electron densities, *total valence* electron densities, or the densities associated with one or more individual *orbitals*. It is possible to compare particular orbitals of two different molecules (e.g. their respective HOMOs or localised orbitals describing the same bond), or the densities associated with a particular atom or group of atoms common to the two molecules. A further option is a comparison of two orbitals, such as the HOMO and the LUMO, of the *same* molecule. The purpose of the introduction of the term p^n into the integrand is to weight preferentially different regions of p-space. Typical values of n are −1, 0, 1, and 2. For example, a factor of p^{-1} places particular emphasis on the slowest moving electrons; this corresponds in the more familiar position-space representation to highlighting the long-range slowly-varying valence electron density. Values of $I_{AB}(n)$ depend on the relative orientation of the two molecules. However, $\rho_A(\boldsymbol{p})$ and $\rho_B(\boldsymbol{p})$ are independent of the choice of origin in r-space, and so $I_{AB}(n)$ is completely independent of the separation between the two systems in position space. The relationship between r-space and p-space is such, of course, that it would be possible to transform Equation (3.1) into an equivalent position-space integral. However, it is *much* easier to work directly in p-space with Equation (3.1).

In the present version of our codes, $I_{AB}(n)$ is computed by direct numerical evaluation, typically using standard library routines. It is usually sufficient to integrate out to p_{max} = 5 atomic units. The most appropriate method of integration varies from case to case. Three such schemes are currently in use – all of them yield exactly the same results, but the computational effort can vary significantly, depending on the characteristics of the integrand. The first of these methods uses an adaptive strategy with repeated subdivision of a cube into smaller three-dimensional boxes. In each subregion, the integrand is estimated using a seventh degree rule and the 'error' is estimated by comparison with a fifth degree rule. Hyper-rectangular subregions associated with large errors are subdivided further. The second of our schemes is much the same as the one just described, except that the range of integration is a sphere rather than a cube, and the various subregions are portions of spherical shells rather than boxes. We have found that this approach tends to be particularly efficient when dealing with total densities for molecules or parts of molecules. The third of our algorithms integrates over a cube using the method of Sag and Szekeres [17] which exploits various properties of the shifted p-point trapezoidal rule. This scheme tends to be much more efficient than the other two when dealing with limited summations over individual orbitals.

The integrals $I_{AB}(n)$ can take any non-negative value, and so it is natural to attempt some form of normalisation. The first strategy we employed, inspired by the work of Carbó [2], was to divide $I_{AB}(n)$ by the geometric mean of the "diagonal" terms:

$$R_{AB}(n) = \frac{100 \times I_{AB}(n)}{\sqrt{I_{AA}(n) \times I_{BB}(n)}}, \tag{3.2}$$

The similarity indices $R_{AB}(n)$ lie in the range 0–100%, with higher values implying greater similarity. An obvious alternative to scaling $I_{AB}(n)$ by the geometric mean of $I_{AA}(n)$ and $I_{BB}(n)$ is to use instead the arithmetic mean:

$$S_{AB}(n) = \frac{100 \times I_{AB}(n)}{\frac{1}{2}\,(I_{AA}(n) + I_{BB}(n))}. \tag{3.3}$$

The similarity indices $S_{AB}(n)$ also lie in the range 0–100%, with higher values implying greater similarity. An expression of this general type was first discussed in detail by Hodgkin and Richards [4]. In the special case that $\rho_A(\mathbf{p}) = m \times \rho_B(\mathbf{p})$, $R_{AB}(n)$ is invariant to the choice of non-zero m whereas $S_{AB}(n)$ is not. In this sense, values of $S_{AB}(n)$ are less dominated by the *shape* of the momentum-space electron densities than are the values of $R_{AB}(n)$. This tends to make the Hodgkin-like definition more appropriate than the Carbó-like definition for some p-space applications.

A third definition which has a number of features to recommend it was inspired by a quantity utilised in statistical analysis. Our Tanimoto-like similarity index takes the form:

$$T_{AB}(n) = \frac{100 \times I_{AB}(n)}{I_{AA}(n) + I_{BB}(n) - I_{AB}(n))}. \tag{3.4}$$

As for the other indices, $T_{AB}(n)$ can adopt values in the range 0–100%, with higher values implying greater similarity. The indices $S_{AB}(n)$ and $T_{AB}(n)$ are by no means independent of one another:

$$100/T_{AB}(n) = 200/S_{AB}(n) - 1. \tag{3.5}$$

The variation of $T_{AB}(n)$ with $S_{AB}(n)$ is illustrated in Figure 3. It is clear that the Tanimoto-like index is rather more discriminating than the Hodgkin-like

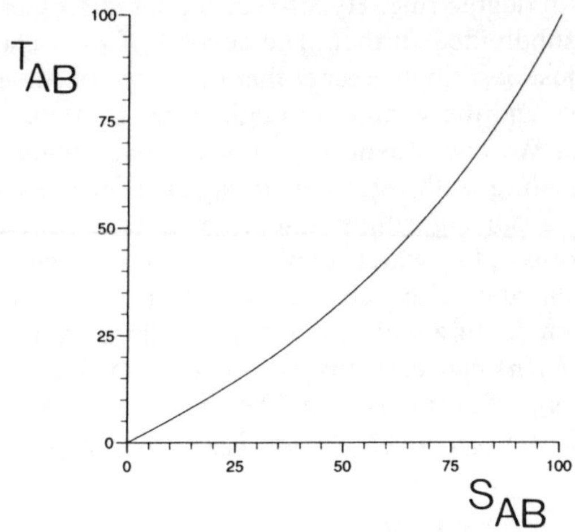

Figure 3. Relationship between values of T_{AB} and S_{AB} (Equation (3.5)).

index when dealing with high values (i.e. with very similar systems). Of course, this conclusion is *not* specific to *p*-space indices. It has been pointed out to us by Carbó [18] that the Hodgkin-like and Tanimoto-like indices can be viewed as special cases ($k = 0$ and $k = 1$, respectively) of a more general definition

$$G_{AB}(n, k) = \frac{100 \times (2 - k)I_{AB}}{I_{AA}(n) + I_{BB}(n) - kI_{AB}}\,. \tag{3.6}$$

We return to this idea later in Section 3b).

When dealing with quantities which show very high similarity, such as the **p**-space electron densities for analogous bonds in closely related molecules, we have found it more appropriate to dispense with the normalisation of $I_{AB}(n)$. A much more discriminating definition for such cases is the family of *p*-space *dissimilarity* indices:

$$D_{AB}(n) = 100 \times (I_{AA}(n) + I_{BB}(n) - 2I_{AB}(n)), \tag{3.7}$$

in which the ratio in $S_{AB}(n)$ has been replaced by a difference. This distance-like index can take any non-negative value such that larger values imply greater dissimilarity, with no upper bound. The magnitude of $D_{AB}(n)$ scales to a large extent with $I_{AB}(n)$, so that it should normally only be employed when $\rho_A(\mathbf{p})$ and $\rho_B(\mathbf{p})$ are very similar indeed, and correspond to closely related systems. This last observation is self-evident given the relationships linking $D_{AB}(n)$ with the normalised quantities $S_{AB}(n)$ and $T_{AB}(n)$:

$$\begin{aligned} D_{AB}(n)/100 &= I_{AB} \times (100/T_{AB}(n) - 1) \\ &= 2 \times I_{AB}(n) \times (100/S_{AB}(n) - 1). \end{aligned} \tag{3.8}$$

3b. Results for model systems

We start by considering the model compounds CH_3OCH_3, CH_3SCH_3 and $CH_3CH_2CH_3$ which have been examined in several previous similarity studies [4, 9–11]. Comparable biological activity is often obtained when a —CH_2— group is replaced by —S—. The same is not usually the case when —O— is substituted for —CH_2—.

Table II lists values of $R_{AB}(-1)$, $S_{AB}(-1)$ and $T_{AB}(-1)$ calculated using total p-space densities computed from *ab initio* wavefunctions (4–31G basis). The appropriate relative orientation of the two molecules is not a problem for this series of molecules – in each case the z-direction was that of the C_2-axis and the y-axis lay in the σ_v' plane. Reference 9 tabulates R_{AB} and S_{AB} for the additional cases $n = 0$, 1 and 2. The same qualitative variation is observed from one molecule to another but the range of values is smaller. In every case the largest similarity indices are between $CH_3CH_2CH_3$ and CH_3SCH_3 and the smallest between CH_3SCH_3 and CH_3OCH_3, consistent with the relative biological activity generally observed. The difference between the largest and the smallest values of $R_{AB}(-1)$, which depends particularly on the shapes of the two electron densities, is only 0.1. S_{AB} is less dominated by considerations of shape, but the range of values of S_{AB} (2.4) is still small. As expected, the range of T_{AB} values (4.6) is larger. For these systems, where all the molecules have much in common, none of the p-space similarity indices is particularly discriminating.

TABLE II
Values of the similarity indices $R_{AB}(-1)$, $S_{AB}(-1)$, $T_{AB}(-1)$ and the dissimilarity index $D_{AB}(-1)$ for the model series, calculated using *ab initio* wavefunctions

	$R_{AB}(-1)$	$S_{AB}(-1)$	$T_{AB}(-1)$	$D_{AB}(-1)$
$CH_3CH_2CH_3/CH_3OCH_3$	99.8	98.4	96.8	29.4
$CH_3CH_2CH_3/CH_3SCH_3$	99.9	99.8	99.6	4.0
CH_3OCH_3/CH_3SCH_3	99.7	97.4	95.0	49.4

As mentioned in Section 3a), the Hodgkin-like and Tanimoto-like indices belong to the same clan, with a direct relationship between them (Equation 3.5) and also the possibility of constructing more general definitions (e.g. Equation 3.6) which include them as special cases. We have considered several such definitions, including variants which incorporate trigonometric and exponential functions. The simplest of the functions we have examined in detail is

$$G_{AB}(n, k, a, b) = \frac{100 \times (2 - k^a) I_{AB}(n)}{I_{AA}(n) + I_{BB}(n) - k^b I_{AB}(n)} \cdot \tag{3.9}$$

which reduces to $S_{AB}(n)$ and $T_{AB}(n)$ for $k = 0$ and $k = 1$, respectively. We show in Figure 4 values of $G_{AB}(0, k, a, a)$ for the model series (*ab initio*)

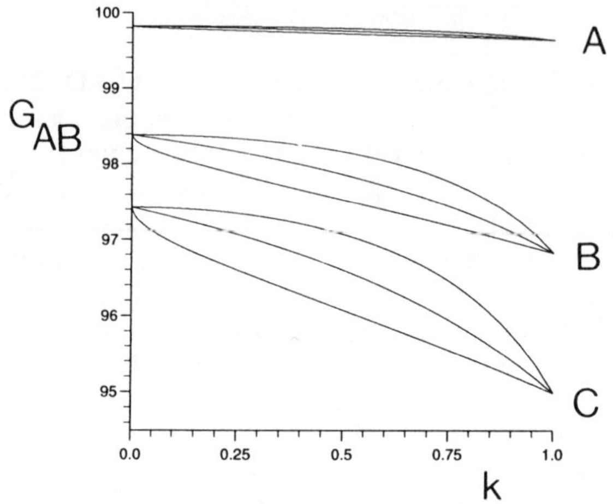

Figure 4. Variation of $G_{AB}(0, k, a, a)$ (Equation (3.9)) with k, computed using *ab initio* wavefunctions for the model systems. A $CH_3CH_2CH_3/CH_3SCH_3$, B $CH_3CH_2CH_3/CH_3OCH_3$, and C CH_3OCH_3/CH_3SCH_3. In each group of curves, the three cases are (top to bottom) $a = 2$, $a = 1$ and $a = 0.5$.

with $a = 0.5$, 1 and 2. At least in the present case, the choice of $a = b = 0.5$ yields an almost linear dependence of $G_{AB}(n, k, a, b)$ on k, except in the neighbourhood of $S_{AB}(n)$.

The distance-like index D_{AB} has no upper limit, unlike R_{AB}, S_{AB} and T_{AB}. Dissimilarity indices, $D_{AB}(-1)$, for the model series, computed from the same *ab initio* wavefunctions, are also listed in Table II. The same conclusions are obtained from this index as with the three similarity indices, but there is a very much larger variation (45.4). We have found that D_{AB} is most useful in such situations.

The calculation of *ab initio* wavefunctions for the large molecules associated with applications such as those discussed later in this paper is computationally very expensive. Accordingly we have also examined the model systems using semi-empirical wavefunctions taken from the well-established MOPAC program [13]. These can be obtained relatively cheaply. Table III collects together R_{AB}, S_{AB}, T_{AB} and D_{AB} for the model series evaluated from

TABLE III
Values of the similarity indices $R_{AB}(-1)$, $S_{AB}(-1)$, $T_{AB}(-1)$ and the dissimilarity index $D_{AB}(-1)$ for the model series, calculated using MNDO wavefunctions

	$R_{AB}(-1)$	$S_{AB}(-1)$	$T_{AB}(-1)$	$D_{AB}(-1)$
$CH_3CH_2CH_3/CH_3OCH_3$	99.8	97.6	95.4	137
$CH_3CH_2CH_3/CH_3SCH_3$	99.9	99.9	99.9	5
CH_3OCH_3/CH_3SCH_3	99.6	97.3	94.7	161

MNDO wavefunctions. Once more the conclusions regarding the relative similarities are unchanged, with CH_3SCH_3 and $CH_3CH_2CH_3$ the most similar pair of molecules. It is particularly useful that the same trends are evident in the similarity indices whether the electron densities are calculated from *ab initio* or semi-empirical techniques, since it allows the use of the cheap semi-empirical methods for series of large molecules. Of course, the *ab initio* and semi-empirical numbers are not the same. This reflects the different qualities of the two types of wavefunction. In addition, the MOPAC calculations generate "valence-electron" densities whereas the *ab initio* densities are "all-electron". Since I_{AB}, for the reasons discussed in the last section, emphasises the valence region of the electron density in any case, this last effect should be small.

It is easy to envisage situations in which it would be more appropriate to compare only certain parts of a series of molecules, where these might be fragments containing the active centres. One way of proceeding is simply to compare the electron densities associated with the atoms of interest. This requires the use of some "atoms-in-a-molecule" method, by means of which the electron density is assigned to individual atoms. The approach of Cioslowski [19], working with similarity measures based on r-space electron densities, has been to use the Bader zero-flux partitioning scheme [20]. We have implemented a much more primitive and considerably cheaper method, in which, for any given atom(s), we consider only the basis functions centred on the nucleus of the atom(s) concerned. Returning to the model series, it is possible in this way to compare just the parts of the molecules that are not in common, i.e., the —O—, —S— and —CH_2— linkages. Table IV lists the values of $R_{AB}(-1)$, $S_{AB}(-1)$, $D_{AB}(-1)$ and $T_{AB}(-1)$ so obtained using MNDO wavefunctions. The same pair of molecules (CH_3SCH_3 and $CH_3CH_2CH_3$) is once again the most similar (the relative order of the other two pairs is now reversed) and the range of values of each index is larger than in Table III.

TABLE IV

Values of the similarity indices $R_{AB}(-1)$, $S_{AB}(-1)$, $T_{AB}(-1)$ and the dissimilarity index $D_{AB}(-1)$ for the model series, ignoring the contribution to the electron density from the methyl groups, calculated using MNDO wavefunctions

	$R_{AB}(-1)$	$S_{AB}(-1)$	$T_{AB}(-1)$	$D_{AB}(-1)$
$CH_3CH_2CH_3/CH_3OCH_3$	88.2	56.2	39.0	170
$CH_3CH_2CH_3/CH_3SCH_3$	98.9	98.7	97.5	8
CH_3OCH_3/CH_3SCH_3	90.9	60.8	43.7	136

An attractive alternative strategy is the use of a reliable and computationally cheap scheme to localise the occupied molecular orbitals. This has enabled us to compare the C—H and C—F bonds in the fluoromethanes CH_xF_{4-x} $(0 \leq x \leq 4)$, a series of molecules which Cioslowski has also considered [19]. Here the semi-empirical method AM1 [13] was used instead of MNDO since the latter tends to produce large errors for highly fluorinated compounds [21]. For the hydrofluoromethanes and CF_4, AM1 wavefunctions were generated, with the z-axis lying along a C—F bond. Localised molecular orbitals were then produced by the Mulliken population procedure due to Pipek and Mezey [22]. In this method, the quantity

$$Z = \sum_A \sum_i |P_A(i)|^2, \tag{3.10}$$

is maximised, where $P_A(\mathrm{i})$ is the contribution from each electron in MO ψ_i to the Mulliken population on centre A.

In a rather different context [23] we have discussed how primary C—F bonds in fluoroalkanes differ little from molecule to molecule. The C—F bond length and the Mulliken charges on F, calculated by the AM1 method, varies only very slightly. The calculated charge on the F atoms in CF_4 is the only slight exception, where the Mulliken charge on F is a little less negative than in the other molecules. Table V lists the values of $D_{AB}(-1)$ calculated for the localised orbitals describing the C—F bonds. All these bonds are very similar. Values of $D_{AB}(-1)$ calculated for orbitals rather than total densities are always smaller than those for total densities, simply because this index involves a difference and not a ratio. Nevertheless the C—F bonds in "adjacent" molecules in the series CH_3F, CH_2F_2 and CHF_3 are the most similar. The values of $D_{AB}(-1)$ are slightly larger when CF_4 is involved in the comparison. It is evident that $D_{AB}(-1)$ can differentiate between the C—F bonds in this series.

TABLE V
Values of the dissimilarity index $D_{AB}(-1)$ for the C—F bonds in the fluoromethanes CH_3F, CH_2F_2, CHF_3 and CF_4

	CH_2F_2	CHF_3	CF_4
CH_3F	0.2	0.4	1.5
CH_2F_2		0.2	1.2
CHF_3			0.8

To consider the variation of C—H bonds, AM1 wavefunctions were calculated for methane and the hydrofluoromethanes, orientating each molecule such that a C—H bond lay along the z-axis. Localised molecular orbitals were then produced using the population localisation method and values of $D_{AB}(-1)$ calculated (Table VI). Both AM1 and experiment indicate that the C—H bond length is insensitive to increasing fluorination. Nevertheless, in these molecules, the C—H bonds show a marked variation in chemical reactivity. This is clear, for instance, from the rate constants and activation energies for H-atom abstraction by free radicals. In this context the reaction with OH is particularly significant because this process is the dominant loss process for hydrofluorocarbons in the troposphere. Several models [24, 25] predict that the tropospheric lifetime of CHF_3 is much longer than that of CH_2F_2

or CH_3F. If we assume that there is a correlation between the properties of the molecules themselves, and those of the corresponding transition states, we would anticipate the $D_{AB}(-1)$ values for the C—H bonds to be larger than those for the C—F bonds. This is evident from a comparison of Tables V and VI. The largest $D_{AB}(-1)$ occurs with CHF_3. The rate constant of the reaction with the OH radical at 298 K is much smaller, and the activation energy much greater, for CHF_3 than for any of the other hydrofluoromethanes [26]. This much lower reactivity of CHF_3 is highlighted by the dissimilarities calculated using localised molecular orbitals.

TABLE VI
Values of the dissimilarity index $D_{AB}(-1)$ for the C—H bonds in the series CH_4, CH_3F, CH_2F_2 and CHF_3

	CH_3F	CH_2F_2	CHF_3
CH_4	1.1	3.9	13.1
CH_3F		1.9	9.2
CH_2F_2			4.5

4. Anti-HIV phospholipids

Our next example [27] relates to the virology data for a particular group of phospholipids [28, 29]. Those considered here have the general structural formula:

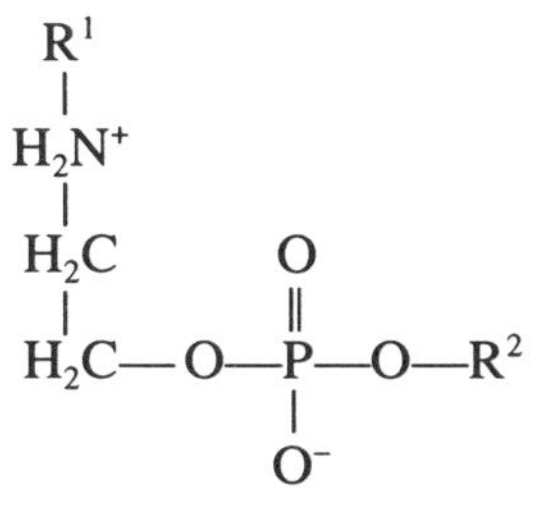

in which R^2 represents the lipid tail and R^1 is a relatively small alkyl substituent on the amine group. Molecules with different choices of R^2 and R^1 give 50% inhibition of HIV1 in C8166 T-lymphoblastoid cells with the ED_{50} (μM) values listed in Table VII. These show no *obvious* correlation between HIV inhibition and the identity of the lipid tail or, indeed, of other alkyl substituents on the amine group. It does not appear to be possible to make much sense of the various trends by consideration of chemical structure alone. For example, the replacement by t-butyl of the methyl group at R^1 results in greatly reduced ED_{50} values in the most inactive compounds (HX1 and DD1), and in some enhancement in the activity of two others (OD1 and OL1), whereas it is clear

TABLE VII

ED_{50} values for the inhibition of HIV1 in C8166 T-lymphoblastoid cells, and two calculated similarity indices $T_{AB}(-1)$. $T_{AB}^{(HX1)}$ is the similarity of the given molecule to the *inactive* compound HX1 and $T_{AB}^{(OL2)}$ is the similarity to the *active* compound OL2

Mnemonic	ED_{50} (μM)	R^1	R^2	$T_{AB}^{(HX1)}$	$T_{AB}^{(OL2)}$
HX1	> 200	methyl	n-hexyl	100	61.0
DD1	> 200	methyl	n-dodecyl	86.5	87.0
OD1	25	methyl	n-octadecyl	67.6	98.9
EG1	110	methyl	ethyl glycolate	97.4	51.4
OL1	10	methyl	oleyl	68.8	98.4
HX2	40	t-butyl	n-hexyl	95.4	75.5
DD2	10	t-butyl	n-dodecyl	76.6	94.8
OD2	3	t-butyl	n-octadecyl	59.9	99.9
EG2	200	t-butyl	ethyl glycolate	99.0	67.0
OL2	3	t-butyl	oleyl	61.0	100
HX3	> 200	hydrogen	n-hexyl	99.2	55.3
DD3	4	hydrogen	n-dodecyl	89.6	83.2
OD3	3.5	hydrogen	n-octadecyl	70.6	97.6
OL3	0.5	hydrogen	oleyl	71.7	97.4

from the data in Table VII that EG2 is *less* active than EG1. The series of compounds in which the amine group is free (R^1 = H) tends to have very low ED_{50} values, except that HX3 is *inactive*, unlike HX2.

The mechanism by which these phospholipids inhibit the virus is not completely understood, although it is thought that they first insert into the membranes of the virus. The concept of molecular similarity is particularly appropriate here where the molecular processes involved are extremely complex or even unknown. Bearing in mind that it is likely that the phospholipids must be able to mimic the membrane lipids, we have chosen to compare total densities for the complete molecules, rather than the densities for individual orbitals or for molecular fragments.

In view of the size of the phospholipids, we chose to obtain computationally inexpensive position-space wavefunctions from semi-empirical (MNDO) geometry optimisations. $T_{AB}(-1)$ was computed for each pair of molecules, with the phospholipids aligned so as to match as closely as possible the positions of those atoms common to all systems, and the values are listed in Table VII. Results are presented here only for the zwitterions, although it should be stressed that all the key features were also exhibited when the neutral molecules were compared.

We display in Figure 5 a representative subset of our results – comparisons relative to the *inactive* compound HX1 and comparisons relative to the highly *active* compound OL2. The corresponding ED_{50} values from Table VII are also shown, with an arbitrary value of 200 μM assigned to the most inactive compounds (HX1, DD1 and HX3). It is clear from the figure that active compounds have a high similarity to OL2 and a low similarity to HX1,

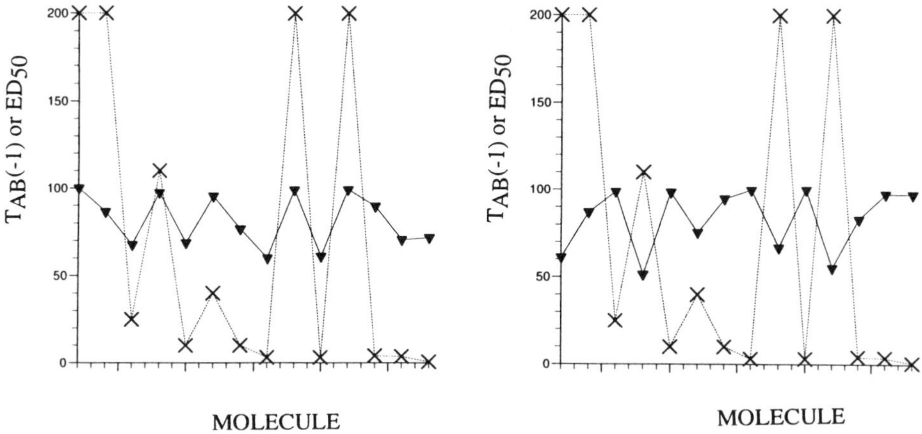

Figure 5. Values of the molecular similarity index $T_{AB}(-1)$ (▼) for various phospholipids relative to (a) the *inactive* compound HX1 and (b) the highly *active* compound OL2. Also shown in each frame are the corresponding ED_{50} (μM) values (X), with an arbitrary value of 200 assigned to the most inactive compounds (HX1, DD1 and HX3). All the values are taken from Table VII.

and *vice versa*. The way in which the various maxima and minima coincide is particularly striking in both cases.

We were initially informed that OD2 was insoluble/inactive, whereas we expected from the various $T_{AB}(-1)$ values that this phospholipid should have a very low ED_{50} value. The most recent data (Table VII) suggests ED_{50} = 3 μM. It is also important to point out that the T_{AB} values for the series with a free amine group were computed *before* the virology data were available to us. In this way we were able to *predict* that OD3 and OL3 would both have low ED_{50} values, but that HX3 would be inactive. For DD3 we observed moderately high T_{AB} values relative to *both* HX1 and OL2, and so it was not possible to make a definitive prediction in this case.

It is clear that the particular approach to molecular similarity employed here can be used to rationalise HIV virology data for the families of phospholipids considered and even to make some successful predictions of active compounds. Preliminary results for various AZT derivatives (reverse transcriptase inhibitors) are also encouraging. The active compounds all feature an —N_3 group. Nonetheless, comparisons of the HOMOs, which are localised in the thymine rings, can be used to distinguish easily between active and inactive compounds.

5. Hyperpolarisabilities

We have found that momentum-space concepts can be particularly beneficial in the field of materials with non-linear optical properties. In particular we describe here structure-activity relations involving the hyperpolarisabilities, β, of conjugated molecules. The *direct* calculation of hyperpolarisabilities is not only difficult, but also very sensitive to the quality of the wavefunction. An alternative approach is *not* to attempt to calculate β *a priori*, but instead to look for an empirical correlation involving β and some quantity which can be evaluated much more easily and which is relatively insensitive to the quality of the wavefunction.

The molecules of interest here can be represented

D—(π)—A

in which D—(π)—A denotes a conjugated π-electron system (cf. the polyenes in Section 2b)), and D and A are electron-donor and acceptor groups, respectively. It is conventional to rationalise the response of such systems to external electric fields by invoking a two-state model [30, 31] involving

D^{+}—(π)—A^{-}

At the simplest level, this second state can be envisaged as arising via excitation from the highest occupied molecular orbital (HOMO) to the lowest unoccupied molecular orbital (LUMO) of the ground state. As a consequence, it is natural to compare the HOMOs and LUMOs of such systems in *p*-space.

Results for just one family of molecules are presented here. These are all 1,4-benzene derivatives of the type,

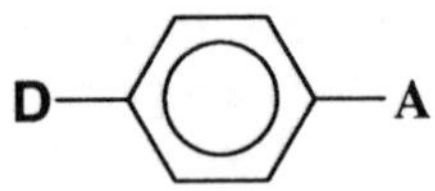

All the wavefunctions were generated using MNDO geometry optimisations. However, the calculated MNDO geometries exhibited some serious errors for a few molecules. For example, MNDO predicts incorrectly that the coordination at the nitrogen atom in Ph—NMe_2 is pyramidal, rather than planar. In such cases, the appropriate experimental values were substituted for the relevant parameters.

For *each* molecule we have calculated the similarity index $R_{AB}(-1)$ between its HOMO and its LUMO, considering only contributions from basis functions centred on atoms in the benzene ring (the structural unit common to all the molecules). Accordingly the problem of relative orientation does not arise. Figure 6 shows a plot of the experimental values of β, taken from ref. 30, vs. $R_{AB}(-1)$. All the values are listed in Table VIII. It is clear that there is a reasonable correlation which could now be used predictively.

Further series of non-centrosymmetric conjugated molecules are currently under investigation [32]. In keeping with the known failure of the two-state model for the family of stilbenes, Ar^1—CH=CH—Ar^2 [30], our preliminary results indicate that such systems do not follow the same correlation observed for the benzene derivatives.

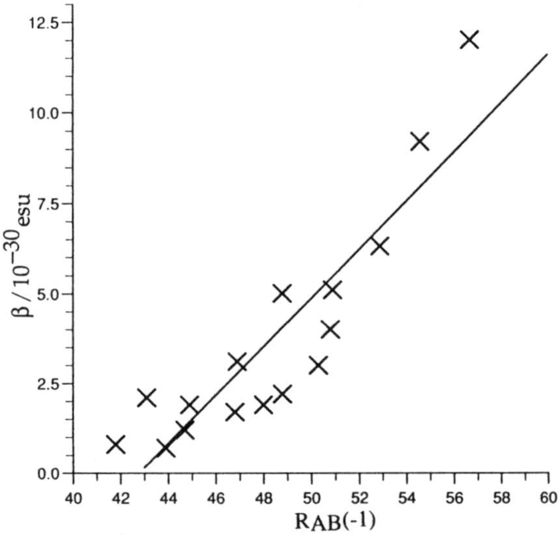

Figure 6. Experimental values of the hyperpolarisability β (10^{-30} esu) vs. the calculated orbital similarity $R_{AB}(-1)$ for a series of 1,4-benzene derivatives. All the data are listed in Table VIII.

TABLE VIII
Experimental values [30] of the hyperpolarisability β (10^{-30} esu) and the calculated orbital similarity index $R_{AB}(-1)$ for a series of non-centrosymmetric 1,4 benzene derivatives. The columns headed A and D list the acceptor and donor groups respectively

A	D	$R_{AB}(-1)$	β(10^{-30} esu)
CN	Cl	41.8	0.8
CN	Me	43.9	0.7
CN	OPh	44.7	1.2
CN	OMe	44.9	1.9
CN	NH_2	46.9	3.1
CN	NMe_2	48.8	5.0
COH	Me	46.8	1.7
COH	OPh	48.0	1.9
COH	OMe	48.4	2.2
COH	NMe_2	52.9	6.3
NO_2	Me	43.1	2.1
NO_2	OH	50.3	3.0
NO_2	OPh	50.8	4.0
NO_2	OMe	50.9	5.1
NO_2	NH_2	54.6	9.2
NO_2	NMe_2	56.7	12

6. Discussion and conclusions

Momentum-space similarity and dissimilarity indices tend to be especially useful in situations for which the bonding topology is less important than the long-range valence electron density. In addition to model systems, the approach has now been applied successfully to a number of "real" problems, such as anti-HIV phospholipids, reverse transcriptase inhibitors, and conjugated systems with high hyperpolarisabilities. In general, the method can be particularly effective for situations in which conventional chemical intuition has failed.

In future applications, not only the relative orientation but also the conformational flexibility of the molecules may prove to be more of a problem than for the cases described here. Of course, it is not hard to anticipate situations in which the *p*-space approach is unlikely to be useful. Obvious examples include cases in which the observed activity is largely dependent on the values of bulk physical properties or on well-defined steric effects. We have stressed previously [9] that *no one* approach to molecular similarity can be expected to be a universal panacea. In this context, it is worthwhile to mention two series of molecules we have been asked to examine. The first of these consists of a range of molecules which have an odour reminiscent of popcorn, or freshly-baked bread. Further examination of the molecular structures of those systems which possess the so-called 'popcorn-nutty factor' indicate that the most important consideration is the relative position of an aromatic nitrogen atom and a carbonyl group [33]. The second example relates to a series of attractants for the gravid cabbage root fly [34]. These attractants are all isothiocyanates of the form R—NCS, in which R is an aliphatic organic group such as ethyl or isopropyl. The naturally occurring compound in bruised leaves of brassicas is allyl isothiocyanate. According to our various *p*-space indices, all of these molecules are rather similar. A further examination of the experimental data available to us suggests that a key consideration when comparing the efficacies of the different solutions of these molecules is their relative volatility. It is important to apply our momentum-space approach to problems for which it is appropriate.

In summary, molecular similarity indices based on momentum-space electron densities appear to show considerable promise for a wide range of applications. A further possibility is the generation of families of *p-s*pace *and r*-space similarity indices, and other properties, followed by the use of appropriate statistical techniques. In this context, we are currently considering the use of neural nets. It may also be worthwhile to investigate the properties of higher-order momentum-space density matrices rather than $\rho(\mathbf{p})$. Work on a wide range of applications, both biological and non-biological, is currently in progress and the results of these studies will be reported in due course.

References

1. Johnson, M. A. and Maggiora, G. M., eds., *Concepts and Applications of Molecular Similarity* (Wiley, New York, 1990); *J. Chem. Inf. Comput. Sci* **32**(6) (1992).
2. See for example, Carbó, R., Leyda, L., and Arnau, M., *Int. J. Quant. Chem.* **17**, 1185 (1980); Carbó, R. and Domingo, Ll., *Int. J. Quant. Chem.* **32**, 517 (1987); Carbó, R. and Calabuig, B., *Int. J. Quant. Chem.* **42**, 1681 (1992); *ibid.* 1695.
3. Ponec, R. and Strnad, M., *J. Phys. Org. Chem.* **4**, 701 (1991); *Int. J. Quant. Chem.* **42**, 501 (1992).
4. Hodgkin, E. E., and Richards, W. G., *Int. J. Quantum Chem. Quantum Biology Symp.* **14**, 105 (1987); Richards, W. G. and Hodgkin, E. E., *Chem. Br.* **24**, 1141 (1988); see also Burt, C. and Richards, W. G., *J. Comput. -Aided Mol. Design* **4**, 231 (1990).
5. Duane-Walker, P., Artera, G. A., and Mezey, P. G., *J. Comput. Chem.* **12**, 220 (1991).
6. See, for example, Brion, C. E., *Int. J. Quantum Chem.* **29**, 1397 (1986).
7. Coulson, C. A. and Duncanson, W. E., *Proc. Camb. Philos. Soc.* **37**, 55 (1941); *ibid.* 67; 74; 397; 406.
8. For example, Epstein, I. R. and Tanner, A. C., in *Compton Scattering*, ed. B. G. Williams (McGraw-Hill, New York, 1977), Chapter 7; Rawlings, D. C. and Davidson, E. R., *J. Phys. Chem.* **89**, 969 (1985) and references therein.
9. Cooper, D. L. and Allan, N. L., *J. Comput. -Aided Mol. Des.* **3**, 253 (1989).
10. Cooper, D. L. and Allan, N. L., *J. Am. Chem. Soc.* **114**, 4774 (1992).
11. Allan, N. L. and Cooper, D. L., *J. Chem. Info. Comp. Sci.* **32**, 587 (1992).
12. Guest, M. F. and Sherwood, P., *GAMESS-UK User's Guide and Reference Manual, Revision B.0*; SERC Daresbury Laboratory (UK) (1992).
13. Stewart, J. J. P., *J. Comput-Aided Mol. Des.* **4**, 1 (1990).
14. Kaijser, P. and Smith, V. H., Jr., *Adv. Quant. Chem.* **10**, 37 (1977).
15. Cooper, D. L., Allan, N. L., and Grout, P. J., *J. Chem. Soc., Faraday Trans. 2* **85**, 1519 (1989).
16. Cooper, D. L. and Allan, N. L., *J. Chem. Soc., Faraday Trans. 2* **83**, 449 (1987).
17. Sag, T. W. and Szekeres, G., *Math. Comput.* **18**, 24553 (1964).
18. Carbó, R., private communication.
19. Cioslowski, J. and Nanayakkara, A., *J. Am. Chem. Soc.* **115**, 11213 (1993).
20. Bader, R. F. W., *Atoms in a molecule: A Quantum Theory* (OUP, 1990).
21. Dewar, M. J. S and Rzepa, H. S., *J. Am. Chem. Soc.* **100**, 58 (1978).
22. Pipek, J. and Mezey, P. G., *J. Chem. Phys* **90**, 4916 (1989).
23. Cooper, D. L., Allan, N. L., and Powell, R. L., *J. Fluorine Chem.* **46**, 317 (1990); *ibid.* **49**, 421 (1990).
24. Derwent, R. G., Volz-Thomas, A., and Prather, M. J., UNEP/WMO Scientific Assessment of Stratospheric Ozone: Appendix. AFEAS Report (AFEAS, 1989), Chapter 5, p. 123.
25. Hampson, R. F., Kurylo, M. J., and Sander, S. P., UNEP/WMO Scientific Assessment of Stratospheric Ozone: Appendix. AFEAS Report (AFEAS, 1989), Chapter 3, p. 47.
26. Cooper, D. L., Allan, N. L., and McCulloch, A., *Atmos. Environ.* **24A**, 2417, 2703 (1990); Cooper, D. L., Cunningham, T. P., Allan, N. L., and McCulloch, A., *Atmos. Environ.* **26A**, 1331 (1990).
27. Cooper, D. L., Mort, K. A., Allan, N. L., Kinchington, D., and McGuigan, C., *J. Am. Chem. Soc.* **115**, 12615 (1993).
28. McGuigan, C., O'Connor, T. J., Swords, B., and Kinchington, D., *AIDS* **5**, 1536 (1991)
29. McGuigan, C. and Kinchington, D., (submitted for publication).
30. Cheng, L., Tam, W., Stevenson, S. H., Meredith, G. R., Rikken, G., and Marder, S. R., *J. Phys. Chem.* **95**, 10631 (1991)
31. Matsuzawa, N. and Dixon, D. A., *Int. J. Quantum Chem.* **44**, 497 (1992).
32. Measures, P. T., Mort, K. A., Allan, N. L., and Cooper, D. L., manuscript in preparation.
33. Southwick, E. and Schiffman, S. S., *Chemical Senses* **5**, 343 (1980); Southwick, E., private communication (1990).
34. Harwood, D., private communication (1992).

Molecular Similarity Measures for Assessing Reactivity

PAUL G. MEZEY
Mathematical Chemistry Research Unit, Department of Chemistry and Department of Mathematics and Statistics, University of Saskatchewan, Saskatoon, Canada, S7N 0W0

1. Introduction

The concept of similarity in chemistry serves in many roles. However, two aspects of similarity appear to dominate in chemistry and in most of natural scences: similarity is a key element in both understanding and prediction [1–10]. Chemical similarity can be analyzed in terms of molecular properties which are *preserved*, and in terms of similarities in the processes (reactions, conformational processes) which *change* the molecules. The interrelations between these two aspects of chemical similarities are of special importance in assessing chemical reactivities.

Similar molecules usually react similarly: molecular similarity implies some degree of similarity in their reactivities. According to one view, molecular similarity is interpreted in terms of properties of isolated molecules, such as the molecular skeleton of the formal bonding pattern, three-dimensional stereochemical features, the shape of the electronic density clouds [11], the electrostatic potential, or spin density distribution, where only limited variations of the nuclear arrangements are considered. Alternatively, a broader range of molecular deformations and all the actual reactivities of a given molecule towards potential reaction partners may be regarded to be integral components of the evaluation of molecular similarity. According to this second view, molecules are regarded as similar *because* they react similarly. In this chapter, the connections between some aspects of these two views are discussed.

It is somewhat easier to follow approaches suggested by the first view: the properties of isolated molecules are simpler to model by quantum chemical and alternative computational techniques. Some but not all aspects of reactivity are evident from isolated molecules; for example, the presence of polar molecular regions, or exposed, vulnerable π-electron densities in double bonds, provide indications of specific reactivities. However, chemical reactivity is a concept that assigns information manifested in various *interactions between molecules* to *individual molecules*. The question is, how much of the information concerning all possible reactions of a molecule can be assigned to the individual molecule? If two molecules participate in a reaction, then the

R. Carbó (ed.), Molecular Similarity and Reactivity: From Quantum Chemical to Phenomenological Approaches, 57–76.

information content of the interacting system is more than that of the two isolated molecules, and if one is to describe reactivity by assigning all information to the individual molecules, then the choice for the formal distribution of the excess information between the two molecules is not unique.

Another approach is suggested by taking the view that not individual molecules but their interactions are of primary importance. One extreme model can be formulated by taking an overall stoichiometry of a fixed set of nuclei, and considering all possible interactions among them. In principle, this model describes all the reactions between molecules involving various sub-families of nuclei. A nuclear configuration space can be assigned to the overall stoichiometry, where each point of this space corresponds to a nuclear arrangement K. The nuclear configuration space can be chosen as a metric space M, where the distance $d(K, K')$ for any two nuclear arrangements K and K' in M corresponds to dissimilarity of the nuclear arrangements, providing an *intrinsic similarity measure* [12]. If the overall stoichiometry involves N nuclei, then the dimension of M is $3N - 6$, whenever $N \geq 3$. Energy relations among these nuclear configurations describe all chemical reactions, conformatinal changes and various deformations of all molecular families generated within the given stoichiometry. For each formal electronic state, these energy relations are described by a function $E(K)$, called a potential energy hypersurface [12]. The family $\{E_0(K), E_1(K), E_2(K), \ldots, E_k(K), \ldots\}$ of potential energy hypersurfaces for all electronic states describes all the reactions within the given stoichiometry, and contains all the relevant reactivity information.

In this chapter, a short review of similarity analysis of reaction paths and reaction mechanisms is presented, based on the underlying topological properties of potential energy hypersurfaces. Relations between molecular shape similarity and potential surface similarity will be used to define similarity measures for reactivity. Simplified shape representations, using generalizations of convexity, will be used for the exploration of some of the connections between algebraic-geometrical properties of molecular shape and reactivity.

2. Reactivity and similarity in the context of potential energy surfaces

2.1. Relations between molecular shape similarity and reaction path similarity

The shapes of formal molecular "bodies" can be described by molecular isodensity contour (MIDCO) surfaces [11]. The density domains (DD) are the formal bodies [13, 14] enclosed by such MIDCO surfaces. These density domains provide more detailed information on chemical bonding than the conventional "skeletal model" of stereochemical bond diagrams, and also better represent the actual shape of the fuzzy electron density clouds. The shapes of these fuzzy molecular bodies are described in detail by the Shape Group Methods (SGM) [11, 15–20].

For a given electronic state, a *molecular isodensity contour surface*, MIDCO $G(K, a)$ of a nuclear configuration K and density threshold a is the collection of all points $\mathbf{r}$ of the three-dimensional space where the electronic density $\rho(K, \mathbf{r})$ is equal to the threshold value a:

$$G(K, a) = \{\mathbf{r} : \rho(K, \mathbf{r}) = a\}. \tag{1}$$

Most of the electronic charge cloud is localized near the nuclei of the molecule. With increasing distance from the nearest nucleus the electronic charge density function $\rho(K, \mathbf{r})$ converges rapidly to zero, becoming negligible at about 8–10 Å from the nearest nucleus, however, in the strict sense, this function becomes zero only at infinite distance from the nuclei. For each threshold value a, an open set, regarded as an approximate *molecular body*, can be defined as the *level set* $F(K, a)$ of all those points $\mathbf{r}$ of the 3D space where the electronic density is greater than the value a,

$$F(K, a) = \{\mathbf{r} : \rho(K, \mathbf{r}) > a\}. \tag{2}$$

Note that negative charge means a positive value for the density function $\rho(K, \mathbf{r})$.

For a closer analogy with macroscopic bodies of well-defined boundaries, it is useful to represent formal molecular bodies by closed sets. For each threshold value a, an approximate molecular body can be defined as the *density domain* [13, 14], denoted by $DD(K, a)$, that includes *all the points and all the boundary points* of the corresponding level set $F(K, a)$ of the same density threshold a:

$$DD(K, a) = \{\mathbf{r} : \rho(K, \mathbf{r}) \geq a\}. \tag{3}$$

The density domain of nuclear configuration K and density threshold a is the union of the corresponding level set $F(K, a)$ and MIDCO $G(K, a)$:

$$DD(K, a) = F(K, a) \cup G(K, a). \tag{4}$$

These density domains provide a natural representation of the three-dimensional pattern of chemical bonding in molecules. Chemical bonding within a molecule is described by the interfacing and mutual interpenetration of local, fuzzy charge density clouds, as represented by the density domains.

The shape and size of each density domain $DD(K, a)$ depend on the nuclear configuration K and density threshold value a. However, for a fixed nuclear configuration K of an isolated molecule, within the entire range of density thresholds a, there are only a finite number of topologically different bodies of density domains. A set of representative density domains $DD(K, a)$ and the associated families of MIDCO's $G(K, a)$ can be used in the analysis of molecular shapes and molecular similarity. For the study of molecular similarity and for the generation of numerical similarity measures, the general GSTE principle (Geometrical Similarity as Topological Equivalence) [11] is applied. The Shape Group Methods (SGM) have been specifically designed [15–20] for this purpose, leading to nonvisual, algebraic-topological shape analysis techniques for molecules. The shape groups are the homology groups of molecular MIDCO surfaces truncated according to some local curvature criteria. The ranks of shape groups, their Betti numbers, can be used as numerical shape codes; shape similarities can be expressed as numerical differences between these shape codes [11]. The GSHAPE 90 computer program [21] evaluates the shape groups and shape codes of *ab initio* or semiempirical electron densities, and also evaluates nonvisual, numerical similarity measures of MIDCO sequences. For macromolecular systems, such as proteins, *ab initio* quality electron density calculations have become possible only recently by the introduction of the MEDLA (Molecular Electron Density Lego Assembler) approach [22]. The new MEDLA 93 computer program [23] extends the applicability of electron density shape modeling to macromolecules (see comments in Section 4), and provides alternatives to earlier approaches [24–32].

For each conformation *K* of a molecule, there are only a finite number of topologically different density domains, consequently, the three-dimensional shape and the bonding pattern of each conformer can be described by a finite sequence of density domains. This sequence can be represented by the shape groups, and the corresponding shape codes. The topological properties of density domains are invariant within small domains of the nuclear configuration space [33], hence the shape groups, shape codes, and alternative shape representations are also invariant within small segments along formal reaction paths [33–47]. However, the density domain sequences eventually change as the change in the nuclear configuration progresses along a formal reaction path. Nevertheless, along each path one finds only a finite number of different shape codes. The comparison of shape codes provides a *molecular similarity measure* for individual nuclear arrangements, whereas a comparison of the *sequences of shape codes* occurring along reaction paths provides a *similarity measure between two reaction paths*. The shape groups and the associated shape codes provide a link between the two levels of similarities. For the study of sequences of topologically different shape patterns the methods of "iterated similarity analysis" can be applied [48].

2.2. Similarity of Reaction Paths and Reaction Mechanisms

In this section a special but rather common family of reaction paths and reaction mehanisms are discussed: those confined to a single potential energy hypersurface. A continuous change of the nuclear arrangement of a molecule corresponds to a formal path in the nuclear configuration space, often called the reaction path. The first precise definition of reaction paths has been given by Fukui [49] in terms of the concept of intrinsic reaction coordinate (IRC), based on the differential geometrical features of the potential energy hypersurface associated with the given electronic state. This approach has been generalized by Tachibana and Fukui to include the meta-IRC [50–52]. These methods, the force method of Pulay, and alternative techniques for the study of isoenergy contours and energy gradients [53–59] have led to the topological description of reaction paths and reaction mechanisms [12, 60–74]. Reaction path techniques can also be used for the study of interrelations between various chirality problems [75, 76].

A classical mechanical, geometrical model of a reaction path is not fully satisfactory, and topology offers alternative models. Molecules, as quantum mechanical objects with wave-packet properties, obey the Heisenberg uncertainty relation, hence, their rearrangements cannot follow strictly any specific path in the classical sense. The concept of *reaction channel* describes better the quantum mechanical reality. Such a reaction channel can be represented by a family of reaction paths which are *similar* in some sense. This similarity can be expressed by topological means. Topological equivalence classes of classical reaction paths, fulfilling some similarity condition, may be taken to represent reaction mechanisms in a topological model. For example, reaction paths p and p' that are deformable into one another below some energy bound A are homotopically equivalent, that is, they belong to the same homotopy equivalence class $[p]_A$ of energy bound A, and can be regarded as *similar*. The corresponding equivalence classes define *energy-dependent reaction mechanisms* [12]. Clearly, at higher energy bound, more paths become homotopically equivalent.

It is convenient to describe the above model of reaction mechanisms using the concepts introduced for the analysis of potential energy hypersurfaces. Several topological models of potential energy hypersurfaces have been introduced. One such topological model, based on the *catchment regions* [12, 61], relies on the differential geometric model of reaction paths [49–52] and on tools and concepts originally proposed for geographical terrains [77, 78]. Consider a potential energy hypersurface $E(K)$. A point K of the nuclear configuration space M where the energy gradient of $E(K)$ vanishes is a *critical point* of $E(K)$. The catchment regions of $E(K)$ are defined in terms of the critical points. Each critical point $K(\lambda, i)$ is characterized by two indices: by the critical point index λ, defined as the number of negative eigenvalues of the Hessian matrix of second derivatives of the function $E(K)$ at point $K(\lambda, i)$, and by a serial index i. The catchment region $C(\lambda, i)$ of critical point $K(\lambda, i)$ is the collection of all points K of the configuration space M from where an infinitely slow relaxation (a steepest descent path in a mass-weighted coordinate system) leads to $K(\lambda, i)$.

Consider two reaction mechanisms, both leading from a given set of reactants to a given set of products; that is, only the actual routes of the chemical transformations are different at the given energy bound A. The similarity between two such reaction mechanisms, $[p_1]_A$ and $[p_2]_A$ of energy bound A, can be characterized by various means. For example, an energy-based measure of their similarity is expressed by the minimum energy increase ΔA needed to reach an energy bound $A + \Delta A$ where the two homotopy classes $[p_1]_A$ and $[p_2]_A$ merge into a single homotopy class $[p_3]_{A+\Delta A}$.

Alternative similarity measures are based on molecular geometry. The differences between the transition structure nuclear arrangements ($K(\lambda, i)$ with $\lambda = 1$) for the two reaction mechanisms provides one such measure. Take the highest transition structure saddle point nuclear arrangements $K(1, 1)$ and $K(1, 2)$ required for interconversion between the reactants and products for the two reaction mechanisms $[p_1]_A$ and $[p_2]_A$, respectively. The

$$\Delta K_{tr} = d(K(1, 1), K(1, 2)), \tag{5}$$

nuclear configuration space distance between these two saddle points characterizes the dissimilarity between the two reaction mechanisms.

Chirality is a special aspect of shape. The changes in chirality along reaction paths can be used for the classification of paths, based on relations between molecular shape properties and shape properties of the paths. A special question of some interest is the following: what are the conditions and probabilities for the occurrence of chiral and achiral nuclear arrangements along interconversion paths between enantiomers? A simple transformation of a formal chiral center, for example, a change of the bonding pattern about the carbon atom in CHDFCl, can always be selected so that it leads through an achiral arrangement, for example, a planar arrangement. However, the question is more general. An early example of transformations between enantiomers by reaction paths along which *all* nuclear arrangements are chiral has been given by Mislow and Bolstad [75]. In fact, this example represents the majority of chemical reactions. Based on a global approach to molecular chirality, general properties of chirality and achirality preserving reaction paths have been discussed [76]. As it can be shown, for any chiral molecular structure of five nuclei or more, almost all interconversion paths between enantiomers have the property that no achiral nuclear arrangement is found along them. (In a formal, mathematical sense, paths with achiral arrangements form a set of measure zero.) There are, of course, exceptional paths with some achiral nuclear arrangements, for example, those which involve planar nuclear arrangements. By contrast, for any four-nucleus chiral structure, all interconversion paths between the enantiomers *must* encounter at least one achiral arrangement. These results can be extended to other dimensions, for example, if one is interested in the interrelations and shapes of *sets* of configurations in a multidimensional nuclear configuration space M, or in nuclear arrangements along a planar catalytic surface. For any chiral set of $m = n + k$ points, where n is the dimension of the space, and $k \geq 2$, almost all interconversion paths between the enantiomers involve only chiral arrangements, as long as $n \geq 2$. If $k = 1$, and $n \geq 2$, then all interconversion paths between the enantiomers *must* involve at least one achiral arrangement. For $k < 1$, no chiral structure exists. The dimension $n = 1$ is special, since all paths must occur along a single line, that in the case of $m = 3$ implies either a formal "collision" between some points, or the occurrence of achiral structures in all rearrangements interconverting enantiomers. Even if coincident points are allowed, chirality preserving interconversion paths for $n = 1$ and $m = 3$ are possible only if all three points have distinct labels.

2.3. SHAPE SIMILARITY OF POTENTIAL ENERGY SURFACES: SIMILARITY IN REACTIVITY OF SINGLE SURFACE REACTIONS

The topological tools of shape similarity analysis are applicable to multidimensional objects such as potential energy hypersurfaces. In principle, the shape of an energy hypersurface $E(K)$ describes the reactivities of all the corresponding chemical species, as long as only reactions confined to the given surface are considered.

For each electronic state, the catchment regions $C(\lambda, i)$ of a nuclear configuration space M represent chemical species, such as stable conformers of molecules or exciplexes ($\lambda = 0$), including the stable relative arrangements of two or more molecules or exciplexes with a specified overall stoichiometry. Lower dimensional ($n = 3N - 7$) catchment regions describe transition structures ($\lambda = 1$) as formal "species" [12, 61]. The distribution and the pattern of arrangement of the catchment regions can be specified by topological tools, for example, by the homology groups of the catchment region pattern restricted to a subset of the nuclear configuration space and constrained by an energy bound [12]. Similarities between domains of two potential energy hypersurfaces, and similarities between the reactivities of the chemical species associated with these surfaces, can be characterized by a comparison of their energy-dependent homology groups [12]. The ranks of these groups, their Betti numbers, can be used as shape codes for various domains of the potential energy hypersurfaces, and these shape codes can be compared directly, leading to numerical similarity measures.

Lower bounds for the number of critical points of various types, *e.g.*, minima, and various saddle points of one or more negative canonical curvatures are given by the Morse inequalities of algebraic topology [79–81]. In many, common chemical problems, both lower and upper bounds on the number of critical points along potential energy hypersurfaces can be given by alternative relations proposed in ref. [62]. Since there is a one-to-one correspondence between the critical points and the catchment regions representing stable and unstable chemical species, these relations also provide bounds for the number of chemical species of various types.

The energy and the relative locations of critical points within the nuclear configuration space M are of special importance in the study of reactivity, and the pattern of reaction paths and reaction mechanisms. By considering an upper bound A for energy, and the family of all possible "loop" paths within the accessible regions of the potential energy hypersurface below the bound A, the paths which are continuously deformable into (that is, "homotopic" to) one another, are members of various energy-dependent homotopy equivalence classes. These homotopy classes form an energy dependent, algebraic group, the *fundamental group of reaction mechanisms* [12, 71–73]. The shape similarities of potential surface domains of various energy bounds A can be expressed in terms of their fundamental groups of reaction mechanisms [12]. The energy dependence can be correlated with temperature dependence. These energy-dependent groups provide an algebraic description of reactivity, and can be used as tools for assessing similarities between potential energy hypersurfaces.

2.4. Shape Similarity of Stacks of Potential Energy Surfaces: Similarity of Reactivities in Multi-Surface Reactions and Photochemistry

If reactions involving several potential energy hypersurfaces are considered, such as photochemical reactions, then the shape characterization problem becomes more complicated. The local shapes of all potential energy hypersurfaces involved influence the reaction. There are, however, important, simplifying relations among potential surfaces corresponding to different electronic states and defined over the same nuclear configuration space. One family of such relations is based on symmetry.

The *vertical point symmetry theorem of nuclear configuration spaces*, proven in [82], interrelates the location and the point symmetries of nuclear configurations of critical points found on different potential energy hypersurfaces. Consider another hypersurface B in the nuclear configuration space M that divides M into two parts, M_1 and M_2. We assume that part M_1 contains the hypersurface B as its boundary. Let $\boldsymbol{R}' = \{R'_1, R'_2, \ldots, R'_p\}$ be a family of symmetry elements $R'_1, R'_2, \ldots, R'_p$ which are present for *each* nuclear configuration K' along the boundary hypersurface B, and let $\boldsymbol{R} = \{R_1, R_2, \ldots, R_q\}$ be a family of symmetry elements $R_1, R_2, \ldots R_q$ which are present at some nuclear configuration K from set M_1. If no configuration along B possesses the entire family $\boldsymbol{R}$ of symmetry elements, or, if configuration K does not have all the symmetry elements of family $\boldsymbol{R}'$, then the family M_1 of configurations must contain at least one critical point for the potential energy surface of each electronic state (and also of each possible overall electronic charge).

The shapes of potential energy hypersurfaces influence the changes of molecular shapes in electronic excitations. Usually, a closed shell molecule with a singlet ground state electronic configuration has a different three-dimensional shape than any of the excited electronic states of the same molecule. These molecular shape differences are partly due to the fact that the potential energy hypersurfaces of different electronic states are usually rather different: local minimum points may occur at different nuclear arrangements K, and the point group symmetries of the most stable nuclear arrangements are also often different.

Some of the shape changes in electronic excitation processes follow a simple rule, analogous to the Quantum Chemical le Chatelier Principle [83]. The original Quantum Chemical le Chatelier Principle (QCLCP) describes some of the regularities of the changes of energy components in various chemical processes. In many reactions and conformational changes, various components of the molecular total energy undergo changes which show characteristic interrelations. For two components, their changes often appear to have opposite phase: one increases when the other decreases. For example, in a nuclear arrangement, the change of the nuclear repulsion energy component is usually accompanied by a change of opposite sign in the electronic energy component of the total energy. According to the QCLCP, the nuclear geometry change applied to the molecular system is regarded as a formal "stress", and as the overall change progresses, the system readjusts by changing the electronic energy in the opposite sense, thereby reducing the total energy change of the molecule. A formal justification of the QCLCP was derived from the variational principle of quantum mechanics [83].

The analogous principle for shape changes in electronic excitations is based on a simple model. In a semiclassical model of an electronic excitation or de-excitation process, the electron distribution changes rapidly, followed by a slower nuclear rearrangement. Using this model, the molecular shape changes associated with an electronic excitation can be formally decomposed into two separate steps as follows:

(i) In the first step, the formal nuclear geometry is preserved and the shape of the electron distribution corresponds to the new electronic state at the old nuclear geometry. The shape change of the electron distribution associated with this step is called the *vertical shape change.*

(ii) In the second formal step, the nuclear arrangement relaxes to some energy minimum of the new electronic state. The electron distribution readjusts as it follows the nuclear rearrangement. The corresponding change in the shape of the electron distribution is called the *shape change due to relaxation.*

The Quantum Chemical le Chatelier Principle for Molecular Shapes (QCLCP-MS) appears to apply to many shape changes in electronic excitations [39]: the *shape change due to relaxation tends to reduce the effect of the initial vertical shape change.*

In a formal sense, the initial vertical shape change in Step 1 may be regarded as a "stress" applied to the molecular system, and the system readjusts in Step 2 by a shape change due to relaxation. The overall shape change is usually less than the vertical shape change. A justification of this trend has been given by another trend: the shape complexity is often a monotonic function of the energy [11].

3. The T-hull method for similarity and reactivity studies

3.1. The convex hull of nuclear arrangements K for shape and similarity measures

The concept of convexity is fundamental to many branches of mathematics [84], as well as to many techniques of molecular shape characterization and the assessment of shape similarity [11]. A body B is a convex subset of the three-dimensional space, if for any two points $\mathbf{r}_1$ and $\mathbf{r}_2$ of B all points $\mathbf{r}$ of the straight line segment between $\mathbf{r}_1$ and $\mathbf{r}_2$ fall within B. Usually, a formal molecular body $B(a)$ enclosed by the corresponding MIDCO $G(a)$ is a convex set only for a low electron density threshold a, that reveals only little about the detailed shape features at higher densities. For more detailed shape and similarity studies, the methods of local convexity analysis and their various generalizations have been introduced [11]. In this chapter, some alternative generalizations will be described, with special emphasis on the nuclear arrangements.

The convex hull $\langle B \rangle$ of a set B is the smallest convex set that contains B. Consider the three-dimensional nuclear arrangement of configuration K, and generate its convex hull, denoted by $\langle K \rangle$. Note that, the same notation K is used for the three-dimensional nuclear arrangement and for the corresponding point K in the (3N – 6)-dimensional configuration space M. The convex hull $\langle K \rangle$ describes some of the aspects of molecular shape. Since the number N of nuclei is finite, $\langle K \rangle$ is a convex polyhedron. Some of the nuclei form the vertices of this polyhedron, whereas some other nuclei may be "buried" within the convex hull $\langle K \rangle$. The polyhedral mobility of vertices [85] is a concept describing the freedom of deformations which do not introduce essential shape changes. The combinatorial type [86, 87] of the polyhedron is invariant to some small variations of the locations of the nuclei, however, essential changes may occur for most larger changes in K.

Consider two convex polyhedra, V and W. The combinatorial type of V can be turned into that of W by adding and deleting vertices, edges, and faces. Let d_v, d_e, and d_f denote the cumulative number of changes in the number of vertices, edges, and faces, respectively, during such a conversion of the combinatorial type. The combinatorial distance $cd(V, W)$ between two convex polyhedra V and W can be defined as

$$cd(V, W) = \min\{d_v + d_e + d_f\}. \tag{6}$$

Let $n_{\text{vef}}(V)$ denote the sum of the number of vertices, edges, and faces of polyhedron V, and define

$$m_{\text{vef}}(V, W) = \max\{n_{\text{vef}}(V), n_{\text{vef}}(W)\}. \tag{7}$$

Considering two nuclear configurations K_1 and K_2, and their convex hulls $\langle K_1 \rangle$ and $\langle K_2 \rangle$, respectively, the combinatorial distance $cd(\langle K_1 \rangle, \langle K_2 \rangle)$ between the two convex hulls provides a combinatorial similarity measure $\text{csm}(K_1, K_2)$ between the two molecular conformations:

$$\text{csm}(K_1, K_2) = (m_{\text{vef}}(\langle K_1 \rangle, \langle K_2 \rangle) - cd(\langle K_1 \rangle, \langle K_2 \rangle))/m_{\text{vef}}(\langle K_1 \rangle, \langle K_2 \rangle). \tag{8}$$

These measures can be calculated easily, since their computation involves only the nuclear coordinates for the two nuclear configurations K_1 and K_2.

3.2. Generalized convexity: α-hulls and T-hulls

In this section we review the concepts of α-hull and T-hull, regarded as generalizations of the convex hull. We shall follow the terminology of the original references for the two-dimensional case of the α-hull [88], and for the generalization to T-hulls [89].

A *generalized disc of radius* $1/\alpha$ is defined as a disc of radius $1/\alpha$ if $\alpha > 0$, as the complement of a disc of radius $-1/\alpha$ if $\alpha < 0$, and as a half-plane if $\alpha = 0$.

The *α-hull* $\langle S\rangle_\alpha$ of a point set S in the plane is the intersection of all closed generalized discs of radius $1/\alpha$ which contain the point set S. In the special case of $\alpha = 0$, the α-hull $\langle S\rangle_\alpha$ of S is the ordinary convex hull $\langle S\rangle$ of the point set S. If the empty intersection is defined as the entire space, then for any set S of the plane the α-hull exists for any α value. For any finite point set S and a sufficiently small negative α, the α-hull $\langle S\rangle_\alpha$ of S is the finite point set S itself.

In spaces of higher dimensions, the α-hull is defined similarly. For our purposes, the three-dimensional case is of major importance, where the definition is entirely analogous. A *generalized ball of radius* $1/\alpha$ is defined as a ball of radius $1/\alpha$ if $\alpha > 0$, as the complement of a ball of radius $-1/\alpha$ if $\alpha < 0$, and as a halfspace if $\alpha = 0$. The *α-hull* $\langle S\rangle_\alpha$ of a finite point set S in a 3D Euclidean space is defined as the intersection of all closed generalized balls of radius $1/\alpha$ which contain S.

The T-hull is a generalization of the α-hull, by replacing disks and balls by more general objects T.

Consider some, closed, three-dimensional set T. This set T will be used as a reference object. The shapes of various other sets S will be described *relative* to the reference object T. The closure $\mathrm{clos}(E^3\backslash T)$ of the relative complement $E^3\backslash T$ of T is required for some of the definitions; for sake of simplicity, $-T$ is written for the closure of the relative complement of T:

$$-T = \mathrm{clos}(E^3\backslash T). \tag{9}$$

In some instances, the set $T' = -T$ is used as a reference object.

A *version* of T is a set obtained by translation and rotation of T. In special applications, the test object T is subject to orientation constraints. If orientation constraints apply then a version of T is a set obtained from T by translation.

The *T-hull* $\langle S\rangle_T$ of a point set S has been defined [89] as the intersection of all versions of T which contain S. If no version of T contains the point set S then the T-hull of S is the entire space. The T-hull of a set S depends on the relative shapes of the objects S and T.

If orientation constraints are considered, then the reference set T is provided with some fixed orientation with respect to the cartesian axes of the three-dimensional Euclidean space. The *oriented T-hull* $\langle S\rangle_{T,\,o}$ of a point set S has been defined [89] as the intersection of all those translated versions of the oriented set T which contain point set S.

Note that the α-hull is a special T-hull where the reference object T is taken as a ball of radius $1/\alpha$. For a ball, all orientations are equivalent.

The ordinary convex hull $\langle K \rangle$ of a nuclear configuration K provides a rather limited shape description. If a long chain molecule is considered in a conformation K that is bent considerably, then the ordinary convex hull $\langle K \rangle$ contains a large volume of essentially empty space, where the electronic density is negligibly small. Hence, the convex hull $\langle K \rangle$ gives a rather poor description of shape. By contrast, if an α-hull or an appropriate T-hull is used, then the new object provides a more faithful representation of the main shape features of the nuclear arrangement of the molecule.

The concept of *T-front* of a point set S interrelates the various T-hulls with polyhedra [89].

T-fronts are defined in terms of *T-extreme points*. A point p of S is a *T-extreme point* in S if there exists a version of reference set T such that p lies on its boundary and it contains S. Similarly, a point p of S is a *T_o-extreme point in S* if there exists a version of the reference set T_o of orientation o such that p lies on its boundary and it contains S.

If for three T-extreme points (or, in the oriented case, T_o-extreme points) p, p', and p'' there exists a version of the 3D set T (or, a version of set T_o of a specified orientation o) such that it has all three points p, p', and p'' on its boundary, and it contains set S, then points p, p', and p'' are said to form a *T-frontal triple of S* (a *T_o-frontal triple of S*, respectively).

The set of all frontal points for all versions of set T (oriented set T_o of orientation o) which contain set S is the *T-front of* set S (the *T_o-front of S*, respectively). The higher-dimensional definitions are analogous.

In the following we shall consider finite point sets S, such as the set of nuclei of a specified configuration K.

A *T-frontal set* (a *T_o-frontal set*) of point set S is a family of all frontal points of S falling on the boundary of a single version of reference object T (oriented reference object T_o, respectively). A frontal set is *maximal* if it is not a subset of a frontal set of more points. Consider all maximal T-frontal sets (T_o-frontal sets) of S, and all their intersections. These intersections are denoted by $f_i(S, T, k)$ (by $f_i(S, T_o, k)$, respectively), where i is a serial index and $k + 1$ is the number of elements in the intersection. A set $f_i(S, T, 0)$ (or $f_i(S, T_o, 0)$ is a *T-frontal vertex* (a *T_o-frontal vertex*) of set S. A set $f_i(S, T, 1)$ (or $f_i(S, T_o, 1)$ is a *T-frontal edge* (a *T_o-frontal edge*) of set S. The single-member intersections, that is, the maximal frontal sets $f_i(S, T, k)$ (or $f_i(S, T_o, k)$) themselves are the *ordinary T-frontal faces* (*ordinary T_o-frontal faces*) of set S. If an intersection $f_i(S, T, k)$ (or $f_i(S, T_o, k)$) is not a maximal frontal set and if $k > 1$, then it is called an *extraordinary T-frontal face* (*extraordinary T_o-frontal face*) of set S. In general, an intersection $f_i(S, T, k)$ (or $f(S, T_o, k)$) is called a k-face of the *generalized T-frontal polyhedron* (*generalized T_o-frontal polyhedron*) of set S. The incidence relations between these k-faces characterize the *combinatorial type* of the generalized T-frontal polyhedron (generalized T_o-frontal polyhedron) of set S.

By analogy with the ordinary convex hull $\langle K \rangle$ of a nuclear arrangement K, similarity measures can be defined using the generalized T-frontal polyhedra (generalized T_o-frontal polyhedra) $G(K)$ of nuclear configurations K.

Consider two nuclear configurations K_1 and K_2. Let $n_{\mathrm{vef}}(G(K))$ denote the sum of the numbers $n_k(G(K))$ of k-faces of generalized polyhedron $G(K)$,

$$n_{\mathrm{vef}}(G(K)) = \Sigma_k\, n_k(G(K)), \tag{10}$$

and define

$$m_{\mathrm{vef}}(G(K_1),\, G(K_2)) = \max\{n_{\mathrm{vef}}(G(K_1)),\, n_{\mathrm{vef}}(G(K_2))\}. \tag{11}$$

Let $d_1(G(K_1), G(K_2))$, $d_2(G(K_1), G(K_2))$, . . . , $d_k(G(K_1), G(K_2))$, . . . , denote the cumulative number of changes in the numbers $n_k(G(K))$ of various k-faces during a conversion of the combinatorial type of generalized T-polyhedron (generalized T_o-polyhedron) $G(K_1)$ into the combinatorial type of $G(K_2)$. The combinatorial distance $cd(G(K_1), G(K_2))$ between these generalized polyhedra can be defined as

$$cd(G(K_1),\, G(K_2)) = \min\{\Sigma_k\, d_k(G(K_1),\, G(K_2))\}. \tag{12}$$

The combinatorial distance $cd(G(K_1), G(K_2))$ between the two generalized polyhedra $G(K_1)$ and $G(K_2)$ provides a combinatorial similarity measure $\mathrm{csm}(K_1, K_2, T)$ (or, $\mathrm{csm}(K_1, K_2, T_o)$) between the two molecular conformations:

$$\begin{aligned}&\mathrm{csm}(K_1,\, K_2,\, T) = \\ &\quad (m_{\mathrm{vef}}(G(K_1), G(K_2)) - cd(G(K_1), G(K_2)))/m_{\mathrm{vef}}(G(K_1), G(K_2)).\end{aligned} \tag{13}$$

(The T_o case is entirely analogous). These measures are the simplest for ordinary α-hulls, that is, if reference object T is chosen as a ball.

3.3. Reactivity and the vertex mobility of T-hulls

Reactivity of molecules depends on the facility of nuclear rearrangements. The mobility of nuclei is also manifested in vibrational properties. In a model relying on T-hulls of nuclear configurations, mobility can be interpreted in terms of invariance and changes of the combinatorial types of generalized polyhedra.

For a given nuclear configuration K_1, a T-frontal polyhedron (T_o-frontal polyhedron) $G(K_1)$ of minimum volume is called a *T-polyhedron* (*T_o-polyhedron*) of configuration K_1. Two configurations K_1 and K_2 are said to be *TG-similar* (*T_oG-similar*) if both have T-frontal polyhedra (T_o-frontal polyhedra) and if the polyhedra $G(K_1)$ and $G(K_2)$ are of the same combinatorial type. If a nuclear configuration K_1 has no T-frontal polyhedron (T_o-polyhedron) $G(K_1)$, then we say that K_1 is not TG-representable (not T_oG-representable, respectively).

If either one of nuclear configurations K_1 and K_2 is not TG-representable (not T_oG-representable) then we say that these nuclear arrangements are not TG-comparable (not T_oG-comparable, respectively).

As follows from the definition, TG-similarity (T_oG-similarity) is an equivalence relation. The corresponding equivalence classes are defined by generalized polyhedral shape representations of nuclear arrangements K with respect to the test object T (oriented test object T_o). Consequently, we may refer to these classes as the TG-shape classes (T_oG-shape classes, respectively). If one regards the non-TG-representable (non-T_oG-representable) nuclear arrangements to belong to a special equivalence class, then the above similarity classification is applicable to all nuclear arrangements K of the configuration space M. For a given nuclear configuration K, non-TG-representability (non-T_oG-representability) indicates that the test object T (oriented test object T_o) is not a suitable shape criterion for a simple, polyhedral shape representation of K.

3.4. Schlegel diagrams of T-hulls as shape descriptors

The Schlegel diagram of the T-hull (T_o-hull) of a finite point set S (*e.g.*, nuclear configuration K) is defined by direct analogy with the Schlegel diagrams of convex polyhedra. For simplicity, we assume that the actual T-frontal generalized polyhedron (T_o-frontal generalized polyhedron) $G(K)$ of configuration K has no extraordinary faces. The graph $g(G(K))$ of $G(K)$ is obtained by adopting the vertices and edges of $G(K)$ as graph vertices and graph edges, respectively. Pictorially, a Schlegel diagram of $G(K)$ can be visualized as the view of the graph obtained when one ordinary face of $G(K)$ is stretched so that the cycle formed by its incident edges surrounds all non-incident vertices of the graph. The Schlegel diagram serves as a graphical shape descriptor of the original nuclear configuration K.

If configuration K has extraordinary faces, then the incidence relations between these and all other k-faces must be accounted for. This requires the introduction of labelled vertices, labelled edges, and labelled 2-faces, where the extraordinary faces are assigned an artificial dimension based on their incidence relations with other k-faces. The labels serve to distinguish ordinary and extraordinary faces of various actual or assigned dimensions.

3.5. T-hull volumes and T-hull densities and reactivity

Consider two nuclear configurations K_1 and K_2, and their convex hulls $\langle K_1\rangle$ and $\langle K_2\rangle$, respectively. If a reference object T is selected, then additional similarity measures are obtained by direct comparisons of the volumes of the convex hulls and the T-hulls obtained. If $V(A)$ denotes the volume of an object A, then the volume ratios

$$r(\langle K_1\rangle_T) = (V(\langle K_1\rangle_T) - V(\langle K_1\rangle))/V(\langle K_1\rangle) \tag{14}$$

and

$$r(\langle K_1\rangle_{T,o}) = (V(\langle K_1\rangle_{T,o}) - V(\langle K_1\rangle))/V(\langle K_1\rangle) \tag{15}$$

describe the relative volume change of the T-hulls as compared to the volume of the ordinary convex hull. Note that the $r(\langle K_1\rangle_T)$ and $r(\langle K_1\rangle_{T,o})$ relative volume changes may take both positive and negative values.

The relative volume change itself can be used as a similarity criterion; for two nuclear configurations, K_1 and K_2, the smaller the difference between their respective r values $r(\langle K_1\rangle_T)$ and $r(\langle K_2\rangle_T)$,

$$\Delta r(K_1, K_2)_T = |r(\langle K_1\rangle_T) - r(\langle K_2\rangle_T)|, \tag{16}$$

the more similar their relative space demand in an environment dominated by the shape conditions specified by the reference object T. The analogous definition applies to the oriented T-hulls:

$$\Delta r(K_1, K_2)_{T,o} = |r(\langle K_1\rangle_{T,o}) - r(\langle K_2\rangle_{T,o})|. \tag{17}$$

The actual mass density (md), as well as the simpler nucleus count density (Nd) within T-hulls provide some correlations with reactivity. Both highly crowded and too sparsely populated molecular moieties show general trends of reactivity. The following quantities are of relevance:

$$md(\langle K_1\rangle_T) = (\Sigma_j m_j)/V(\langle K_1\rangle_T), \tag{18}$$

$$md(\langle K_1\rangle_{T,o}) = (\Sigma_j m_j)/V(\langle K_1\rangle_{T,o}), \tag{19}$$

$$Nd(\langle K_1\rangle_T) = N/V(\langle K_1\rangle_T), \tag{20}$$

and

$$Nd(\langle K_1\rangle_{T,o}) = N/V(\langle K_1\rangle_{T,o}), \tag{21}$$

where m_j is the mass of the j-th nucleus. The changes of mass densities of T-hulls along chemical transformations are characteristic features of reactions. In many instances, a lowering of mass density is expected near the saddle point geometry of the transition structure of the reaction.

3.6. Reactivity and configuration space distribution of T-hulls; T-hull shape invariance domains of the configuration space

Although the discussion in this section is valid for any reference object T, for simplicity and visualization, it is useful to relate the concepts and tools to a fixed reference object T, for example, to a ball of radius $1/\alpha$. For any nuclear configuration K the generalized T-polyhedron $G(K)$ is characterized by its combinatorial type. The combinatorial type can be described by the family of incidence relations between k-faces of different dimensions. Here we shall focus on the incidence relations between the 0-faces (vertices) and 1-faces (edges) of the generalized T-polyhedra $G(K)$. Other incidence relations, for example, those between the 1-faces and 2-faces, can be described similarly.

The elements of the $N \times N$ binary matrix $\mathbb{G}(G(K))$ are defined by the incidence relations between the vertices and edges of the generalized T-polyhedron $G(K)$, as expressed by pairs of T-extreme points p_j and p_k of set K falling on edges $f_i(K,T,1)$ of $G(K)$ where the maximum possible dimension N is used:

$$\mathbb{G}(G(K))_{jk} = \begin{cases} 1 & \text{if exists } f_i(K, T, 1) \text{ such that } p_j, p_k \in f_i(K, T, 1) \\ 0 & \text{otherwise.} \end{cases} \tag{22}$$

The matrix representations of the incidence relations of the $G(K)$ generalized polyhedra are suitable to study the topologically significant shape changes of T-hull sequences along reaction paths.

In chemical reactions and conformational changes the matrix $\mathbb{G}(G(K))$ remains invariant for most small deformations, but it changes as the nuclear configuration K undergoes significant changes. A reaction path p is regarded as a formal mapping from the unit interval I to the metric nuclear configuration space M,

$$p : I = [0, 1] \to M, \tag{23}$$

using a parametrization,

$$p = p(u) = K(u), \quad 0 \le u \le 1, \tag{24}$$

where $u = 0$ corresponds to the initial point (reactant configuration)

$$K_0 = p(0) \in M, \tag{25}$$

and $u = 1$ corresponds to the "final" (product) nuclear configuration

$$K_f = p(1) \in M. \tag{26}$$

Most small displacements Δu do not necessarily alter the topological pattern of the T-hull as described by $\mathbb{G}(G(K(u)))$, and one finds

$$\mathbb{G}(G(K(u))) = \mathbb{G}(G(K(u + \Delta u))). \tag{27}$$

Along the entire path $p(u)$, there are only a *finite number* of different $\mathbb{G}(G(K))$ matrices, as follows from the discrete nature of the matrix representation. The path $p(u)$ can be decomposed into a finite number w of invariance intervals of these matrices,

$$p_{G,1}, p_{G,2}, \ldots, p_{G,w}. \tag{28}$$

The corresponding sequence

$$\mathbb{G}(p, 1), \mathbb{G}(p, 2), \ldots, \mathbb{G}(p, w) \tag{29}$$

of matrices along path p characterizes the topological patterns of T-hulls and their variations in the corresponding reaction.

The matrix sequences $\mathbb{G}(p, 1), \mathbb{G}(p, 2), \ldots, \mathbb{G}(p, w)$ are suitable for studying the similarities of reaction paths and some aspects of reactivity. Two reaction paths p_1 and p_2 are regarded to be T-shape equivalent within the above context (that is, $\mathbb{G}$-shape equivalent) if and only if

(i) the numbers w_1 and w_2 of their shape invariance intervals agree,

$$w_1 = w_2 = w,$$

and

(ii) the two matrix sequences $\mathbb{G}(p_1, 1), \mathbb{G}(p_1, 2), \ldots, \mathbb{G}(p_1, w)$ and $\mathbb{G}(p_2, 1), \mathbb{G}(p_2, 2), \ldots, \mathbb{G}(p_2, w)$ are the same,

$$\begin{aligned}\{\mathbb{G}(p_1, 1), \mathbb{G}(p_1,2), \ldots, \mathbb{G}(p_1, w)\} = \\ \{\mathbb{G}(p_2, 1), \mathbb{G}(p_2, 2), \ldots, \mathbb{G}(p_2, w)\}.\end{aligned} \tag{31}$$

The above T-shape equivalence of two reaction paths p_1 and p_2 is denoted by

$$p_1 \,\mathbb{G}\, p_2. \tag{32}$$

This equivalence relation generates equivalence classes PT of reaction paths.

One may study the T-shape variations of nuclear configurations K within conformational domains instead of along reaction paths. Take a subclass $PT_1(C(\lambda, i), C(\lambda', i'))$ of a T-shape equivalence class PT_1 of reaction paths, defined by the following condition: $PT_1(C(\lambda, i), C(\lambda', i'))$ is the family of all paths from class PT_1 which start at the catchment region $C(\lambda, i)$, end at the catchment region $C(\lambda', i')$, and are homotopic to one another (continuously deformable into one another) while preserving the properties which qualify them for the equivalence class PT_1. The new set of conditions correspond to a new equivalence relation among paths, consequently, $PT_1(C(\lambda, i), C(\lambda', i'))$ is an equivalence class. Such equivalence classes $PT_1(C(\lambda, i), C(\lambda', i'))$ represent formal *reaction mechanisms defined in terms of the T-shape* ($\mathbb{G}$-shape) *of nuclear configurations.*

For each reaction mechanism defined by such conditions, one can use a representative reaction path p. The similarities between various reaction mechanisms can be analyzed and quantified by direct, numerical comparisons of the corresponding matrix sequences $\mathbb{G}(p, 1), \mathbb{G}(p, 2), \ldots, \mathbb{G}(p, w)$.

In a more general context, the entire nuclear configuration space M can be partitioned into invariance domains according to the $\mathbb{G}(K)$ matrices. A matrix $\mathbb{G}(K)$ can be defined for each nuclear configuration K, even for those not compatible with the reference object T; for these latter families of nuclear configurations the $N \times N$ zero matrix is assigned. These matrices are invariant for most small variations of the nuclear configuration K. Except for some pathological cases, there are only a finite number of $\mathbb{G}(K)$ matrices associated with the nuclear configuration space M. The invariance domains of these matrices are the T-shape ($\mathbb{G}$-shape) invariance domains of the metric nuclear configuration space M, where in the actual example only the incidence relations of the 0-faces and the 1-faces of the generalized polyhedra $G(K)$ of the T-hulls are considered.

4. Summary

The role of shape characterization in the study of molecular similarity and reactivity is discussed, with special emphasis on the changes of topological patterns of electron density and nuclear distributions along reaction paths and in more general chemical processes. The *ab initio* quality electron densities calculated for proteins by the MEDLA method will extend reactivity studies to the shapes of macromolecules [90, 91].

Acknowledgments

This work was supported by both strategic and operating research grants from the Natural Sciences and Engineering Research Council of Canada, and by additional financial assistance from the Computational Chemistry Laboratory of the Upjohn Company.

References

1. Johnson, M. A., and Maggiora, G. M., eds., *Concepts and Applications of Molecular Similarity* (Wiley, New York, 1990).
2. Carbó, R., Leyda, L., and Arnau, M., *Int. J. Quantum Chem.* **17**, 1185 (1980).
3. Carbó, R. and Domingo, Ll., *Int. J. Quantum Chem.* **32**, 517 (1987).
4. Carbó, R. and Calabuig, B., *Comput. Phys. Commun.* **55**, 117 (1989).
5. Carbó, R. and Calabuig, B., *Int. J. Quantum Chem.* **42**, 1681 (1992).
6. Carbó, R. and Calabuig, B., *Int. J. Quantum Chem.* **42**, 1695 (1992).
7. Hodgkin, E. E. and Richards, W. G., *J. Chem. Soc. Chem. Commun.* **1986**, 1342.
8. Hodgkin, E. E. and Richards, W. G., *Int. J. Quantum Chem.* **14**, 105 (1987).
9. Burt, C., Richards, W. G., and Huxley, P., *J. Comput. Chem.* **11**, 1139 (1990).
10. Good, A. and Richards, W. G., *J. Chem. Inf. Comp. Sci.* **32**, 112 (1992).
11. Mezey, P.G., *Shape in Chemistry: An Introduction to Molecular Shape and Topology*, (VCH Publishers, New York, 1993).
12. Mezey, P. G., *Potential Energy Hypersurfaces*, (Elsevier, Amsterdam, 1987).
13. Mezey, P. G., *J. Chem. Inf. Comp. Sci.* **32**, 650 (1992).
14. Mezey, P. G., *Canad. J. Chem.* **72**, 928 (1994). (Special issue dedicated to Prof. J. C. Polanyi).
15. Mezey, P. G., *Int. J. Quant. Chem. Quant. Biol. Symp.* **12**, 113 (1986).
16. Mezey, P. G., *J. Comput. Chem.* **8**, 462 (1987).
17. Mezey, P. G., *J. Math. Chem.* **2**, 325 (1988).
18. Mezey, P. G., "Three-Dimensional Topological Aspects of Molecular Similarity", in *Concepts and Applications of Molecular Similarity*, eds. M. A. Johnson and G. M. Maggiora (Wiley, New York, 1990).
19. Mezey, P. G., "Molecular Surfaces", in *Reviews in Computational Chemistry*, eds. K. B. Lipkowitz and D. B. Boyd (VCH Publishers, New York, 1990).
20. Mezey, P. G., "Topological Quantum Chemistry", in *Reports in Molecular Theory*, eds. H. Weinstein and G. Náray-Szabó (CRC Press, Boca Raton, 1990).
21. Walker, P. D., Arteca, G. A., and Mezey, P. G., *Program GSHAPE 90*, (Mathematical Chemistry Research Unit, University of Saskatchewan, Saskatoon, Canada, 1990).
22. Walker, P. D. and Mezey, P. G., *J. Am. Chem. Soc.* **115**, 12423 (1993).

23. Walker, P. D. and Mezey, P. G., *Program MEDLA 93*, (Mathematical Chemistry Research Unit, University of Saskatchewan, Saskatoon, Canada, 1993).
24. Mezey, P. G., *J. Math. Chem.* **3**, 407 (1989).
25. Maggiora, G. M., Mezey, P. G., Mao, B., and Chou, K. C., *Biopolymers* **30**, 211 (1990).
26. Arteca, G. A. and Mezey, P. G., *J. Mol. Graphics* **8**, 66 (1990).
27. Arteca, G. A. and Mezey, P. G., "A Topological Analysis of Macromolecular Folding Patterns", in *Theoretical and Computational Models for Organic Chemistry*, eds. S. J. Formosinho, I. G. Csizmadia, and L. G. Arnaut (Kluwer Academic Publishers, Dordrecht, 1991).
28. Arteca, G. A., Tapia, O., and Mezey, P. G., *J. Mol. Graphics* **9**, 148 (1991).
29. Arteca, G. A. and Mezey, P. G., "Algebraic Approaches to the Shape Analysis of Biological Macromolecules", in *Computational Chemistry, Structure, Interactions and Reactivity, Part A*, ed. S. Fraga (Elsevier, Amsterdam, 1992).
30. Mezey, P. G., "Dynamic Shape Analysis of Biomolecules Using Topological Shape Codes", in *The Role of Computational Models and Theories in Biotechnology*, ed. J. Bertran (Kluwer Academic Publishers, Dordrecht, 1992).
31. Arteca, G. A. and Mezey, P. G., *Biopolymers* **32**, 1609–1621 (1992).
32. Mezey, P. G., *J. Math. Chem.* **12**, 365 (1993).
33. Mezey, P. G., *J. Math. Chem.* **2**, 299 (1988).
34. Arteca, G. A. and Mezey, P. G., *J. Phys. Chem.* **93**, 4746–4751 (1989).
35. Mezey, P. G., *J. Math. Chem.* **4**, 377–381 (1990).
36. Arteca, G. A., Heal, G. A., and Mezey, P. G., *Theor. Chim. Acta* **76**, 377 (1990).
37. Arteca, G. A. and Mezey, P. G., *Int. J. Quantum Chem.* **38**, 713–726 (1990).
38. Arteca, G. A. and Mezey, P. G., *Int. J. Quantum Chem. Symp.* **24**, 1 (1990).
39. Mezey, P. G., "Non-Visual Molecular Shape Analysis: Shape Changes in Electronic Excitations and Chemical Reactions", in *Computational Advances in Organic Chemistry (Molecular Structure and Reactivity)*, ed. C. Ogretir and I. G. Csizmadia, Nato ASI Series (Kluwer Academic Publishers, Dordrecht, 1991).
40. Mezey, P. G., "New Symmetry Theorems and Similarity Rules for Transition Structures", in *Theoretical and Computational Models for Organic Chemistry*, ed. S. J. Formosinho, I. G. Csizmadia, and L. G. Arnaut (Kluwer Academic Publishers, Dordrecht, 1991).
41. Arteca, G. A. and Mezey, P. G., *J. Mol. Structure Theochem* **230**, 323 (1991).
42. Luo, X., Arteca, G. A., and Mezey, P. G., *Int. J. Quantum Chem. Symp.* **25**, 335 (1991).
43. Luo, X., Arteca, G. A., and Mezey, P. G., *Int. J. Quantum Chem.* **42**, 459 (1992).
44. Arteca, G. A. and Mezey, P. G., *J. Math. Chem.* **10**, 329–371 (1992).
45. Mezey, P. G., *Canad. J. Chem.* **70**, 343 (1992). (Special issue dedicated to Prof. S. Huzinaga).
46. Mezey, P. G., *J. Math. Chem.* **13**, 59 (1993).
47. Mezey, P. G., *J. Math. Chem.* **7**, 39 (1991).
48. Mezey, P. G., *J. Chem. Inf. Comp. Sci.* **34**, 244 (1994).
49. Fukui, K., *J. Phys. Chem.* **74**, 4161 (1970).
50. Tachibana, A. and Fukui, K., *Theor. Chim. Acta* **49**, 321 (1978).
51. Tachibana, A. and Fukui, K., *Theor. Chim. Acta* **51**, 189 (1979).
52. Tachibana, A. and Fukui, K., *Theor. Chim. Acta* **51**, 275 (1979).
53. Pulay, P., *Mol. Phys.* **17**, 197 (1969).
54. Pulay, P., *Mol. Phys.* **18**, 473 (1970).
55. Pulay, P., "Direct use of gradients for investigating Molecular Energy Surfaces", in *Applications of Electronic Structure Theory*, ed. H. F. Schaefer (Plenum, New York, 1977).
56. Mezey, P. G., *Progr. Theor. Org. Chem.* **2**, 127 (1977).
57. Basilevsky, M. V., *Chem. Phys.* **24**, 81 (1977).
58. Garrett, B. C., Truhlar, D. G., and Grev, R. S, in *Potential Energy Surfaces and Dynamic Calculations*, ed. D. G. Truhlar (Plenum, New York, 1981).
59. Quapp, W. and Heidrich, D., *Theor. Chim. Acta* **66**, 245 (1984).
60. Mezey, P. G., *Theor. Chim. Acta* **54**, 95 (1980).
61. Mezey, P. G., *Theor. Chim. Acta* **58**, 309 (1981).

62. Mezey, P. G., *Chem. Phys. Letters* **82**, 100 (1981); **86**, 562 (1982).
63. Mezey, P. G., *Theor. Chim. Acta* **60**, 97 (1981).
64. Mezey, P. G., *Internat. J. Quantum Chem., Quant. Biol. Symp.* **8**, 185 (1981).
65. Mezey, P. G., *Theor. Chim. Acta* **62**, 133 (1982).
66. Mezey, P. G., *Theor. Chim. Acta* **63**, 9 (1983).
67. Mezey, P. G., *J. Chem. Phys.* **78**, 6182 (1983).
68. Mezey, P. G., *Can. J. Chem.* **61**, 956 (1983). (Volume dedicated to Prof. H. Gunning).
69. Mezey, P. G., "The Topological Model of Non-Rigid Molecules and Reaction Mechanisms", in *Symmetries and Properties of Non-Rigid Molecules: A Comprehensive Survey*, eds. J. Maruani and J. Serre (Elsevier Sci. Publ. Co., Amsterdam, 1983).
70. Mezey, P. G., *Internat. J. Quantum Chem.* **26**, 983 (1984).
71. Mezey, P. G., *Internat. J. Quantum Chem. Symp.* **18**, 77 (1984).
72. Mezey, P. G., *Theor. Chim. Acta* **67**, 43 (1985).
73. Mezey, P. G., *Theor. Chim. Acta* **67**, 91 (1985).
74. Mezey, P. G., *Theor. Chim. Acta* **67**, 115 (1985).
75. Mislow, K. and Bolstad, R., *J. Amer. Chem. Soc.* **77**, 6712 (1955).
76. Mezey, P. G., "A Global Approach to Molecular Chirality", in *New Developments in Molecular Chirality*, ed. P. G. Mezey (Kluwer Academic Publishers, Dordrecht, 1991).
77. Cayley, C. A., *Philos. Mag.* **18**, 264 (1859).
78. Maxwell, J. C., *Philos. Mag.* **40**, 233 (1870).
79. Morse, M., *Calculus of Variation in the Large*, (Am. Math. Soc. Colloq. Publ., Vol. 18, 1934).
80. Morse, M. and Cairns, S. S., *Critical Point Theory in Global Analysis and Differential Topology*, (Academic Press, New York, 1969).
81. Milnor, J., *Morse Theory*, (Annals of Math. Studies, Vol. 51, Princeton Univ. Press, Princeton, 1973).
82. Mezey, P. G., *J. Am. Chem. Soc.* **112**, 3791 (1990).
83. Mezey, P. G., *Internat. J. Quantum Chem.* **25**, 853 (1984).
84. Berger, M., *Amer. Math. Monthly* **97**, 650 (1990).
85. Luo, X., Taylor, K., and Mezey, P. G., *Bull. Math. Biol.* **55**, 131 (1993).
86. Grünbaum, B., *Convex Polytopes*, (John Wiley and Sons, New York, 1967).
87. Federico, P. J., *Geometriae Dedicata* **3**, 469 (1975).
88. Edelsbrunner, H., Kirkpatrick, D. G., and Seidel, R., *IEEE Trans. Inform. Theor.*, **IT-29**, 551 (1983).
89. Mezey, P. G., *J. Math. Chem.* **8**, 91 (1991).
90. Walker, P. D. and Mezey, P. G., *Canad. J. Chem.* **72**, 2531 (1994).
91. Walker, P. D. and Mezey, P. G., *J. Amer. Chem. Soc.* **116**, 12022 (1994).

Electron Density Approximations for the Rapid Evaluation of Quantum Molecular Similarity Measures

J. MESTRES, M. SOLÀ, E. BESALÚ, M. DURAN and R. CARBÓ
Institut de Química Computacional, Universitat de Girona, Plaça de l'Hospital 6, 17071 Girona, Catalonia, Spain

Key words: *ab initio* similarity, density fitting, Quantum Molecular Similarity Measures

Abstract. An approximation to the exact density function has been devised by expanding the density function in a set of gaussian functions. This approximation allows the fast computation of Quantum Molecular Similarity Measures from any kind of wave function, the only requirement being the need of a density matrix. Wave functions obtained by means of techniques ranging from semiempirical to high ab initio methodologies are tested and the use of this approximate density is justified from the large computing time savings and the results of Quantum Molecular Similarity Measures obtained.

1. Introduction

One of the basic concepts of human knowledge is that of comparing how similar is a new concept to any other concept previously acquired. In a first stage, new concepts are entered in our brain as related to older concepts before a more concrete concept is gained when dissimilarities are found [1]. In Chemistry we can profit from these basic concepts of similarity and dissimilarity in order to compare series of molecules, study the evolution of a reaction, analyze the effect of substitution in a given molecule, design new drugs, *etc*. All we need is a well defined chemical property to quantify similarities among chemical systems. In this sense, the density function can be considered as the natural candidate to be used in the evaluation of the similarity.

In fact, one of the most used expression to compute the Quantum Molecular Similarity Measure (QMSM) [2] between two chemical systems $\{I, J\}$ uses their first-order electron density functions $\{D_I, D_J\}$ as:

$$Z_{IJ}(\Theta) = \int D_I(\mathbf{r}_1)\, \Theta(\mathbf{r}_1, \mathbf{r}_2)\, D_J(\mathbf{r}_2)\, d\mathbf{r}_1\, d\mathbf{r}_2. \quad (1)$$

$\Theta(\mathbf{r}_1, \mathbf{r}_2)$ being a positive definite operator depending on two-electron coordinates. Overlap-like QMSM are obtained when the $\Theta(\mathbf{r}_1, \mathbf{r}_2)$ operator is chosen as the Dirac Delta function $\delta(\mathbf{r}_1 - \mathbf{r}_2)$. Use of the operator $1/r_{12}$ or $1/r_{12}^2$ gives raise to Coulomb-like QMSM and Gravitational-like QMSM, respectively.

The main drawback of the definition of QMSM given in Equation (1) is found in the need to evaluate and process the four-index integrals involved, whose number scales as N^4, being N the number of basis functions. This

R. Carbó (ed.), Molecular Similarity and Reactivity: From Quantum Chemical to Phenomenological Approaches, 77–85.

problem constitutes the most significant bottleneck in QMSM evaluation as well as in many other types of *ab initio* calculations.

As yet, different methododogies have been proposed in order to circumvent this common N^4 scaling problem. For instance, pseudospectral Hartree-Fock methodology expand the density in an auxiliary basis set and solves the integrals by the generation of a molecular grid [3]. More recently, Vahtras, Almlöf and Feyereisen, [4] have developped a new promising methodology in which the products of basis functions are expanded in an auxiliary basis set. Finally, in the method proposed in a previous publication [5] we have followed the initial work by Baerends *et al.* [6] using Slater type orbitals and by Sambe and Felton [7] and also by Dunlap and coworkers [8] using Gaussian type functions, and we have approximated the electronic first order density function by a linear combination of a gaussian function set $\Gamma(\boldsymbol{r}) = \{g_i(\boldsymbol{r})\}$:

$$D(\boldsymbol{r}) \simeq \rho(\boldsymbol{r}) = \sum_i ci \; g_i(\boldsymbol{r}). \tag{2}$$

Use of $\rho(\boldsymbol{r})$ instead of $D(\boldsymbol{r})$ in Equation (1) eliminates the need to evaluate costly four-centered integrals in the QMSM evaluation, and reduces the order of the problem to N_f^2, being N_f the number of auxiliary functions used in the density expansion. In this previous work [5] we have analyzed and discussed various different fits to the exact electron density in terms of computational time used and ability to reproduce *ab initio* QMSM. We have concluded that analytical integration is preferred to numerical integration in order to avoid the large dependence on the grid used attached to numerical fitting. Furthermore, this work shows that excellent Overlap-like QMSM results can be obtained from fitted densities when compared to the exact Overlap-like QMSM.

The main goal of the present work is to extend our previous analysis [5] to semiempirical and correlated wave functions and compare the exact QMSM values to the QMSM values obtained from the different fitted densities.

2. Methodology

The methodology used here in order to compute QMSM has been previously reported [5]. Basically, an approximate density ($\rho(\boldsymbol{r})$) is least square fitted to the exact charge density ($D(\boldsymbol{r})$) by minimizing the so-called quadratic error integral:

$$S(\Theta) = \int (D(\boldsymbol{r}_1) - \rho(\boldsymbol{r}_1))\; \Theta(\boldsymbol{r}_1, \boldsymbol{r}_2)(D(\boldsymbol{r}_2) - \rho(\boldsymbol{r}_2))\; d\boldsymbol{r}_1\, d\boldsymbol{r}_2. \tag{3}$$

where $\Theta(\boldsymbol{r}_1, \boldsymbol{r}_2)$ is, again, a positive definite operator depending on two-electron coordinates. Here the $\Theta(\boldsymbol{r}_1, \boldsymbol{r}_2)$ operator appearing in Equation (3) has been chosen to be the Dirac delta function $\delta(\boldsymbol{r}_1 - \boldsymbol{r}_2)$.

In this work we have analyzed the performance of four different sets of $\{g_k\}$ functions which corresponds to the so-called BSA, PSA, PSA + IBS and PSA-MIN fitting methodologies [5]. In all these fitting procedures *s*-type functions have been used in order to preserve the rotational invariance, which is of paramount importance when one is maximizing the similarity. To obtain s-type functions from the basis functions, all $l > 0$ functions have been transformed into $l = 0$ functions by setting their angular quantum numbers to zero and renormalizing. Linear dependency problems (as those due to the presence of sp shells) have been avoided by using a set of independent functions. Integrations have been performed analitically over all space.

In the BSA fit, the set of $\{g_k\}$ functions has been chosen to be the same as the squared molecular s-type basis functions, whereas in the PSA fit the set of $\{g_k\}$ functions has been obtained from the squared s-type primitive functions. So the only difference between these two fitting procedures is the presence in the BSA method of the squared basis functions leading to crossed terms between primitives. Inter-bonding functions have been included in the PSA + IBS fitting methods to analyze the effect of these functions in the fit and in the calculation of molecular similarities from an approximate density. For each bond, two s-type gaussian functions have been added at one third and two thirds of the bond length. We have taken as the exponents of these inter-bonding functions the polarization exponents given by Pople *et al.* [9c] corresponding to the nearest atom to the function, multiplied by two. Finally, when large basis sets have been used in the calculation of the molecular wave function, a minimal set of gaussian functions coming from the STO-3G basis set [9d] is proposed as auxiliary functions for the fitting method named PSA-MIN.

The wave functions from which the electron density is fitted have been calculated by means of the GAUSSIAN-90 system of programs [10]. The program MESSEM [11] extracts the geometry, basis set and one-electron density matrix from the GAUSSIAN system of programs, performs a least-squares fitting to the exact electron density and computes exact and fitted QMSM using diferent $\Theta(\boldsymbol{r}_1, \boldsymbol{r}_2)$ operators in Equation (1). Semiempirical wave functions have been obtained using the AM1 method [12]. *Ab initio* calculations have been carried out using the common STO-3G, 3-21G and 6-31G** basis sets from Pople and coworkers [9]. The electron correlation effect into the one-electron density matrix (at the MP2 and CISD levels of theory) has been introduced by using the generalized corrections at each level [13]. All calculations have been performed on an IBM RISC/6000-355 workstation.

3. Results and discussion

Three different molecules have been considered in our study, namely, water, formaldehide, and furan in their C_{2v} symmetry. Tables I–III collect the overlap self-QMSM values obtained for the different fit methods and levels of theory used, together with the quadratic error (Equation (3)) with respect to the exact self-QMSM values.

TABLE I

Overlap-QMSM values and quadratic error (S) for H_2O at the different levels of theory

METHOD[a]	AM1/VSTO-3G		RHF/STO-3G		RHF/3-21G		RHF/6-31G**		MP2/6-31G**		CISD/6-31G**	
	QMSM	S	QMSM	S	QMSM	S	QMSM	S	QMSM	S	QMSM	S
BSA	3.12434	0.0572	79.45811	0.1161	80.21731	0.0425	81.00364	0.0248	81.01984	0.0210	81.01474	0.0217
PSA	3.13998	0.0552	79.19956	0.2304	80.18224	0.0910	81.00969	0.0227	81.02600	0.0193	81.02024	0.0200
PSA + IBS	3.13136	0.0483	79.18932	0.2037	80.16154	0.0794	81.01678	0.0134	81.03160	0.0114	81.02617	0.0119
PSA-MIN	–	–	–	–	80.17357	0.0980	80.92206	0.1404	80.93739	0.1369	80.93201	0.1376
EXACT	3.20917	–	79.41220	–	80.24519	–	81.03520	–	81.04600	–	81.04111	–

[a] See text for fit method description.

TABLE II
Overlap-QMSM values and quadratic error (S) for H_2CO at the different levels of theory

METHOD[a]	AM1/VSTO-3G		RHF/STO-3G		RHF/3-21G		RHF/6-31G**		MP2/6-31G**		CISD/6-31G**	
	QMSM	S	QMSM	S	QMSM	S	QMSM	S	QMSM	S	QMSM	S
BSA	3.67860	0.0511	110.32843	0.1135	111.40123	0.0644	112.54072	0.0526	112.57247	0.0431	112.56962	0.0459
PSA	3.54847	0.0538	109.82887	0.2855	111.37317	0.1317	112.63804	0.0441	112.66072	0.0359	112.65979	0.0384
PSA + IBS	3.55924	0.0444	109.84076	0.2691	111.40304	0.1168	112.64121	0.0245	112.66252	0.0206	112.66173	0.0222
PSA-MIN	–	–	–	–	111.41574	0.1354	112.46492	0.2127	112.49186	0.2041	112.49005	0.2067
EXACT	3.66067	–	110.18264	–	111.55010	–	112.66061	–	112.67687	–	112.67760	–

[a] See text for fit method description.

TABLE III
Overlap-QMSM values and quadratic error (*S*) for Furan at the different levels of theory. Fitting times (in seconds) are added in parenthesis

METHOD[a]	AM1/VSTO-3G		RHF/STO-3G		RHF/3-21G		RHF/6-31G**		MP2/6-31G**		CISD/6-31G**	
	QMSM	S	QMSM	S	QMSM	S	QMSM	S	QMSM	S	QMSM	S
BSA	5.80655 (10.37)	0.0671	202.81060 (21.31)	0.1570	204.77216 (17.41)	0.1026	207.01070 (213.45)	0.0842	207.16120 (213.45)	0.0663	207.14264 (213.45)	0.0741
PSA	5.50856 (5.29)	0.0751	201.90350 (10.66)	0.4797	204.71808 (10.66)	0.2337	207.19869 (94.33)	0.0658	207.33350 (94.33)	0.0510	207.32101 (94.33)	0.0576
PSA + IBS	5.53330 (9.40)	0.0557	201.91671 (15.79)	0.4461	204.79301 (15.79)	0.1912	207.18504 (117.96)	0.0246	207.31734 (117.96)	0.0204	207.30559 (117.96)	0.0223
PSA-MIN	–	–	–	–	204.81004 (12.37)	0.2304	206.81746 (55.18)	0.3914	206.96156 (55.18)	0.3740	206.94529 (55.18)	0.3820
EXACT	5.66799	–	202.47714	–	205.02216	–	207.17714	–	207.30527	–	207.29455	–

[a] See text for fit method description.

From the results collected in these three tables one can extract several general trends:

1) Regarding the results at the SCF level we can observe that improving the basis set translates into larger QMSM values. This can be explained by the fact that addition of more gaussian functions, to obtain a better description of the core orbitals, increases the core electron density which in turn produces larger values of the QMSM.

2) The small QMSM value obtained at the AM1 level is due to the fact that we are using a valence basis set. To test the consistency of these values we perform two calculations from the SCF/STO-3G wave function of the water molecule using the BSA fit method. First, we perform a fitting of the total density and then we obtain the QMSM value only by taking into account the valence orbitals. This gives us a value of 2.93179. Second, only a fitting of the valence density was performed and then the total QMSM value obtained was of 3.05188. As can be seen, both numbers clearly agree with that obtained from the AM1 wave function (3.12434).

3) Including the effect of the electron correlation into the one-electron density matrices gives also larger QMSM values and slightly smaller S values. A possible explanation can be obtained if one takes into account that QMSM values reflect, in fact, how localized is the distribution of the total electron density in the molecule. For instance, if we take two different molecules with the same number of electrons, as H_2O and CH_4, we obtain exact QMSM values of 79.4123 and 31.0998, respectively. As a consequence, the trend of the QMSM values when adding the electron correlation effect, will give us an idea about the concentration or depletion of the total electron density distribution.

4) The quadratic error value (S) reflects the quality of the fit. In this sense, the fitted semiempirical AM1 density is of quality enough in the three molecular systems although the auxiliary basis set used in the fit is poor. This is due to the fact that valence electron density is smoother and easier to be well-fitted than the core electron density. This fact is reflected in the large S values obtained at the SCF/STO-3G level. As the auxiliary basis set used in the fit is augmented, the S value decreases dramatically. In poor basis sets, fit BSA becomes better adapted than fits PSA and even PSA + IBS. When the basis set is large enough the PSA fit is better than BSA, although in this case the best fit is obtained when inter-bonding functions are added in the fit procedure (PSA + IBS). The S values obtained using the STO-3G basis set as a minimal auxiliary basis set for the fit, are not of good enough quality although use of the PSA-MIN procedure has the advantage of being less time consuming. Fitting times (in seconds) have been included in Table III for comparison between all fitting methods. As can be seen, the PSA fit method joins a good performance together with a small CPU computing time. However, when thinking of real large molecular systems, the fitting time becomes of paramount importance. This fact let us to think in the possibility of constructing optimized minimum auxiliary basis sets for each atom, corrected with inter-

bonding functions in the molecular system, to obtain really small S values using minimal computing time. Research in this direction is under way in our laboratory.

5) Another important point to remark is that, whatever the fit method used, the relationship between values obtained at the different levels of theory agrees with these observed from an exact computation of the QMSM values. The only case in which this tendence is not followed, appears in the relationship between the MP2 and CISD levels for H_2CO (Table II), due to the fact that exact values have only a difference of ca. 0.001 a.u.

To have an idea of where the most important differences between the exact and fitted densities are located, we present a density difference map between the exact density and the fitted density obatined from PSA method for the furan molecule (Figure 1). This picture clearly illustrates that inter-bonding areas are poorly described by the fitting procedure and justifies the fact that PSA + IBS fitting method gives the smallest S values.

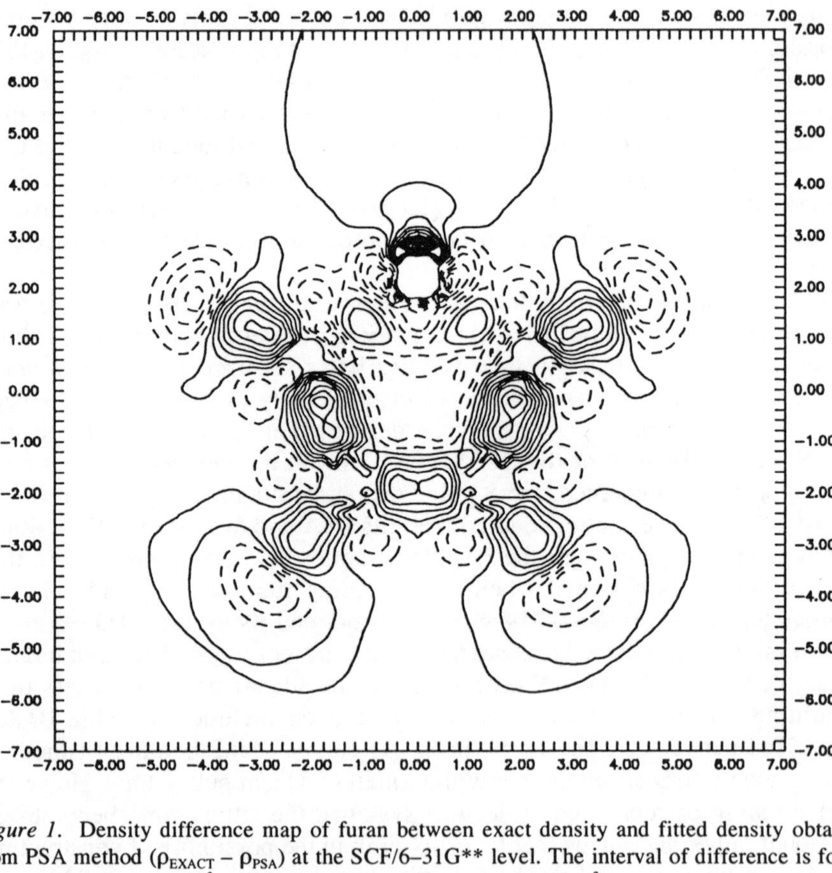

Figure 1. Density difference map of furan between exact density and fitted density obtained from PSA method ($\rho_{EXACT} - \rho_{PSA}$) at the SCF/6–31G** level. The interval of difference is found to be (–0.07, 0.13) $bohr^{-3}$ and lines depict contours of 0.01 $bohr^{-3}$.

Acknowledgments

This work has been founded through Spanish DGICYT Project No. PB92-0333. J.M. acknowledges the financial help provided by the Generalitat de Catalunya through CIRIT Project No. AR91-425. One of us (E.B.) benefits from a grant from the "Departament d'Ensenyament de la Generalitat de Catalunya".

References

1. Rouvray, D. H., in *Concepts and Applications of Molecular Similarity*, eds. M. A. Johnson and G. M. Maggiora (John Wiley & Sons, Inc., New York, 1990).
2. a) Carbó, R., Arnau, M., and Leyda, L., *Int. J. Quantum Chem.* **17**, 1185 (1980). b) Carbó, R. and Calabuig, B., *Int. J. Quantum Chem.* **42**, 1681 (1992). c) Carbó, R. and Calabuig, B., *Comp. Phys. Commun.* **55**, 117 (1989). d) Carbó, R. and Calabuig, B., *J. Mol. Struct. (Theochem)* **254**, 517 (1992). e) Carbó, R. and Calabuig, B., *Int. J. Quantum Chem.* **42**, 1695 (1992). f) Carbó, R. and Domingo, Ll., *Int. J. Quantum Chem.* **32**, 517 (1987).
3. a) Friesner, R., *J. Phys. Chem.* **92**, 3091 (1988). b) Ringnalda, M. N., Won, Y., and Friesner, R. A., *J. Chem. Phys.* **92**, 1163 (1990).
4. Vahtras, O., Almlöf, J., and Feyereisen, M. W., *Chem. Phys. Lett.* **213**, 514 (1993).
5. Mestres, J., Solà, M., Duran, M., and Carbó, R., *J. Comp. Chem.* **15**, 1113 (1994).
6. Baerends, E. J., Ellis, D. E., and Ros, P., *Chem. Phys.* **2**, 41 (1973).
7. Sambe, H. and Felton, R. H., *J. Chem. Phys.* **62**, 1122 (1975).
8. a) Dunlap, B. I., Connolly, J. W. D., and Sabin, J. R., *J. Chem. Phys.* **71**, 3396 (1979). b) Dunlap, B. I., Brenner, D. W., Mintmire, J. W., Mowrey, R.C., and White, C. T., *J. Phys. Chem.* **95**, 5763 (1991).
9. a) Hehre, W. J., Ditchfield, R., and Pople, J. A., *J. Chem. Phys.* **56**, 2257 (1972). b) Francl, M. M., Pietro, W. J., Hehre, W. J., Binkley, J. S., Gordon, M. S., Frees, D. J., and Pople, J. A., *J. Chem. Phys.* **77**, 3654 (1982). c) Hariharan, P. C. and Pople, J. A., *Theor. Chim. Acta* **28**, 213 (1973). d) Hehre, W. J., Stewart, R. F., and Pople, J. A., *J. Chem. Phys.* **51**, 2657 (1969). e) Binkley, J. S., Pople, J. A., and Hehre, W. J., *J. Am. Chem. Soc.* **102**, 939 (1980).
10. Frisch, M. J., Head-Grodon, M., Trucks, G. W., Foresman, J. B., Schlegel, H. B., Raghavachari, K., Binkley, J. S., Gonzalez, C., Defrees, D. J., Fox, D. J., Whiteside, R. A., Seeger, R., Melius, C. F., Baker, J., Martin, R. L., Kahn, L. R., Stewart, J. J. P., Topiol, S., and Pople, J. A., *GAUSSIAN 90, Revision H* (Gaussian Inc., Pittsburgh PA, 1990).
11. Mestres, J., Solà, M., Besalú, E., Duran, M., and Carbó, R., *MESSEM version 1.0* (Girona CAT, 1993).
12. Dewar, M. J. S., Zoebisch, E. G., Healy, E. F., and Stewart, J. J. P., *J. Am. Chem. Soc.* **107**, 3902 (1985).
13. Wiberg, K. B., Hadad, C. M., LePage, T. J., Breneman, C. M., and Frisch, M. J., *J. Phys. Chem.* **96**, 671 (1992).

Section B

Computational Methods

General Suggestions and Applications of Quantum Molecular Similarity Measures from *ab initio* Fitted Electron Densities

J. MESTRES, M. SOLÀ, M. DURAN and R. CARBÓ
Institut de Química Computacional, Universitat de Girona, Plaça de l'Hospital 6, 17071 Girona, Catalonia, Spain

Key words: *ab initio* similarity, Computer-Aided Drug Design, maximizing the similarity, Quantum Molecular Similarity Measures

Abstract. Fast calculations of Quantum Molecular Similarity Measures from *ab initio* wave functions are performed through use of an approximate first-order electron density, which is obtained by a least-squares fit to the exact electron density. The purpose of fitting the density is to eliminate the need of calculating costly four-centered integrals as appear in the exact framework. This procedure can be applied to large systems in order to reproduce exact Quantum Molecular Similarity Measures at a reasonable computing time cost. Several practical advices are discussed concerning the process of maximizing the Similarity measure between two molecules. Finally, various applications of Quantum Molecular Similarity Measures computed within the present methodology are presented.

1. Introduction

Pharmacological research has as one of its goals the design and synthesis of new drugs with increased activity and reduced toxicity. Since production of a new marketable compound requires enormous human and economic efforts, many researchers in the field are becoming interested in existing and new Computer-Aided Drug Design facilities [1] which making use of the structure-activity relationships allow chemists to estimate the biological activity of a compound series.

The number of available structure-activity methods has greatly increased in the last years, in such a way that nowadays these techniques are an almost everyday practice in basic and applied pharmacological research. In the past decade, as a consequence of an ever growing interest, new ideas and concepts on the subject of molecular properties estimation have emerged. The main characteristic of the new methods can be found in the fact that they are not based on statistical QSAR techniques, but on less restrictive mathematical backgrounds.

Among the most promising new methodologies are those based on the concept of Similarity between molecules. In this case, it is assumed that the larger the Similarity between two compared molecules, the more similar the biological activity of these two molecules as well as other physicochemical characteristics may be. Therefore, one can start from a known molecule, which

R. Carbó (ed.), Molecular Similarity and Reactivity: From Quantum Chemical to Phenomenological Approaches, 89–111.

is commonly referred to as a lead structure, and a series of possible drugs which can be used as a substitute of the lead compound. Then, performing calculations of Molecular Similarity among them, it is quite feasible to screen a large number of these compounds and select only those that will likely bear a biological activity similar to that presented by the model drug. The same process can be applied to any molecular property in general. Moreover, we can profit from this basic concept of Similarity to compare series of molecules, study the evolution of a reaction, or analyze the effect of different substituents on the reactivity of a given molecule.

Molecular Similarity Measures can be divided into three main categories depending on the source the measure is obtained from, namely, Three-dimensional Molecular Shape Similarity based on topological approaches [2], Graphical Similarity based on graph theory [3], and Quantum Similarity based on quantum mechanical calculations [4, 5]. All these approaches have been used to quantify Molecular Similarity and have been successfully applied to the design of new powerful drugs [1b]. Among these techniques, probably, the most widely used are the Molecular Similarity Measures computed in a Quantum Mechanical framework, which one known as to Quantum Molecular Similarity Measures (QMSM).

Since the pharmacological activity of a molecule is due to both the spatial arrangement and the electronic nature of its atoms, one can assume that similar molecules with similar biological activity must have similar electron distributions. In fact, the most widespread definition of QMSM [5a, b] is based on the aforementioned assumption, and consequently, the similarity between two chemical systems $\{I, J\}$ is defined from their first-order electron density functions $\{D_I, D_J\}$ as:

$$Z_{IJ}(\Theta) = \iint D_I(\mathbf{r}_1)\, \Theta(\mathbf{r}_1, \mathbf{r}_2)\, D_J(\mathbf{r}_2)\, d\mathbf{r}_1\, d\mathbf{r}_2. \tag{1}$$

where $\Theta(\mathbf{r}_1, \mathbf{r}_2)$ is a positive definite operator depending on two-electron coordinates. In the particular case that $\Theta(\mathbf{r}_1, \mathbf{r}_2)$ is chosen as the Dirac delta function $\delta(\mathbf{r}_1 - \mathbf{r}_2)$ the formula to calculate the Overlap-like Similarity arises:

$$Z_{IJ} = \int D_I(\mathbf{r}_1)\, D_J(\mathbf{r}_2)\, d\mathbf{r}. \tag{2}$$

and by the same token the Coulomb-like Similarity is given by:

$$Z_{IJ} = \int D_I(\mathbf{r}_1) \frac{1}{\mathbf{r}_{12}}\, D_J(\mathbf{r}_2)\, d\mathbf{r}_1\, d\mathbf{r}_2, \tag{3}$$

Other operators and more general definitions can be used depending on the information one wishes to obtain from the Similarity Measures [5b].

Once a Similarity Measure is known, a Similarity Index (SI) [5a] can be calculated as a generalized cosine of the angle between functions D_I and D_J as:

$$I_{I,J} = Z_{I,J}(Z_{I,I}\, Z_{J,J})^{-1/2}. \tag{4}$$

This cosine-like index ranges, for any pair of compared systems, from 0 (total dissimilarity) to 1 (complete similarity) depending on the Similarity Measure associated to the two molecules.

Likewise an euclidean distance between molecular electronic distributions I and J can be defined as [5a]:

$$d_{I,J} = [Z_{I,I} + Z_{J,J} - 2Z_{I,J}]^{-1/2}. \tag{5}$$

Much work has been carried out on QMSM using a rotationally invariant ZDO approach [5c–e]. Furthermore, some effort has been done towards the description of *ab initio* QMSM from Equation (1) [5f]. A strong drawback of such an *ab initio* QMSM lies in the need of calculating the integrals involved in Equation (1), whose number scales as N^4. This problem is especially cumbersome in maximizing the Similarity between two large molecules, because a lot of QMSM calculations must be carried out.

In order to circumvent this N^4 scaling problem, a computational device has been designed [6]. In this method, the electronic first order density function is approximated by a linear combination of a gaussian function set $\Gamma(\boldsymbol{r}) = \{g_i(\boldsymbol{r})\}$:

$$D(\boldsymbol{r}) \simeq \rho(\boldsymbol{r}) = \sum_i c_i\, g_i(\boldsymbol{r}). \tag{6}$$

This substitution eliminates the need to evaluate costly four-centered integrals in such a way that the Similarity evaluation process becomes an N^2-depending process, as opposed to the N^4-dependence of the exact computation. Further, in Equation (1) if one substitutes $D_I(\boldsymbol{r})$ by:

$$\rho_I(\boldsymbol{r}) = \sum_{i \in I} c_i^I\, g_i^I(\boldsymbol{r}), \tag{7}$$

and $D_J(\boldsymbol{r})$ by:

$$\rho_J(\boldsymbol{r}) = \sum_{j \in J} c_j^J\, g_j^J(\boldsymbol{r}), \tag{8}$$

a new expression involving two function integrals is found:

$$Z_{I,J} = \sum_{i \in I} \sum_{j \in J} c_i^I\, c_j^J \iint g_i^I(\boldsymbol{r}_1)\, \Theta(\boldsymbol{r}_1, \boldsymbol{r}_2)\, g_j^J(\boldsymbol{r}_2)\, d\boldsymbol{r}_1\, d\boldsymbol{r}_2. \tag{9}$$

In previous work [6] we analyzed and discussed various different fits to the exact electron density in terms of computational time used and ability to reproduce *ab initio* QMSM. In the same study, QMSM have also been carried out for a set of small molecules in order to show that QMSM, as obtained from fitted densities, yield small mean deviation errors from exact values. This methodology has also been applied successfully to large systems. It has been shown that the so-called PSA and BSN methods are those most suitable when performing QMSM from fitted densities. In the particular case of the PSA method, the gaussian function set $\Gamma(\boldsymbol{r})$ has been chosen to be the same as the squared s-type renormalized primitive basis functions. Integrations are then performed analytically. For the BSN method, the gaussian function set $\Gamma(\boldsymbol{r})$

elements are the squared *s*-type basis functions, the main two differences with the PSA method being the existence, in the BSN case, of crossed terms among primitive functions, and the integrations being performed numerically using an atom-centered grid.

The main goal of the present study is to provide several examples showing how fitted densities can be used to study large systems whose size prevents calculations of exact QMSM. In the first section of this work, we discuss the full mathematical implications of the methodology used, together with all possible new definitions arising from the substitution of the exact density by a fitted density. Practical suggestions in the process of maximizing QMSM are also discussed. In the second section, we discuss the application to a series of amino acids and the use of these measures as an interpretative tool in chemical reactivity.

2. Methodology

2.1. General computational aspects

Fitting of first-order density functions can be designed in a general way akin to the Similarity definition given in Equation (1).

Let us define the function difference:

$$P(\mathbf{r}) = D(\mathbf{r}) - \rho(\mathbf{r}) \tag{10}$$

and take $\Theta(\mathbf{r}_1, \mathbf{r}_2)$ as a positive definite operator. Then, one can fit optimally $\rho(\mathbf{r})$ to $D(\mathbf{r})$ provided that the quadratic error integral

$$\Delta(\Theta) = \iint P(\mathbf{r}_1)\, \Theta(\mathbf{r}_1, \mathbf{r}_2)\, P(\mathbf{r}_2)\, d\mathbf{r}_1\, d\mathbf{r}_2. \tag{11}$$

is minimized according to the least-squares technique. Definition (11) allows us to see that the quadratic error in generalized least-squares sense is nothing but a Self-Similarity measure associated to function $P(\mathbf{r})$. If the coefficients of the expansion and the gaussian function set $\Gamma(\mathbf{r})$ in Equation (6) are collected into the column vectors $\mathbf{c}$ and the row vector $\mathbf{G}(\mathbf{r}) = (g_1(\mathbf{r}), g_2(\mathbf{r}), \ldots)$, respectively, that is:

$$\rho(\mathbf{r}) = \mathbf{G}(\mathbf{r})\mathbf{c}, \tag{12}$$

then $\mathbf{c}$ is given in a least squares sense by the matrix expression:

$$\mathbf{c} = \mathbf{S}^{-1}\mathbf{t}, \tag{13}$$

where the elements of the vector $\mathbf{t}$ are defined by:

$$t_i = \iint D(\mathbf{r}_1)\, \Theta(\mathbf{r}_1, \mathbf{r}_2)\, g_i(\mathbf{r}_2)\, d\mathbf{r}_1\, d\mathbf{r}_2, \tag{14}$$

and the metric matrix $\boldsymbol{S}$ is constructed as:

$$S_{ij} = \iint g_i(\boldsymbol{r}_1)\ \Theta(\boldsymbol{r}_1, \boldsymbol{r}_2)\ g_j(\boldsymbol{r}_2)\ d\boldsymbol{r}_1\ d\boldsymbol{r}_2. \tag{15}$$

If $\Delta(\Theta)$ is minimized subject to the constraint that the equality:

$$\int \rho(\boldsymbol{r})\ d\boldsymbol{r} = N_e. \tag{16}$$

holds, then the expansion coefficients are given by the expression:

$$\boldsymbol{c} = \boldsymbol{S}^{-1}(\boldsymbol{t} + \lambda\boldsymbol{n}), \tag{17}$$

where

$$n_i = \int g_i(\boldsymbol{r})\ d\boldsymbol{r}, \tag{18}$$

and

$$\lambda = \frac{(N_e - \boldsymbol{n}^T\boldsymbol{S}^{-1}\boldsymbol{t})}{\boldsymbol{n}^T\boldsymbol{S}^{-1}\boldsymbol{n}}\ , \tag{19}$$

N_e being the total number of electrons and λ a Lagrange multiplier, which is necessary to guarantee charge conservation.

As suggested by the work of Sambe and Felton [7] and by that of Dunlap *et al.* [8], two kinds of fitting procedures can be analyzed at once; namely, the Overlap fitting: $(\Theta(\boldsymbol{r}_1, \boldsymbol{r}_2) = \delta(\boldsymbol{r}_1 - \boldsymbol{r}_2))$, and the Coulombic fitting: $((\Theta(\boldsymbol{r}_1, \boldsymbol{r}_2) = 1/r_{12})$. The same program can be used to carry out Overlap and Coulomb fittings, the only difference being the calculation of integrals involved in Equations (14) and (15). It is generally found that Coulomb fitting is more expensive than Overlap fitting, even though it does not yield a better description of the electron density from a QMSM point of view, as it will be shown below.

2.2. Practical details

The wave functions whose corresponding electron densities are fitted have been calculated by means of the GAUSSIAN-90 system of programs. [9] The program MESSEM [10] extracts the one-electron density matrix from the GAUSSIAN-90 output file, performs a least-squares fitting to the exact electron density (different fitting basis sets are available from numerical and analytical integration), and computes QMSM. It allows also for the optimization of the mutual orientation between the two systems studied, in order to maximize their Similarity by commonplace steepest-descent, Newton and quasi-Newton algorithms [11]. Overlap-like and Coulomb-like QMSM from both densities, exact and fitted, can be calculated and restart calculations are permitted if the number of cycles during the optimization are exceeded. In order to account for Similarities between particular regions of two molecules, QMSM between sets of atoms and/or between sets of orbitals can also be performed. Other capabilities of the program involve searching of bond critical points, construction of electron density maps, and calculation of gradient norm and laplacian of the charge density maps.

2.3. Comparison with exact similarities

In order to show how fitted and exact QMSM compare, Table I collects the values of Self-QMSM for different systems at the STO-3G-optimized geometries. QMSM calculations reported in this table have been performed with the help of a PC-486/66 Mhz with 8Mb of RAM. From these values one can conclude that both the BSN and the PSA methods of fitting electron densities yields quite satisfactory results. In fact, for the systems studied the relative error given by the BSN method is never larger than 0.49%, whereas the PSA method yields error percentages which never exceeds a 0.32%. As the computational cost of calculating the exact QMSM is very expensive when compared to the same cost using fitted densities, calculation of fitted QMSM instead of exact QMSM is highly recommended for systems of medium to large size. A clear example in Table I of this statement is the case of glycine: here computing QMSM from a fitted density requires only 0.3 seconds of CPU time, whereas the calculation of the exact Self-Similarity for the same molecule demands 4705 seconds of CPU time. Although this last value could be reduced by improving the four-centered integrals algorithm, the N^4 scaling in the CPU time will remain.

TABLE I
Self-QMSM (in a.u.) for selected molecules at the STO-3G-optimized geometries. The computer time (in seconds) needed to calculate the Similarity value is given in parenthesis[a]

Species	BSN	PSA	EXACT
H_2O	79.3427 (0.1)	79.2010 (0.1)	79.4123 (17)
CH_4	31.2506 (0.1)	31.0527 (0.1)	31.0998 (41)
CH_3F	146.8569 (0.2)	146.7585 (0.2)	147.2043 (171)
CH_3OH	110.3398 (0.2)	109.9367 (0.2)	110.2460 (223)
CH_2O	110.5076 (0.2)	109.8301 (0.2)	110.1825 (160)
Glycine[b]	271.9905 (0.3)	270.6555 (0.3)	271.3874 (4705)

[a] Computed with a PC 486/66 Mhz machine.
[b] Result obtained at the STO-3G//AM1 level.

So far we have dealt with Similarities between two identical chemical systems perfectly superimposed, the so-called Self-Similarity. However, our ultimate goal is to derive similarities among a series of chemical systems. Table II gathers QMSM and the corresponding Similarity indices for methane, methyl fluoride and methanol at the STO-3G-optimized geometries. As one can see,

TABLE II
Exact and fitted (PSA procedure) QMSM (in a.u.) and Similarity Indices (SI) corresponding to the series of methane, methyl fluoride and methanol at the STO-3G-optimized geometries. The mutual orientation of the two molecules has been optimized for the fitted QMSM in order to maximize the QMSM. The same orientation has been used to compute exact QMSM

$Z_{I,J}$	Exact		Fitted	
	QMSM	SI	QMSM	SI
CH_4/CH_3F	31.7026	0.4685	31.7218	0.4699
CH_4/CH_3OH	31.5042	0.5380	31.5243	0.5395
CH_3OH/CH_3F	121.5329	0.9540	121.1627	0.9539

there are small differences between the obtained indices from the exact density and those obtained from fitted densities, showing that this method gives results which are also highly reliable when comparing Similarity Indices.

Since we have tested two methodologies for fitting the electron density, namely, the Overlap and Coulomb fits, the question regarding which of those fits are more adequate for calculating QMSM immediately arises. Table III gathers Overlap-like QMSM and Coulomb-like QMSM obtained from Overlap and Coulomb fits, together with the exact QMSM for various small molecules. It can be seen from these values that QMSM obtained from both Overlap and Coulomb fittings give results close to the exact ones. This is especially remarkable in the case of Overlap-like QMSM, although Coulomb-like QMSM obtained from a Coulomb fit are closer to the exact values than those obtained from an Overlap fit. In any event, and given that the Coulomb fit is compu-

TABLE III
Comparison between fitted QMSM, from an Overlap fit or a Coulomb fit, and exact QMSM, from an Overlap-like (Ol) or a Coulomb-like (Cl) QMSM calculation

Molecule	QMSM	Exact	PSA Fit	
			Overlap Fit	Coulomb Fit
H_2	Ol QMSM	0.15422	0.14839	0.15258
	Cl QMSM	2.51315	2.43871	2.48482
CH_4	Ol QMSM	31.09995	31.05239	31.25065
	Cl QMSM	65.73317	66.50281	65.82838
NH_3	Ol QMSM	51.39196	51.28402	51.60633
	Cl QMSM	78.27782	79.26548	78.35072
H_2O	Ol QMSM	79.41220	79.19946	79.75903
	Cl QMSM	94.06899	95.00788	93.93506
HF	Ol QMSM	116.39891	116.06363	116.99695
	Cl QMSM	112.91307	113.91078	112.61703

tationally more expensive than the Overlap fit, for large molecules the Overlap fit is recommended since differences in QMSM obtained from both kind of fits are very small.

2.4. Avoiding the core dependence: Valence and Coulomb-like similarity measures

One of the main drawbacks of Overlap-like QMSM is their important core dependence due to the existence of cusps of electron density at nuclei. This fact means that once the mutual orientation of the two compared systems has been optimized to maximize the Overlap-like QMSM, very small displacements of one system with respect to the other carries about radical changes in the value of the Overlap-like QMSM. This is especially noticeable in the case where the two compared systems have only one heavy atom, where optimization process becomes fairly complicated, because large gradient components arise in the neighborhood of the optimal mutual orientation. This problem can be circumvented through use of Coulomb-like QMSM, because presence of Coulomb operator smoothes the electron density surface.

Table IV shows the differences between Overlap-like and Coulomb-like Self-QMSM of water as obtained from Overlap and Coulomb fittings. The most striking aspect of Table IV turns out to be the large percentages of Similarity owing merely to the two oxygens of the two superimposed water molecules. This is especially noticeable in the case of the Overlap-like QMSM, where more than 99% of the total Similarity is due to the particular contribution of the two oxygens. From these values one can expect that a slight displacement of one water molecule, leading to a mutual orientation where the two oxygens of water do not completely overlap, will reduce drastically the value of the QMSM obtained.

TABLE IV

Values of the Overlap-like and Coulomb-like QMSM obtained from Overlap and Coulomb fitted densities for the case of H_2O at the STO-3G optimized structure with an exact Overlap-like QMSM of 79.4123 a.u. Values in parenthesis are the percentage of similarity due to the mere contribution of the two oxygens of the two water molecules under comparison

	Overlap-like QMSM	Coulomb-like QMSM
Overlap Fit	79.2010 (99.72%)	95.0081 (89.96%)
Coulomb Fit	79.7605 (99.64%)	93.9352 (86.93%)

In Table V we gather the values of Self-QMSM and Self-QMSM gradient norm appeared when a water molecule (located at the xz plane with the z-axis bisecting the two hydrogens) has been displaced 0.01 and 0.1 Å along the z-axis from the position of the maximum Self-QMSM. It is clearly seen that the Coulomb-like QMSM changes are smaller than Overlap-like QMSM changes. In the case of a 0.1 Å displacement, Coulomb-like QMSM changes

TABLE V

Overlap-like (Ol) and Coulomb-like (Cl) Quantum Molecular Self-Similarity Measures (QMSM) and gradient norm (GN) for the water molecule, when a water molecule has been displaced 0.01 and 0.1 Å along the z-axis. The PSA method has been used in the process of fitting the density, and the STO-3G basis set has been used in *ab initio* calculations

Displacement		QMSM	GN
0. Å	Ol	79.2010	0.00
	Cl	95.0081	0.00
0.01 Å	Ol	78.9215	31.63
	Cl	94.9907	2.74
0.1 Å	Ol	58.2442	243.96
	Cl	93.5021	24.17

less than 2 a.u., whereas Overlap-like QMSM reduces its value more than 20 a.u., while the gradient norm increases up to 244 a.u.

Another way to reduce the core dependence problem involves the use of valence density functions in the calculation of QMSM. Avoiding the core basis functions in the calculation dramatically reduces density cusps and increases the importance of bond regions, which are the most relevant regions in chemical reactions. Using the present methodology, four possible ways of performing Valence QMSM can be described. In the first one (method A), an *ab initio* pseudopotential or semiempirical calculation can be carried out instead of an *ab initio* all-electron calculation. The exact density obtained from the wave function can then be fitted, so Overlap-like and Coulomb-like QMSM can be calculated from this fitted density. In the second method (method B) an all-electron calculation is carried out, and then the exact density is calculated using only valence orbitals. In this way, a valence density is obtained and fitted with the aforementioned methods to obtain QMSM.

A third possibility (method C) is based on the fact that fitted densities have been expanded as a linear combination of a set of gaussian functions. Let us write Equation (9) in matrix form, that is:

$$\begin{aligned} Z_{I,J}(\Theta) &= \sum_{i\in I}^{n} \sum_{j\in J}^{n} c_i^I c_j^J \langle i, I | j, J\rangle \\ &= \mathbf{C}_I^T \mathbf{Z}_{IJ} \mathbf{c}_J, \end{aligned} \tag{20}$$

where $\mathbf{Z}_{IJ}$ is a matrix collecting the integrals between gaussian functions g_i^I and g_j^J: $\langle i, I|\Theta|j, J\rangle$, and $\{\mathbf{c}_I, \mathbf{c}_J\}$ are vectors corresponding to the expansion coefficients of the density function fit. A general partitioning into p subvectors of the coefficient vector $\mathbf{c} = \{\mathbf{c}^1, \mathbf{c}^2, \ldots, \mathbf{c}^p\}$ in Equation (20) yields,

$$Z_{I,J}(\Theta) = \sum_{\alpha\in I}^{n} \sum_{\beta\in J}^{n} (\mathbf{c}_I^{\alpha})^T \mathbf{Z}_{IJ}^{\alpha\beta} \mathbf{c}_J^{\beta}. \tag{21}$$

Therefore, using the submatrix product:

$$Z_{I,J}^{\alpha\beta} = (\boldsymbol{c}_I^{\alpha})^T \boldsymbol{Z}_{IJ}^{\alpha\beta} \boldsymbol{c}_J^{\beta}, \tag{22}$$

it is always possible to define a set of contributions for every partition pair:

$$Z_{I,J}(\Theta) = \sum_{\alpha \in I}^{n} \sum_{\beta \in J}^{n} Z_{IJ}^{\alpha\beta}. \tag{23}$$

An interesting example permits the definition of Valence-QMSM. Since the gaussian function set $\Gamma(\boldsymbol{r})$ elements are the contracted or uncontracted basis functions of the first-order density function, it is possible to expand the sums over these functions in terms of core and valence functions. By partitioning of the vector coefficients into a two element core-valence partition $\boldsymbol{c} = \{\boldsymbol{c}^{\text{core}}, \boldsymbol{c}^{\text{val}}\}$ use of Equation (23) above brings about

$$Z_{IJ}(\Theta) = Z_{IJ}^{\text{core, core}} + Z_{IJ}^{\text{core, val}} + Z_{IJ}^{\text{val, core}} + Z_{IJ}^{\text{val, val}}. \tag{24}$$

Therefore, it is possible to define the Valence-QMSM as the $Z_{IJ}^{\text{val, val}}$ part of $Z_{IJ}(\Theta)$, which constitutes the method labelled C.

The last method (method D) combines the philosophy of methods B and C: the density is fitted using only valence orbitals, and the similarity is obtained from $Z_{IJ}^{\text{val,val}}$.

The four methods of calculating Valence QMSM are compared for disulphur difluoride (FSSF) and thiothionyl difluoride (F_2SS) in Table VI. Experimental geometries of FSSF and F_2SS have been used in all calculations. The STO-3G basis set [12] has been used for computing all-electron wave functions, whereas Hay and Wadt's ECP [13] and the STO-3G basis set (basis set *lanl1mb* in the GAUSSIAN-90 program) are employed for the calculation of wave functions using pseudopotentials to account for innermost electrons.

The main advantage of using method A in the computation of Valence QMSM is the fact that calculation of the wave function is less expensive: this characteristic must be taken into account when one is interested in comparing Similarities among large systems. Method A can be more directly compared to Method D than to any other because Method D uses only valence

TABLE VI

Coulomb-like QMSM obtained from a PSA Overlap fitting of F_2SS and FSSF wave functions at their experimental geometry. All-electron wave function has been obtained with the STO-3G basis set, and Hay and Wadt ECP's have been used to compute pseudopotential wave functions

	Method A[a]	Method B	Method C	Method D	Full[b]
F_2SS[c]	402.5668	933.0718	586.8433	472.5898	1379.8609
FSSF[c]	388.1935	899.3514	587.8257	472.0516	1378.5680

[a] See text for an explanation of the different methods used.
[b] All basis functions are used for both the density fitting and calculation of QMSM.
[c] Experimental geometry from Kuczkowski, R. L., *J. Am. Chem. Soc.* **86**, 3617 (1964).

functions in the calculation of the exact electron density and the evaluation of QMSM. Differences between methods A and D arise mainly because of the changes in sulphur basis set when using ECP's and the STO-3G basis set, and also by the presence of *1s* orbitals of fluorine atoms in pseudopotential calculations.

Among all methods tested, Method B is the less recommended in the calculation of Valence QMSM, given that it is as expensive as a full Similarity calculation and does not reduce meaningfully the total QMSM. However, an estimate of the importance of core and core-valence cross terms appearing after a valence density fitting can be found from the difference between values obtained with Method B and D. Furthermore, the difference between the values obtained by method C and the full Similarity calculation provides also an estimate of the importance of core basis functions on the total Similarity measure; however, here a fully fitted density is used in the calculation of the Similarity. Interestingly, the full Similarity for the F_2SS can be divided into 586.8433 a.u. corresponding to the valence-valence contribution, 138.6136 a.u. which corresponds to the core-core Similarity and 654.4040 a.u. corresponding to the valence-core contribution. Method C is especially suitable when one is maximizing the full QMSM between two systems. In this case, a Valence QMSM optimization proceeds generally faster than a full QMSM optimization, and results in a very accurate starting point for a full QMSM optimization, yet at a very low computational cost.

Comparison between methods C and D shows that there is an important gap between the final value of the Valence Similarity computed from full- or valence-fitted densities. However, we believe that both methods can be used to compute valence QMSM, a key point being the use of the same method along all a given series of calculations.

All methods but Method C result in F_2SS having a larger Self-QMSM than FSSF, although the difference between both values is very small, as expected from the fact that atoms intervening in both compounds are the same.

2.5. Quantum Molecular Similarities Between Fragments

Another appealing aspect of the methodology presented above is that it allows for calculation of Similarities between fragments. Comparison of chemical fragments sharing a common structure is a very interesting feature that permits to emphasize some aspects of a chemical problem that could otherwise be masked by the presence of the common structure. A practical application of the Similarity between fragments is presented later in Section 3.1.

As in the case of methods C and D for calculating valence Similarities, fragment Similarity is based on the fact that the density has been expanded over a gaussian function set $\Gamma(\boldsymbol{r})$, each element of this set, belonging to a given atom. Therefore, once the density has been fitted, a fragment Similarity can be computed through calculation of the Similarity from Equation (1) by taking into account only the gaussian functions that belong to the atoms defined by the fragment. In such a way, QMSM between fragment α of molecule I and fragment β of molecule J can be defined from Equation (23) as the value of $Z_{IJ}^{\alpha\beta}$. Likewise, it is also possible to define valence-QMSM between fragments through consideration of only the functions of the fragments, and by using any of the methods discussed in previous section.

2.6. Maximizing the similarity: Useful suggestions

Once a Similarity calculation between two molecules has been carried out, the next step is to change their relative position in order to maximize their QMSM. Similarity measure optimization is, from a numerical point of view, a problem of unconstrained maximization. Several studies discussing a wide variety of algorithms for constrained and unconstrained optimizations are found in the literature [11]. The goal of this section is to analyze various of these methods and provide a few hints that can be helpful in QMSM optimizations.

Available methods can be divided, following increasing complexity order, into (a) those that need neither the gradient vector nor the Hessian matrix, (b) those that deal with only the gradient vector, and finally (c) those that make use of both the gradient vector and the Hessian matrix.

Among the first methods, the most widely used is the Simplex procedure [14], which has been found to be reliable and robust when maximizing the Electrostatic Potential and Electric Field-based Molecular Similarity [15]. However, in this work methods belonging to the second and third groups have been preferred. Gradient vector and Hessian matrix needed in these cases have been computed numerically, yet work is planned to develop analytical codes.

The steepest-descent method, corresponding to the second group of optimization methods, has been found to be very useful when the systems lie far from their optimum mutual orientation. Nevertheless, near the maximum of Similarity, where all eigenvalues of the Hessian matrix are negative, the typical zig-zag progress of this method prevents fast location of the maximum, so methods of the third group at this point are much more recommendable.

Among those methods belonging to the third group, we have tested the plain Newton method and the Davidon-Fletcher-Powell (DFP) quasi-Newton method. In the Newton method, the gradient vector and the Hessian matrix are computed at each iteration, while in the DFP method, the gradient vector is computed at each optimization step and the Hessian matrix is calculated only in the first iteration or every *n* iterations; for the remaining iteration steps the matrix is updated from the gradient information gathered along the optimization process. We have found that both methods are fast and quite efficient when the optimization is close to the maximum, and if an approximate linear search is used. Therefore, we recommend use of a steepest-descent method until its well-known zig-zag behaviour appears; at that point, optimization proceeds faster using DFP or Newton procedures.

Any of the methods commented here avoids reaching a local maximum. Therefore, beginning the optimization from a good starting point is quite relevant. It has been found that a reasonable initial mutual orientation can be obtained in many cases by placing the center of charges of both molecules, as defined from gross atomic populations, at the origin of coordinates [5].

In the optimization process, one of the molecules is given complete freedom of motion in the three dimensional space, whereas the other system is kept fixed. It has been generally found that optimizations perform usually faster if the larger molecule is that being kept fixed.

Either differences in the norm of the gradient or differences in QMSM between two adjacent cycles can be used to stop optimization. In this work, the mutual orientation of the two molecules has been changed until differences in QMSM between successive iterations are smaller than 10^{-4} a.u. In this case, the gradient norm in a.u. is usually small, with values under unity.

A final remark worth noting is that the optimization process is clearly improved when performed first using Valence QMSM and later, from the optimum mutual orientation achieved in the valence space, the process is continued in the full space. In this way, the number of cycles required to reach the maximal QMSM is indeed reduced, because the QMSM surface is smoother in the valence space than in the full space, and the optimum mutual orientation corresponding to valence space is usually very close to the optimum mutual orientation in the full space.

2.7. Summary

The methodology presented above, which consists of fitting electron densities in order to calculate QMSM, gives results which are very close to the exact QMSM yet can be obtained at a much lower cost.

The main characteristics of the methodology presented in this work can be summarized in the following statements:

a) Similarity measures can be obtained from wave functions of any form (SCF, MP, CI, . . .), the only requirement being the need for the density matrix. This allows also to use densities from Density Functional calculations.
b) It is computationally feasible.
c) The computed Similarity values are only slightly dependent on the quality of the basis sets used in the computation of wave functions.
d) It allows for calculation of Valence QMSM.
e) Fragment comparison or partial comparison between molecules can be performed.
f) It allows for optimization of the mutual orientation of the two molecules under comparison in order to maximize their Similarity.
g) It can be applied to the study of fairly large systems.

3. Applications of quantum molecular similarity measures

3.1. The amino acids: a test case

In this section we present the results of Similarity calculations on fourteen naturally occurring amino acids, namely, Alanine, Arginine, Asparagine, Aspartic Acid, Cysteine, Glutamine, Glutamic Acid, Glycine, Histidine, Isoleucine, Leucine, Lysine, Methionine and Phenylalanine (throughout this section we refer to these molecules by using the usual symbols: Ala, Arg, Asn, Asp, Cys, Gln, Glu, Gly, His, Ile, Leu, Lys, Met and Phe, respectively). Side chains of these amino acids are depicted in Figure 1.

Since the large size of several of those systems prevent us the use of *ab initio* methods everywhere, an AM1 [16] full geometry optimization using the AMPAC program [17] has been carried out for all amino acids considered. Then, *ab initio* wave functions have been computed with the STO-3G basis set [12] at the AM1-optimized geometry. In this study, fitted densities are obtained by means of a Coulomb fitting to the exact STO-3G//AM1 densities. Coulomb-like Similarities have been preferred to avoid the core dependence problem and to improve the maximization processes. All the compared amino acids share the *S* configuration on the asymmetric carbon.

Table VII collects the Similarity Indices obtained through use of Equation

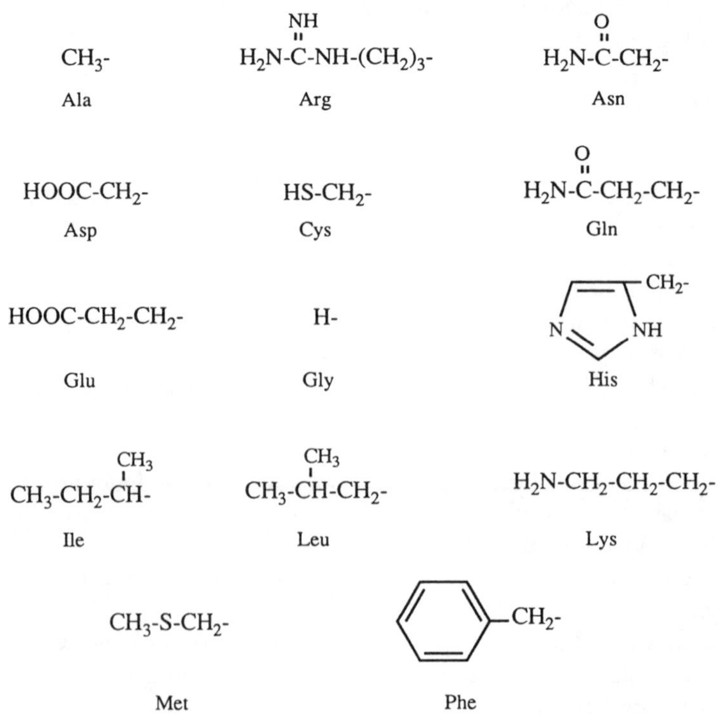

Figure 1.

TABLE VII
Similarity Indices matrix for the studied amino acids computed from Equation (4) at the STO-3G//AM1 level of calculation

	Ala	Arg	Asn	Asp	Cys	Gln	Glu	Gly	His	Ile	Leu	Lys	Met	Phe
Ala	1.0													
Arg	0.7719	1.0												
Asn	0.9058	0.8095	1.0											
Asp	0.9063	0.8279	0.9284	1.0										
Cys	0.8754	0.7802	0.8922	0.8920	1.0									
Gln	0.8686	0.8815	0.8760	0.9052	0.8580	1.0								
Glu	0.8569	0.8869	0.8658	0.8966	0.8493	0.9785	1.0							
Gly	0.9727	0.7338	0.8661	0.8637	0.8339	0.8261	0.8126	1.0						
His	0.8674	0.8723	0.8881	0.9049	0.8707	0.9168	0.9194	0.8214	1.0					
Ile	0.8741	0.8546	0.8663	0.9202	0.8514	0.9073	0.9068	0.8313	0.8907	1.0				
Leu	0.9138	0.8170	0.8954	0.8753	0.8460	0.8790	0.8842	0.8655	0.8880	0.8715	1.0			
Lys	0.8732	0.8717	0.8567	0.8411	0.8103	0.8647	0.8609	0.8267	0.8694	0.8524	0.9381	1.0		
Met	0.8215	0.8138	0.8150	0.7987	0.7693	0.8673	0.8636	0.7750	0.8552	0.8302	0.9025	0.9348	1.0	
Phe	0.8570	0.8665	0.9334	0.9132	0.8644	0.9147	0.9132	0.8142	0.9247	0.8453	0.8826	0.8632	0.8415	1.0

(4) for all systems considered. Euclidean distances between amino acids obtained from Equation (5) are gathered in Table VIII. In general, there is almost a complete parallelism between Similarity Indices and the calculated distances: the higher the Similarity Index, the smaller the distance between the systems compared. This is not at all surprising because cosine-like Similarity Indices and distances are well known to exhibit opposite changes [5a]. This trend can be broken in cases where the distances are very similar (for example Ala-Gly and Glu-Gln).

From the euclidean distances in Table VIII, it is found that the most similar amino acids are Alanine and Glycine, whereas the most dissimilar are Glycine and Arginine. This is not striking if one takes into account the side chains of these amino acids depicted in Figure 1. The only change when going from Glycine to Alanine is the substitution of a hydrogen by a methyl group, thus explaining the large Similarity Index and the small distance between these two amino acids. On the other hand, huge structural differences arise when going from Glycine to Arginine, hence accounting for the large distance between them.

By means of a Kruskal tree [5f] it is possible to connect all amino acids studied by a relation of minimum euclidean distances (Figure 2). This picture obtained from the QMSM among amino acids, has been drawn with the help of the HYPERCLOUD program [5e], a set of graphic design facilities developed in our Laboratory.

Again, in the Kruskal tree Glycine and Alanine are directly connected, whereas Glycine is found far from Arginine. Nevertheless, there is one point

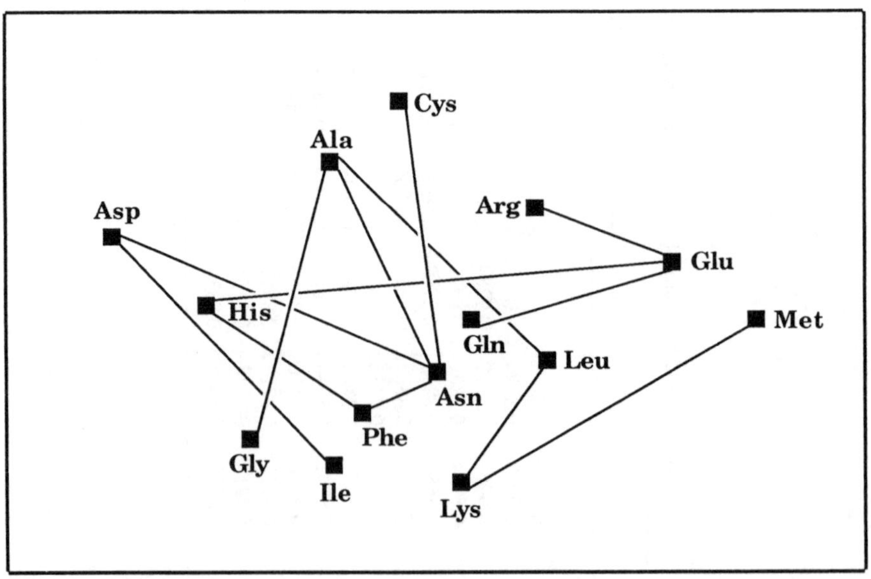

Figure 2.

TABLE VIII
Euclidean distances matrix for the studied amino acids computed from Equation (5) at the STO-3G//AM1 level of calculation

	Ala	Arg	Asn	Asp	Cys	Gln	Glu	Gly	His	Ile	Leu	Lys	Met	Phe
Ala	0.0													
Arg	29.08	0.0												
Asn	17.07	26.31	0.0											
Asp	17.16	25.11	14.36	0.0										
Cys	18.91	28.14	17.58	17.63	0.0									
Gln	20.87	21.13	19.52	17.13	20.88	0.0								
Glu	21.68	20.68	20.39	17.96	21.60	8.34	0.0							
Gly	7.07	31.35	20.41	20.65	21.87	23.96	24.72	0.0						
His	22.36	22.08	19.30	17.87	20.68	16.84	16.60	25.70	0.0					
Ile	18.81	23.24	19.51	15.11	20.57	16.90	17.04	21.79	19.08	0.0				
Leu	16.10	25.80	17.17	18.79	20.84	19.22	18.90	19.93	19.30	18.97	0.0			
Lys	20.20	21.93	20.77	21.90	23.89	20.71	21.08	23.48	20.96	21.03	13.69	0.0		
Met	25.18	26.73	24.89	25.95	27.70	21.44	21.77	28.07	22.77	23.87	18.40	15.14	0.0	
Phe	24.25	22.87	15.94	17.90	21.97	17.66	17.81	27.43	16.70	23.35	20.59	22.13	24.32	0.0

that can be surprising from the results of Tables VII and VIII and Figure 2. From a chemical point of view, one can expect that Asparagine will be closer to Glutamine than to Alanine, since Asparagine and Glutamine share the same functional group. However, distances, Similarity Indices and Kruskal tree show that Asparagine is more similar to Alanine than to Glutamine. This is a clear problem of a chemical question being masked by the presence of a common structure. Valence and fragment QMSM can sometimes help in distinguishing between this kind of compounds. Thus, Table IX gives the Similarity Indices for Ala-Asn and Asn-Gln obtained from Valence Similarities and fragment Similarities.

TABLE IX
Similarity Indices obtained from Equation (4) using Valence Similarity (Method C) and fragment Similarity for Asparagine compared to Alanine and Glutamine computed from STO-3G//AM1 wave functions

	Similarity Indices			
	Valence	Fragment		
		Common	Different	Global
Ala-Asn	0.9063	0.9830	0.7536	0.7408
Asn-Gln	0.8966	0.9746	0.9315	0.9078

It can be seen that Valence Similarities reduce the differences between the Similarity Indices of the two groups of amino acids compared, although the same Similarity ordering is maintained, Asparagine being more similar to Alanine than to Glutamine. At this point we can calculate fragment Similarities for the common fragment share by all amino acids, and for the different fragments which are the side chains drawn in Figure 1. These calculations were performed at the AM1 optimized geometry for each amino acid, and therefore some differences appear in the common fragment, despite having the same structure. The mutual orientation of the two compared systems is optimized in order to maximize the Similarity between fragments. Now, the global fragment Similarity Index, obtained through the product of the common fragment and different fragment Similarity Indices, recovers the expected result: the Similarity between Asparagine and Glutamine is larger than that between Alanine and Asparagine. Therefore, it can be concluded that comparison of chemical fragments sharing a common structure can be very helpful in the process of distinguishing some aspects of a chemical problem that could otherwise be masked by the presence of the common structure.

The drawback of relating Similarity results to experimental data is that, despite the definition of Similarity involving an overlap integral between two observables, it cannot be directly compared with any experimental property. This notwithstanding, various experimental facts that can be fairly explained through QMSM are outlined below.

Site-directed mutagenesis experiments [18] are very useful to determine the role of a particular amino acid in the protein chain. These experiments can shed some light in the process of grouping amino acids. For example, it has been reported [19] that substitution of Met85 by Leu in the yeast transcriptional activator SWI5 does not affect the folding of the protein. Similarity Indices in Table VII, calculated distances in Table VIII and Kruskal tree in Figure 2 show that Leu and Met are not far away from each other, therefore they should share properties, in good agreement with the experimental result.

Other authors have been also found that the enzyme L-Aspartase from *Escherichia Coli* has a great specificity for its L-Aspartic Acid substrate. In fact substitution of Asp by Gly, Ala, Gln, Leu, Phe and His have all failed to show turnover [20]. From the Kruskal tree, it can be found that Asp is not directly related to any of the amino acids tested in the experiment. It can also be predicted that Asn and Ile are the main candidates to show turnover.

A third case involves Carbonic Anhydrase where it has been found that substitution of His64 by Ala inhibits the enzymatic process [21]. The large distance between both amino acids (22.4 a.u.) may explain this experimental fact. Nevertheless, some caution has to be taken into account when this kind of comparisons be made. For instance, in Carbonic Anhydrase the His64 residue acts as a proton shuttle between the active site and the environment. Therefore, even the amino acid most similar to His, which is Phe, will certainly inhibit the enzyme since it lacks the presence a proton shuttle group.

Despite some conclusions being inferred from site-directed mutagenesis experiments, it still remains the unsolved problem of how QMSM can, at least in the case of amino acids, be related to experiment.

3.2. APPLICATION TO CHEMICAL REACTIVITY: QUANTIFYING THE HAMMOND POSTULATE

Lately, use of QMSM has become more widespread as an interpretative tool when comparing wave functions from different systems. For instance, Cioslowski [4d] has proposed the use of NOEL indices [4a] as QMSM in order to quantify the Hammond postulate [22], which basically states that if the transition state is near an adjacent stable complex in the potential energy surface, then it is also close in structure to the same complex. This postulate has been used for predicting the effects of substituent changes and external perturbations on transition-state geometries, and is essentially accomplished if force constant matrices associated to reactants and products are similar [23]. Different methodologies have been developed trying to rationalize this postulate [24], most based on semiquantitative assumptions. This is not the case for the most recent attempt to quantify the Hammond postulate due to Cioslowski [4d]. This author has shown that quantification of the Hammond postulate can be achieved by defining two new parameters, the so-called structural proximity (β) and isosynchronicity (α) parameters between transition state and stable complexes. In particular, β is defined from QMSM euclidean distances in Equation (5) as:

$$\beta = [d_{A,TS} - d_{B,TS}]/d_{A,B}, \tag{25}$$

therefore taking values ranging from −1 to 1. If the reactants (A) and transition state (TS) are closer in the potential energy surface then the β value is negative, and positive otherwise. Furthermore, Cioslowski defined α as:

$$\alpha = [d_{A,TS} + d_{B,TS}]/d_{A,B}. \tag{26}$$

This parameter is never smaller than 1, and takes a value close to 1 only when the distance of either A or B to the TS is small.

The aim of this section is to use these definitions made by Cioslowski together with the QMSM defined in Section 2, as an interpretative tool to (a) understand better the way a series of reactions proceed along the reaction coordinate, and (b) to analyze the changes undergone by the molecular systems when going from reactants to products. This methodology is applied to the quantification of the Hammond postulate in the well-known HNC/HCN [25] rearrangement reaction.

This reaction has been studied at an *ab initio* SCF level using the 6-31G** [26] basis set. Geometrical parameters of optimized stationary points on the potential energy surface are R_{N-C} = 1.15 Å and R_{N-H} = 0.99 Å for the reactant (A), R_{N-C} = 1.17 Å, R_{N-H} = 1.45 Å and angle ∠HCN = 77.5° for the transition state (TS), and R_{N-C} = 1.13 Å and R_{C-H} = 1.06 Å for the product (B). The forward (A → B) energy barrier from is 38.5 kcal/mol, whereas the backward energy barrier is 49.5 kcal/mol.

From those energies and in the light of Hammond postulate, it can be expected that the transition state of the HNC/HCN rearrangement reaction will be closer to HNC than to HCN. From the Similarity Indices and distances of Table X, it can be seen that this result is properly reproduced: the distance of the TS to HNC (0.817) is shorter than that of the TS to the HCN (0.938), thus giving rise to a negative value of β in this reaction, although its small absolute value (0.13) reflects the fact that the TS is found not far from the midpoint of the reaction coordinate. Small distances found between reactant, TS and product can be easily interpreted from the geometric parameters corresponding to these complexes: in this reaction, the C—N bond length does not change meaningfully when going from reactant to product, thus explaining the large values of the Similarity Indices and the small distances between

TABLE X
Similarity Indices (SI), distances (d), and transition state parameters (α and β) for the HNC/HCN rearrangement reaction obtained with the 6-31G** basis set

System	$SI_{A,B}$	$SI_{A,TS}$	$SI_{B,TS}$	$d_{A,B}$	$d_{A,TS}$	$d_{B,TS}$	α	β
HNC/HCN	0.9969	0.9976	0.9968	0.9250	0.8170	0.9381	1.9130	−0.1304

stationary points. The small value of α reflects the fact that, when going from the reactant to the transition state, small structural changes occur.

These results show that taking advantage of the information provided by the QMSM it is possible to draw an approximate shape of the computationally very demanding Intrinsic Reaction Path, independently on the definition of the reaction coordinate being more or less complicated, and also allowing the quantification of the Hammond and anti-Hammond behaviour. Further work is now in progress in our laboratory concerning other rearrangement reactions and the effect of external perturbations on the Walden inversion reaction.

4. Conclusions

This paper has attempted to discuss a new method to calculate QMSM, which essentially consists of the substitution of exact electron densities by fitted densities, expanded over a set of gaussian functions. It has been demonstrated that this methodology gives excellent QMSM results at a low computational cost. Therefore, this technique is highly recommended in the study of large systems, for which computational cost prevents calculation of exact QMSM.

One of the main drawbacks of Overlap-like QMSM, the core dependence problem, can be avoided using Coulomb-like and Valence QMSM. Further, choosing a good starting point by placing the center of charges of the two systems being compared at the origin of coordinates, and using Valence QMSM, are two clues suggested to improve the performance of QMSM maximizations.

Site-directed mutagenesis experiments can be somewhat related to QMSM among naturally occurring amino acids. In this case, comparison of chemical fragments instead of the full structure can be very useful when one tries to differentiate some aspects of a chemical problem that could otherwise be masked by the presence of a common structure.

Finally, it has been shown that QMSM, when used together with structural proximity and isosynchronicity parameters, can be a helpful interpretative tool when studying relationships between minima and maxima of the potential energy surfaces.

Acknowledgments

This work has been founded through Spanish DGICYT Project No. PB92-0333. J. M. acknowledges the financial help provided by the Generalitat de Catalunya through CIRIT Project No. AR91-425.

References

1. a) Franke, R., *Theoretical Drug Design Methods*, (Elsevier, Amsterdam, 1984). b) Sheridan, R. P. and Venkataraghavan, R., *Acc. Chem. Res.* **20**, 322 (1987). c) Richards, W. G., *Computer-Aided Molecular Design*, (IBC Technical Services, London, 1989). d) Mulholland, A. J., Grant, G. H., and Richards, W. G., *Protein Engng.* **6**, 133 (1993). e) Matthews, B. W., *Acc. Chem. Res.* **21**, 333 (1988). f) Johnson, M. A. and Maggiora, G. M., *Concepts and Applications of Molecular Similarity*, (Wiley, New York, 1990). g) Boyd, D. B., in *Reviews in Computational Chemistry*, eds. K. B. Lipkowitz and D. B. Boyd (VCH Publishers, Inc., New York, 1990, p. 355).
2. a) Mezey, P. G., *Shape in Chemistry: An introduction to molecular shape and topology*, (VCH Publishers, New York, 1993). b) Mezey, P. G., ref. 1f, p. 321. c) Mezey, P. G., *J. Chem. Inf. Comp. Sci.* **32**, 650 (1992). d) Mezey, P. G., in *The Role of Computational Models and Theories in Biotechnology*, ed. J. Bertrán (Kluwer Academic Publishers, Dordrecht, 1992, p. 83). e) Mezey, P. G., *J. Math. Chem.* **11**, 27 (1992). f) Mezey, P. G., *J. Comp. Chem.* **8**, 462 (1987).
3. a) Grindley, H. M., Artymiuk, P. J., Rice, D. W., and Willett, P., *J. Mol. Biol.* **229**, 707 (1993). b) Randić, M., *J. Chem. Educ.* **69**, 713 (1992). c) Trinajstić, N., *Chemical Graph Theory*, (CRC Press, Boca Ratón, FL, 1993). d) Mihalić, Z. and Trinajstić, N., *J. Chem. Educ.* **69**, 701 (1992). e) Randić, N., *J. Math. Chem.* **7**, 155 (1991).
4. a) Cioslowski, J. and Fleischmann, E. D., *J. Am. Chem. Soc.* **113**, 64 (1991). b) Cioslowski, J., and Challacombe, M., *Int. J. Quantum Chem. Quantum Chem. Symp.* **25**, 81 (1991). c) Ortiz, J. V. and Cioslowski, J., *Chem. Phys. Lett.* **185**, 270 (1991). d) Cioslowski, J., *J. Am. Chem. Soc.* **113**, 6756 (1991). e) Good, A. C., Hodgkin, E. E., and Richards, W. G., *J. Chem. Inf. Comp. Sci.* **32**, 108 (1992). f) Manaut, F., Sanz, F., Jose, J., and Milesi, M., *J. Comput.-Aided Mol. Des.* **5**, 371 (1991). g) Sanz, F., Manaut, F., Dot, T., and López de Briñas, E., *J. Mol. Struct. (Theochem)* **256**, 287 (1992). h) Ponec, R. and Strnad, M., *J. Phys. Org. Chem.* **4**, 701 (1991). i) Ponec, R. and Strnad, M., *J. Phys. Org. Chem.* **5**, 764 (1992). j) Ponec, R. and Strnad, M., *Int. J. Quantum Chem.* **42**, 501 (1991).
5. a) Carbó, R., Arnau, M., and Leyda, L., *Int. J. Quantum Chem.* **17**, 1185 (1980). b) Carbó, R. and Calabuig, B., *Int. J. Quantum Chem.* **42**, 1681 (1992). c) Carbó, R. and Calabuig, B., *Comp. Phys. Commun.* **55**, 117 (1989). d) Carbó, R. and Calabuig, B., *J. Mol. Struct. (Theochem)* **254**, 517 (1992). e) Carbó, R. and Calabuig, B., *Int. J. Quantum Chem.* **42**, 1695 (1992). f) Carbó, R. and Domingo, Ll., *Int. J. Quantum Chem.* **32**, 517 (1987).
6. Mestres, J., Solà, M., Duran, M., and Carbó, R., *J. Comp. Chem.* **15**, 1113 (1994).
7. Sambe, H. and Felton, R. H., *J. Chem. Phys.* **62**, 1122 (1975).
8. a) Dunlap, B. I., Connolly, J. W. D., and Sabin, J. R., *J. Chem. Phys.* **71**, 3396 (1979). b) Dunlap, B. I., Brenner, D. W., Mintmire, J. W., Mowrey, R. C., and White, C. T., *J. Phys. Chem.* **95**, 5763 (1991).
9. Frisch, M. J., Head-Grodon, M., Trucks, G. W., Foresman, J. B., Schlegel, H. B., Raghavachari, K., Binkley, J. S., Gonzalez, C., Defrees, D. J., Fox, D. J., Whiteside, R. A., Seeger, R., Melius, C. F., Baker, J., Martin, R. L., Kahn, L. R., Stewart, J. J. P., Topiol, S., and Pople, J. A., *GAUSSIAN 90, Revision H* (Gaussian Inc., Pittsburgh PA, 1990).
10. Mestres, J., Solà, M., Duran, M., and Carbó, R., *MESSEM version 1.0* (Girona CAT, 1993).
11. a) Scales, L. E., *Introduction to Non-Linear Optimization* (Springer-Verlag, New York, 1985). b) Fletcher, R., *Practical Methods of Optimization* (Wiley, Chichester, 1981). c) Wolfe, M. A., *Numerical methods for Unconstrained Optimization* (Van Nostrand Reinhold, New York, 1978). d) Beveridge, G. S. G., and Schechter, R. S., *Optimization: Theory and Practice* (McGraw Hill Kogakuska, Tokyo, 1970). e) Pierre, D. A., *Optimization Theory with Applications* (John Wiley, New York, 1969).
12. Hehre, W. J., Stewart, R. F., and Pople, J. A., *J. Chem. Phys.* **51**, 2657 (1969).
13. a) Hay, P. J. and Wadt, W. R., *J. Chem. Phys.* **82**, 270 (1985). b) Wadt, W. R. and Hay,

P. J., *J. Chem. Phys.* **82**, 284 (1985). c) Hay, P. J. and Wadt, W. R., *J. Chem. Phys.* **82**, 29 (1985).

14. Nelder, J. A. and Mead, R., *Comput. J.* **7**, 308 (1965).
15. Hodgkin, E. E. and Richards, W. G., *Int. J. Quantum Chem. Biol. Symp.* **14**, 105 (1987).
16. Dewar, M. J. S., Zoebisch,E. G., Healy, E. F., and Stewart, J. J. P., *J. Am. Chem. Soc.* **107**, 3902 (1985).
17. Liotard, D. A., Healy, E. F., Ruiz, J. M., and Dewar, M. J. S., AMPAC-version 2.1. Quantum Chemistry Program Exchange, program 506, *QCPE Bull.* **9**, 123 (1989).
18. a) Kaiser, E. T., *Angew. Chem. Int. Ed. Engl.* **27**, 913 (1988). b) Gardell, S. J., Craik, C. S., Hilvert, D., Urdea, M. S., and Rutter, W. J., *Nature* **317**, 551 (1985). c) Kolvenbach, C. G., Elliot, S., Sachdev, R., Arakawa, T., and Narhi, L. O., *J. Prot. Chem.* **12**, 229 (1993). d) Ishikawa, K., Kimura, S., Kanaya, S., Morikawa, K., and Nakamura, H., *Protein Engng.* **6**, 85 (1993).
19. Nagai, K., Nakaseko, Y., Nasmyth, K., and Rhodes, D., *Nature* **332**, 284 (1988).
20. Falzone, C. J., Karsten, W. E., Conley, J. D., and Viola, R. E., *Biochemistry* **27**, 9089 (1988).
21. Tu, C., Silverman, D., Forsman, C., Jonsson, B.-H., and Lindskog, S., *Biochemistry* **28**, 7913 (1989).
22. Hammond, G. S., *J. Am. Chem. Soc.* **77**, 334 (1955).
23. Formosinho, S. J., in *Theoretical and Computational Models for Organic Chemistry*, eds. S. J Formosinho, I. G. Csizmadia, and L. G. Arnaut (Kluwer Academic Publishers, Dordrecht, 1991, p. 159).
24. a) Varandas, A. J. C. and Formosinho, S. J., *J. Chem. Soc. Faraday Trans. 2* **82**, 953 (1986). b) Formosinho, S. J., *J. Chem. Soc. Perkin Trans. 2* **2**, 839 (1988). c) Parr, C. and Johnston, H. S., *J. Am. Chem. Soc.* **85**, 2544 (1963). d) Agmon, N. and Levine, R. D., *Isr. J. Chem.* **19**, 30 (1980). e) Agmon, N., *Int. J. Chem. Kinet.* **45**, 343 (1981). f) Lendway, G., *J. Phys. Chem.* **93**, 4422 (1989). g) Miller, A. R., *J. Am. Chem. Soc.* **100**, 1984 (1978). h) Arteca, G. A. and Mezey, P. G., *J. Comp. Chem.* **9**, 728 (1988).
25. a) Müller, K. and Brown, L. D., *Theor. Chim. Acta* **53**, 75 (1979). b) Gray, S. K., Miller, W. H., Yamaguchi, Y., and Schaefer, H. F. III, *J. Chem. Phys.* **73**, 2733 (1980). c) Garret, B. C., Redmon, M. J., Steckler, R., Truhlar, D. G., Baldridge, K. K., Bartol, D., Schmidt, M. W., and Gordon, M. S., *J. Phys. Chem.* **92**, 1476 (1988). d) Bentley, J. A., Bowman, J. M., Gazdy, B., Lee, T. J., and Dateo, C. E., *Chem. Phys. Lett.* **198**, 563 (1992). e) Fan, L. Y. and Ziegler, T., *J. Am. Chem. Soc.* **114**, 10890 (1992). f) Bentley, J. A., Huang, C. N., and Wyatt, R. E., *J. Chem. Phys.* **98**, 5207 (1993).
26. a) Hehre, W. J., Ditchfield, R., and Pople, J. A., *J. Chem. Phys.* **56**, 2257 (1972). b) Francl, M. M., Pietro, W. J., Hehre, W. J., Binkley, J. S., Gordon, M. S., Frees, D. J., and Pople, J. A., *J. Chem. Phys.* **77**, 3654 (1982). c) Hariharan, P. C. and Pople, J. A., *Theor. Chim. Acta* **28**, 213 (1973).

Quantitative Measurement of Molecular Similarity Using Shape Descriptors*

ROBERT P. BYWATER
Novo Nordisk A/S, DK-2880 Bagsværd, Denmark

Abstract. Molecular Shapes can be quantitatively described by a set of Fourier shape descriptors. The most suitable base for calculating these descriptors is spherical harmonics. Different molecules can be compared and their similarity/difference determined quantitatively in terms of the difference in their shape descriptors, expressed as vectors in a multidimensional space. Even very dissimilar molecules may have local regions of highly congruent shape, as in receptor-ligand interactions, for example. The shape descriptor approach can be used to study ligand docking and molecular recognition.

1. Introduction

Chemistry is concerned with the structure and properties of molecules but the subject only acquires its full meaning and usefulness when these are studied quantitatively. Whenever one is dealing with two or more molecules, a natural question is to ask how similar or different the molecules are to oneanother. Although there is a strong link between structure and properties, there is some degeneracy built into the system to the extent that molecules having apparently different structures, as defined by the arrangement of covalently-linked atoms, may have similar properties. It is therefore important to be aware of what it is one is comparing when studying molecular similarity.

Of all the many published methods of studying molecular similarity, there have been two general approaches, one which is *geometrical* in character, represented by the various connectivity and topological indexes [e.g. 1], in which the focus is on molecular structure, and another which is based on *quantum mechanical* considerations [2–6] where the focus is perhaps more on properties, at least in the sense that it is the wavefunction that ultimately dictates these properties.

It may be said of the earlier attempts at describing molecules by connectivity or topological indices [1], that they are perhaps too simplisitic or ambitious. They aimed to describe an object as complex as the geometry of a molecule in terms of *one single number.*[1] This number may be different for two molecules with different molecular skeletons which nonetheless have similar properties. In other cases, there will probably be examples of pairs of molecules with very different properties which have closely similar indices, because of the ways in which the indices are calculated. There is, as it were,

* Paper given at the First Girona Symposium on Molecular Similarity, July 8–10 1993, Girona, Catalunya.

R. Carbó (ed.), Molecular Similarity and Reactivity: From Quantum Chemical to Phenomenological Approaches, 113–122.

a kind of two-way degeneracy built into this approach. This has meant that these methods were confined to certain restricted families of compounds, and indeed in such cases e.g. boiling points of alkanes [7], very good predictions of molecular properties could be made. But for the purposes of calculating similarity between more diverse populations of molecules, these methods are unsatisfactory. However, recent progress involving the development of normalised topological indices, as described elsewhere in this volume [8], have gone a long way to allaying these criticisms.

Quantum mechanical methods [2–6] offer a much more rigorous approach to the problem, and can indeed be very successful. However, it may be that these methods are restricted to cases where molecules in question are "sufficiently similar" to begin with.

This paper describes an approach based on mathematical studies of molecular shape. A such, it would appear to belong to the geometrical category of methods. Whatever definition one may choose to use for molecular shape, one is faced with the need to deal with a complex, irregular object. One way of dealing with this has been to make a simpler version of the object, as for example in the STERIMOL approach [9], in the many QSAR studies where shape is approximated by molecular volume or surface area, or in the topology index methods mentioned above. A disadvantage of all these methods is that one is thereby discarding information that might well be decisive for accurate shape description and comparison.

One representation of molecular structure commonly used in molecular modelling is that of the van der Waals surface. However meaningful this may be in physical terms, it does have the advantage of possessing a well-defined geometry which reflects the physical "space-filling" propensities of the constituent atoms as well as the connectivity of the atoms themselves. From this surface, it is possible to calculate other surfaces, such as the solvent-accessible surface [10], or the interfaces between molecules that are associated, noncovalently, in complexes such as ligand/protein complexes. Anyone who has studied these complexes using molecular graphics can not escape noticing that the surfaces between the associated molecules seem to "fit" prefectly, even though these surfaces are very irregular. It is one of the aims of this work to give a quantitative measure to this degree of "fit".

In this work we have settled on Fourier transform methods for describing these molecular shapes, because they are well suited to handling the small but probably highly significant details of the surfaces that are important for selectivity in molecular recognition.

The mathematical description of complex shapes in 2 dimensions is well established [11], and has culminated in many industrial [12], and military applications [13], and in character recognition [14]. In our case, we of course need a three dimensional equivalent of this method, and for reasons of convenience, we work in a polar coordinate system, and resort to the use of spherical harmonics in order to calculate our shape descriptors.

2. Methods

The method of describing molecular shape in terms of expansions in spherical harmonics has been described in the literature [16, 17]. Briefly, the description of the molecule proceeds from a set of Cartesian coordinates for the constituent atoms in the molecules derived from experiment, theoretical calculation or model-building, from which a numerical description of the molecular "surface" is computed.

There are of course several different types of surface that one might be interested in. The van der Waals surfaces, solvent accessibility surfaces and molecular interfaces described above would be natural choices, but one could easily adapt the method to electrostatic potential or electron density surfaces, if so desired.

In this work we have chosen solvent accessible surface. The MS program [10] was used to calculate the coordinates of points located on this surface. This is achieved by simulating the rolling of a sphere of defined radius (e.g. the radius of a water molecule) over the van der Waals surface, which is in turn calculated from the atomic coordinates. The MS method identifies areas on the surface according to whether they are convex, concave or saddle regions, and displays these as sets of points distributed on these subsurfaces. The parameters of the MS program can be adjusted according to the size of the molecule and the degree of resolution it is intended to preserve from the original structure. (Computation time is approximately proportional to the number of atoms, the $^1/_2$ power of point density and the $^2/_3$ power of the probe radius [10]).

The declared aim of the MS program [10] is that: "the main use of these surface points is to display them on a real-time vector computer graphics system in order to examine the structure and interactions of biological macromolecules". Thus it is a purely numerical/graphical-display algorithm, while we are seeking an algorithm that can lend itself to an analytical treatment of the data, such that *comparisons* between and *classification* of the shapes of many molecules can be made.

The technique requires for its starting point a polar representation of the surface or region of a surface of a molecule. The surface model used in this work takes the molecular surface generated by the MS program. Following that, interpolation calculations are performed to obtain a function defined at integer values of the polar coordinates. Effectively, then the surface description produced by MS is mapped to a unit sphere. Single-valued functions can be ensured by choosing moleclues without severe indentations, or by choosing a Connolly sphere large enough to "smooth out" such indentations. But it may in some cases be required to collect and utilize the information contained in such a reentrant region. The problem of how to do this has been solved [17, 18], and involves making more than one "circuit" of the azimuthal angle θ, in which information omitted on the first pass through 2π radians, is collected in the second and, if necessary, subsequent circuits.

Such a single-valued function can be represented by a convergent series of single valued spherical harmonics [18] which, in complex number representation, is given by:

$$r(\theta, \phi) = \sum_{l=0}^{\infty} \sum_{m=-1}^{+l} a_{lm} Y_l^m(\theta, \phi). \tag{1}$$

From this expression the expansion coefficients or descriptors are given by the double integral over the unit sphere as follows:

$$a_{lm} = \int_0^{2n\pi} \int_0^{\pi} r(\theta, \phi)\, (Y_l^m)^*(\theta, \phi) \sin(\theta)\, d\theta\, d\phi. \tag{2}$$

Notice that the first integration in Equation (1) is taken over $2n\pi$, where n is the number of circuits required to collect all of the shape information.

3. Results

The aim of this work is to be able to quantitatively calculate shape similarities/differences. This can be done from the shape descriptors by treating the set of descriptors for each molecule as a vector in a Euclidean space of dimension l. Then a measure of the similarity between molecule **A** and **B** can be obtained from the Euclidean distance between the vectors $\vec{A}$ and $\vec{B}$:

$$D = |\vec{A} - \vec{B}| = \left\{ \sum_{n=N}^{+N} (a_n - b_n)\, (a_n - b_n)^* \right\}^{\frac{1}{2}}. \tag{3}$$

This is shown schematically in Figure 1, for two "similar molecules" **A** and **B**.

The calculation of coefficients in the spherical harmonics expansion can be pursued to any desired value of l, depending on how much shape information it is required to retain. The coefficient of the first term contains only size information, while true shape information is conveyed in the second and succeeding coefficients. The coefficients eventually become very small so that in practice a cut-off value of l can be chosen, according to the use to which the shape descriptors are to be put. From the shape coefficients the original shape can be reconstructed and, obviously, the accuracy depends on the number of harmonics used in the shape reconstruction. In one published [16] example of this a small protein molecule, haemerythrin B [19] was studied. It was shown that the after inclusion of just 10 harmonics, the reconstructed molecule was almost indistinguishable by visual inspection from the original and with $l = 20$ the error was no more than 1% [16]. For an accurate calculation, values of l up to 50 are recommended. The computer power required for these calculations increases as $l(l + 1)/2$, the number of integrals that need to be calculated.

Apart from the number of harmonics that need to be considered, there is

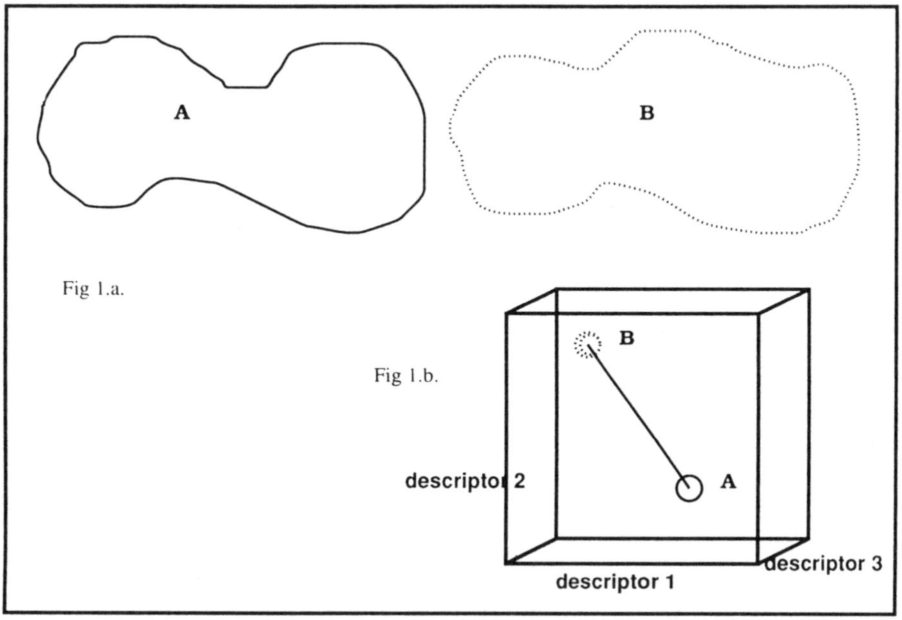

Figure 1a. Showing two "molecules" of different shape with their corresponding vectors displayed in a Fourier space (only 3 dimensions shown). *Figure 1b.* The Euclidean distance between these vectors is a measure of the similarity between the shapes of molecules **A** and **B**.

another question which requires attention, especially when one wishes to use the descriptors for comparison purposes. The values of the shape descriptors for a given molecule will vary according to how the coordinate system was originally set up. This is due to the fact that, for example, a rotation of the contour (or coordinate system) will affect the values obtained for the Fourier descriptors. To overcome this, a procedure referred to as normalisation must be carried out on the descriptors before a comparison can be made. This is is achieved by rotating and translating one molecule with respect to the second such that the distance measure is minimised with respect to coordinate operations. Full details of this procedure have been published [17, 18], and for all work involving comparison of shapes reported below and in references [18 and 20], normalised descriptors were used.

The study of small molecules is illustrated by investigations into the correlation between sweet taste and molecular shape. Typically, a number peptides and peptide analogues possess sweet flavour, although their enantiomers are either tasteless or have a bitter taste. These substances were studied alongside a number of nonpetide sweet-tasting substances. The starting structures were obtained by minimisation using molecular mechanics followed by semiempirical quantum mechanics calculation with geometry optimisation using MOPAC 5.0 [21]. From these, shape descriptors for l values up to

50 were calculated [20] and this data used to predict their index of sweetness. A regression analysis done on this data for a set of seven compounds had a correlation coefficient of 0.91, indicating that there is clearly a strong relationship between molecular shape and sweetness. Indeed, it would seem that one could predict sweetness from molecular shape alone.

Methotrexate is an inhibitor of the enzyme dihydrofolate reductase (DHFR) and a high-resolution crystal structure of the complex between these two exists [22]. This has become one of the most widely studied enzyme-inhibitor systems in drug design and discovery work [e.g. 23].

Methotrexate when bound to DHFR does not have the same conformation as it has in solution. The N—C bond linking the aromatic moieties in the molecule is very flexible and in the crystal structure of the complex with DHFR [20] the molecule is twisted around this bond in a way that is unlikely to resemble the minimum-energy conformation in solution. There are no known rigid analogues of the ligand in its enzyme-bound conformation. It has therefore become the standard practice to study only partial inhibitors which bind to one or other of the subsites [23] on the enzyme, but these are rigid enough for most kinds of investigation involving molecular shape or geometry.

One line of attack on the problem of constructing a QSAR on DHFR ligands has been the so-called "Molecular Shape Analysis" (MSA) method [24]. But, for example, Hopfinger's S_o MSA parameter is defined as representing "the need to have an analogue adopt the $\theta = 310°$ conformer state such that the 3 and/or 4 substituents possess size and/or conformational freedom so that specific spaces are occupied" [24]. Rather than use a parameter with a definition such as this we prefer to use *genuine* shape descriptors [20] when attempting to construct our QSAR model.

Using data obtained from Hopfinger [24], the correlation between DHFR inhibition and and molecular "shape" alone, expressed as the S_o parameter, had a correlation coefficient of no more than 0.448. If one includes hydrophobicity (π) and the MSA $\Delta\theta$ parameter [24] the Hopfinger correlation does improve considerably, to 0.965 [24]. Using our shape descriptors (with $l = 7$) *only*, without invoking any assistance from other parameters such as charge or hydrophobicity, the correlation is high as 0.882 [20]. These results demonstrate that in this case, although shape is not the only determinant of activity, it certainly is the major one, as long as *shape is properly defined.*[2]

In addition to being able to make quantitative descriptions and comparisons of molecular shape, it is possible to examine regions on the surfaces of molecules that may have very different structures and size even, but where each possesses local surface regions that have a mutually similar shape (Figure 2 illustrates this schematically). An example of the use of this approach is the study [17] of trypsin binding to trypsin inhibitor for which an accurate crystal structure exists [25]. This interaction is characterised by an intimate matching of the parts of the surfaces of trypsin and trypsin inhibitor respectively, and in which a lysine side-chain of the inhibitor is docked into a binding pocket on the trypsin surface.

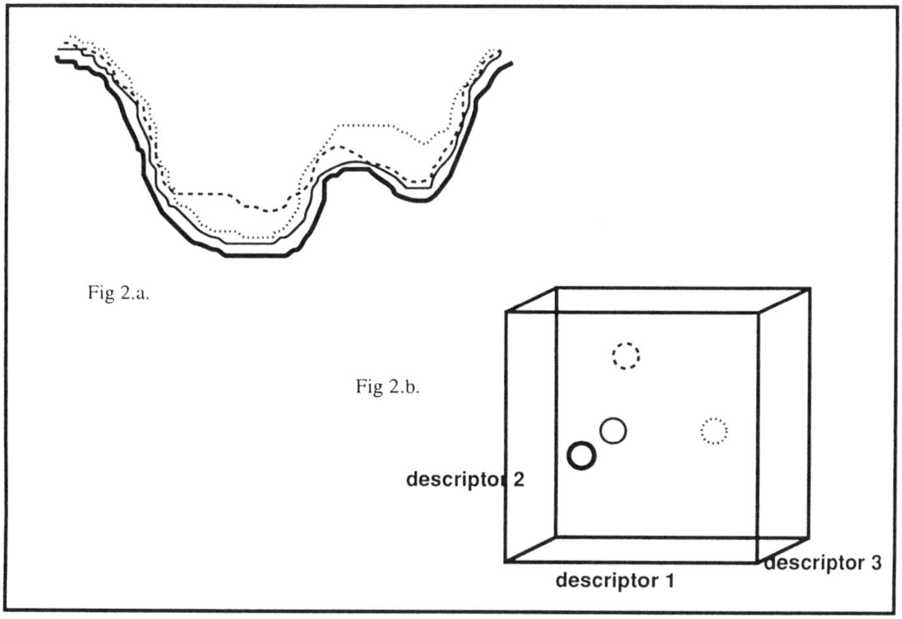

Figure 2a. Showing three "ligands" (faint lines) binding to the surface of a "receptor" (bold line). Their different degree of fit can be clearly demonstrated in a Fourier descriptor space ("only 3 dimensions shown) in *Figure 2b.* where the receptor (bold circle) could be regarded as a "target" e.g. for drug design.

A program was written [18] which identifies the regions on the surfaces of the molecules in contact with each other, defined by whether Connolly surface points on one molecule are within a given cut-off distance of surface points on the other. Connolly surfaces were produced using a surface density of 5 points/Å^2 and then the interface program run to obtain surface regions that were within 1.4 Å of each other. The resulting interface, reconstructed from the surface shape descriptors, together with the trypsin inhibitor docked into the pocket, is shown in Figure 3.

Although the relative positions of the enzyme and inhibitor were already known in this case, it was demonstrated [17], that it is perfectly feasible to devise a fully automatic docking procedure which can be used to find the optimum orientation (translation and rotation) of one surface relative to another, based entirely on shape information.

3. Discussion

The question of similarity has been addressed in this work, but from a somewhat different point of view than that in other work in this area. The methodology is radically different to anything which has been used before,

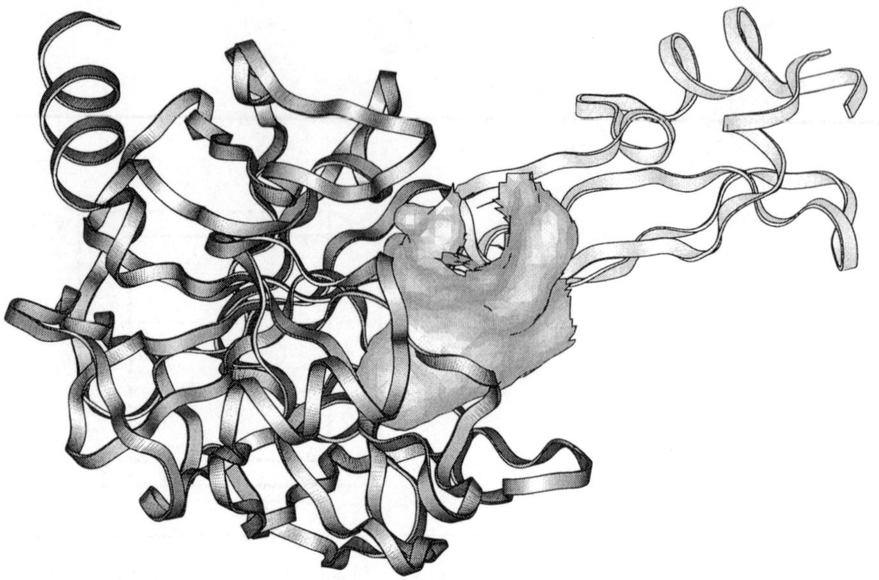

Figure 3. Interface between trypsin and trypsin inhibitor reconstructed from the surface shape descriptors, together with the trypsin inhibitor docked into the binding pocket.

in that attention is paid to surface properties rather than the molecular skeleton.

Many authors have tried to include some measure of "shape" in their attempts to construct a QSAR models for the binding of ligands to their target proteins, but the method reported here is the only one where genuine shape information is used, rather than parameters that are only remotely related to shape.

Furthermore, unlike any of the other published methods, attention can be paid not only to similarities between entire molecules (Figure 1), but also between congruent regions on molecules of different structure (Figure 2) which bind together to form noncovalent complexes (the question of "fit", referred to in the Introduction).

The concept of affinity, meaning the selective binding of one substance to another, has been a recurring theme in chemistry and biochemistry and is central to all modern research in molecular recognition, drug design etc. The idea itself has its origin in antiquity; Titus Lucretius Carus conjectured that substances which bind to each other have a mutual (congruent) pattern of indentations and protrusions of their surfaces [26]. The notion of affinity as arising from the intimate fitting of one surface to another is exactly what is seen in crystal structures of ligand-protein complexes, referred to above.

This definition of affinity has at least one major limitation. It gives only a static picture of the binding of one substance to another, and an important question is that of flexibility of ligands and their corresponding receptors.

Can the shape method cope with that? The answer depends entirely on how flexible these entities are. If there is any appreciable change in the molecular geometry then one must regard the structures corresponding to different energy minima in the overall energy-conformation space as being distinct, and, consequently, one must do separate calculations on these "different" molecules.[3]

Very often, it will be advantageous to do this anyway, as when one is dealing with allosteric proteins where conformational changes concomitant with binding are essential determinants of the mechanism of action. Cases are also known where the ligand undergoes considerable conformational changes on binding. One example of this is the binding of methotrexate to dihydrofolate reductase, referred to above. In the case of flexible ligands it is always advisable to have a knowledge of the active conformation of the bound ligand at the outset, before comparing it with other potential ligands, whether or not the latter are rigid.

One feature of this method which is something of a drawback is that it is a global shape descriptor method so that from the shape descriptors of a particular surface, it is not possible to obtain any information about isolated local features. The only way to achieve this is to identify the interacting interface between the two molecules and then calculate the shape descriptors from the local region at the interface (Figure 2). This has been shown to be quite feasible in practice [17]. However, it does presuppose some experimental data or calculations that give reliable information about the structure of the two molecules at the interface.

Throughout this work and that of other authors in this field, the subject of discussion has been molecular *similarity*. Yet if we consult the experts in the field of geometry, we shall find that this may not be the best choice of word. For example, Coxeter [27] writes: "Euclidean geometry, we might at first expect . . . to be the continuous group of all isometries. But since the propositions remain valid when the scale is altered, . . . , the "principle group" for Euclidean geometry [Klein C. F. 'Elementary Mathematics from an Advanced Standpoint' vol. 2, p. 133, Macmillan New York 1933] includes also *similarities* which may *change the distance* although they preserve the angles)". In the comparison of molecules we are of course only interested in rigid-body transformations and not in dilatations, thus we should really confine our attention to the concept of *congruence* (first defined by Euclid and, in the last century, incorporated by Klein into his generalised system of geometry [28]), in which both distances and angles are preserved.

Notes

1. The only shape which can be encoded by a single number is a sphere; obviously, more complex shapes require a larger set of numbers.
2. In reality, shape is not the underlying determinant of activity, but is, rather, a reflection of those factors which do determine activity. These factors, which may include charge and

"hydrophobicity" and above all, the steric repulsion which limits how close molecules can approach oneanother, are distributed around the molecular surface in a manner which gives rise to the shape as we perceive it using e.g. computer graphics.

3. This objection really applies, or ought to apply, to all other methods of similarity analysis.

References

1. Charton, M. and Motoc, I., Eds., "Steric Effects in Drug Design", Springer-Verlag, Berlin (1983), and references cited therein.
2. Carbó, R., Leyda, L., and Arnau, M., *Int. J. Quantum Chemistry* **17**, 1185 (1980).
3. Carbó, R., *J. Math. Chem.*, this volume (1993).
4. Hodgkin, E. and Richards, W. G., *Int. J. Quant. Chem.* **S14**, 105 (1987).
5. Good, A. C., Peterson, S. J., and Richards, W. G., *J. Med. Chem.* **S14**, 2929 (1993).
6. Burt, C., Huxley, P., and Richards, W. G., *J. Comput. Chem.* **11**, 1139 (1990).
7. Randic, M., *J. Amer. Chem. Soc.* **97**, 6609 (1975).
8. Randic, M., *J. Math. Chem.*, this volume (1993).
9. Verloop, A. Hoogenstraaten, W., and Tipker, J., *Drug Design* **7**, 165 (1976).
10. Connolly, M. L., "Molecular Surface Program", *QCPE Bull.* **1**, 74 (1981) (QCPE No.: 429).
11. Cosgriff, R. L., "Identification of Shape", Ohio State Univ. Res. Found., Columbus Rep. 820-11 ASTIA AD, 254–792 (1960).
12. Etesami, F. and Vicker, J. J., *Computat. Graph. and Image Proc.* **29**, 216 (1985).
13. Wallace, T. and Wintz, P. A., *Computat. Graph. and Image Proc.* **13**, 99 (1980).
14. Fritzsche, D. L., "A Systematic Method for Character Recognition", Ohio State Univ. Res. Found., Columbus Rep. 1222-4 ASTIA AD, 268–360 (1961).
15. Hobson, R. A., "The theory of Spherical and Ellipsoidal Harmonics", Cambridge University Press (1931).
16. Leicester, S. E., Finney, J. L., and Bywater, R. P., *J. Molec. Graphics* **6**, 104 (1988).
17. Leicester, S. E., Finney, J. L., and Bywater, R. P., *J. Math. Chem.* **16**, 315 (1994).
18. Leicester, S. E., PhD thesis, University of London 1989.
19. Klotz, I. M., Klippenstein, G. L., and Hendrickson, W. A., *Science* **192**, 335 (1976).
20. Leicester, S. E., Finney, J. L., and Bywater, R. P., *J. Math. Chem.* **16**, 343 (1994).
21. Stewart, J. J. P., MOPAC 5.0, QCPE 455.
22. Bolin, J. T., Filman, D. J., Matthews, D. A., Hamlin, R. C., and Kraut, J., *Biol. Chem.* **257**, 13650 (1982).
23. Blaney, J. M., Hansch, H., Silipo, C., and Vittoria, A., *Chem Revs.* **84**, 333 (1984).
24. Battershell C., Malhotra D., and Hopfinger, A. J., *J. Med. Chem.* **24**, 812 (1981).
25. Huber, R., Dietmar, K., Bode, W., Schweger, P., Bartels, J. and Steigemann, W., *J. Mol. Biol.* **89**, 73 (1974).
26. Titus Lucretius Carus, *De Rerum Naturae* **V**, 1089 (55**BC**).
27. Coxeter, H. S. M., "Introduction to Geometry", Wiley NY (1969).
28. Klein, C. F., *Das Erlanger Programm*, inaugural lecture at the Univeristy of Nürnberg/Erlangen (1872).

PROTEP: A Program for Graph-Theoretic Similarity Searching of the 3-D Structures in the Protein Data Bank*

PETER J. ARTYMIUK[1], HELEN M. GRINDLEY[2], AMANDA B. MACKENZIE[1], DAVID W. RICE[1], ELIZABETH C. UJAH[2] and PETER WILLETT[2,*]

[1] *Krebs Institute for Biomolecular Research, Departments of Molecular Biology and Biotechnology and of* [2] *Information Studies, University of Sheffield, Sheffield, U.K. (*author for correspondence)*

Abstract. The geometric arrangement of the α-helix and β-strand secondary-structure elements in a three-dimensional protein structure can be described by a labelled graph, in which the nodes represent the secondary-structure elements and the edges the inter-element angles and distances. Pairs of these graphs are compared using a maximal common subgraph isomorphism algorithm, which permits the identification of all areas of structural commonality between the proteins that are being compared and hence of a quantitative measure of the similarity between two protein structures. Given a target protein, the program PROTEP carries out a similarity search by calculating this measure between the target structure and each structure in the Protein Data Bank. The use of the program is exemplified by searches for calcium-binding proteins using troponin-C as the target structure, and by the identification of a previously-unknown structural relationship amongst the β-sandwich domains of prealbumin, protocatechuate 3,4-dioxygenase and thaumatin.

1. Introduction

The increasing amounts of structural data that are becoming available from X-ray and NMR studies has led to the development of a range of systems for the storage and retrieval of information pertaining to the three-dimensional (3-D) structures of proteins [1]. These systems use a range of models from computer science, including relational theory [2], object orientation [3] and logic programming [4]. An alternative model, and one that seems to be particularly well-suited to the processing of macromolecular structural data, is provided by the graph-theoretic methods that are used for the storage and retrieval of information pertaining to both two-dimensional (2-D) and 3-D small molecules [5].

A molecule in a chemical information system is represented by a labelled graph, in which the nodes and edges represent the atoms and bonds, respectively, of a 2-D molecule, i.e., a planar chemical structure diagram [6]. This graph-theoretic representation enables searching operations on databases of chemical structures to be implemented using isomorphism algorithms, which

* Presented at the First International Girona Seminar on Molecular Similarity, 8th–10th July 1993.

R. Carbó (ed.), Molecular Similarity and Reactivity: From Quantum Chemical to Phenomenological Approaches, 123–140.

compare one graph with another to determine the structural relationships that exist between them. Thus, subgraph isomorphism algorithms are used in substructure searching systems, which retrieve all molecules from a database that contain a user-defined partial structure, e.g., all molecules containing a quinazoline ring system. Maximal common subgraph isomorphism algorithms are used in reaction-indexing systems, where they are used to identify those parts of the reactant and product molecules that remain unchanged during a reaction [7], and in similarity searching systems, which permit a user to identify those molecules in a database that are structurally most-similar to an input target molecule [8]. The last few years have seen considerable interest in the extension of such graph-theoretic matching techniques to databases of 3-D structures, where the nodes and edges of a graph describe the atoms and inter-atomic distances, respectively, of a 3-D molecule (see, e.g., [9–12]). An ongoing project in the University of Sheffield is studying the use of these techniques for the representation and searching of 3-D protein structures [13–20], specifically those in the Brookhaven Protein Data Bank [22, 23]. Alternative graph-theoretic approaches to the analysis of protein structure have been reported by Koch *et al.* [24, 25] and by Haneef *et al.* [26, 27].

We have made extensive use of both subgraph and maximal common subgraph isomorphism algorithms, and have used representations that describe proteins at both the amino-acid residue and secondary-structure levels. In this paper, we describe PROTEP (for Protein Topographic Exploration Program) [28]. This is a computer program that has been developed to provide similarity-searching facilities for the 3-D structures in the Protein Data Bank, and that has already proved capable of identifying previously-unknown inter-protein structural similarities. Thus, we have discovered a strong resemblance between leucine aminopeptidase (1LAP) and carboxypeptidase A (5CPA), where there is a common substructure containing no less than eight β-strands in a sheet and five of the ten α-helices in the C-terminal domain of 1LAP [18]. More recently, we have identified a strong similarity between the ribonuclease H domain of HIV-1 reverse transcriptase (RT) and the ATPase folds of hexokinase, heat-shock cognate protein and actin, and an intramolecular resemblance between the same ribonuclease H domain and two other domains in RT [21]. PROTEP uses a maximal common subgraph isomorphism algorithm with a graph representation of a 3-D protein in which the nodes describe the α-helix and β-strand secondary-structure elements and the edges the inter-element angles and distances. The graph representation and the searching algorithm are described in Section 2, and Section 3 then illustrates the use of this approach by means of searches for calcium-binding proteins. Section 4 reports a further, previously-unknown structural relationship between the β-barrel domains of prealbumin, protocatechuate 3,4-dioxygenase and thaumatin. The paper concludes with a summary of our main findings and suggestions for future work.

2. Representation and searching of 3-D protein structures

2.1. Processing of secondary structure information

The main focus of interest to date in our work has been the representation and searching of *tertiary-structure motifs*, i.e., patterns of secondary-structure elements in 3-D space. The graph representation of a protein that we have adopted makes use of the fact that the two most common types of secondary-structure element, the α-helix and the β-strand, are both approximately-linear, repeating structures, which can hence be represented by vectors drawn along their major axes. The set of vectors corresponding to the secondary-structure elements in a protein can then be used to describe that protein's 3-D structure, with the secondary-structure elements and the inter-element angles and distances corresponding to the nodes and to the edges, respectively, of a graph [14, 15]. More precisely, each edge in such a labelled graph is a three-part data element that contains the angle between a pair of vectors describing secondary-structure elements, the distance of closest approach of the two vectors and the distance between their mid-points. Figure 1 is a diagrammatic representation of the protein troponin-C, which contains 12 secondary-structure elements and which was used for the PROTEP searches that are detailed in Section 3. Figure 2 is the corresponding graph representation. This is a 12 × 12 matrix, the *IJ*-th element of which contains the three components that specify the geometric relationship between the *I*-th and *J*-th secondary-structure elements.

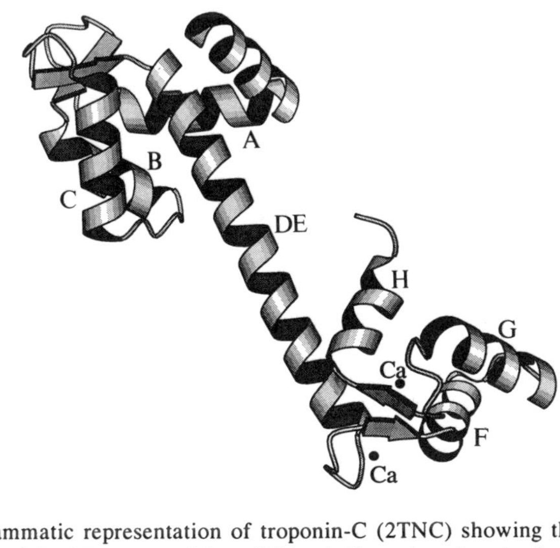

Figure 1. Diagrammatic representation of troponin-C (2TNC) showing the N-terminal and C-terminal domains joined by a central long linking helix and produced using the MOLSCRIPT program (P. J. Kraulis, *Journal of Applied Crystallography* **24**, 946 (1991)). In this diagram, the α-helices are indicated by cylinders and the β-strands by flat ribbons; the numbering of the helices is described in the text.

	4 HEL 12	16 HEL 28	41 HEL 48	55 HEL 65	75 HEL 105	115 HEL 124	131 HEL 140	151 HEL 158	36 STR 38	72 STR 74	112 STR 114	148 STR 150
4	0	117	22	-129	53	62	-165	80	98	-63	40	-127
HEL	0	93	182	139	85	378	372	268	168	122	391	361
12	0	159	202	188	192	440	429	327	190	158	432	408
16	117	0	138	-88	115	96	-52	-112	47	-161	157	-30
HEL	93	0	99	194	117	374	410	306	84	116	417	396
28	159	0	125	211	208	457	480	381	137	147	463	455
41	22	138	0	121	-50	69	-166	76	111	-45	18	-145
HEL	182	99	0	101	112	386	420	313	61	88	378	364
48	202	125	0	124	175	445	478	354	109	124	429	421
55	-129	-88	121	0	153	-74	63	-49	-128	79	-109	59
HEL	139	194	101	0	110	415	399	299	122	88	345	325
65	188	211	124	0	185	461	474	336	142	121	424	403
75	53	115	-50	153	0	115	-120	124	69	-81	54	-145
HEL	85	117	112	110	0	104	174	102	74	29	68	99
105	192	208	175	185	0	284	305	182	256	243	266	253
115	62	96	69	-74	115	0	124	-33	126	-67	-75	79
HEL	378	374	386	415	104	0	100	141	518	495	20	68
124	440	457	445	461	284	0	105	156	536	526	93	127
131	-165	-52	-166	63	-120	124	0	112	-68	131	-150	49
HEL	372	410	420	399	174	100	0	100	522	485	105	81
140	429	480	478	474	305	105	0	147	556	538	130	126
151	80	-112	76	-49	124	-33	112	0	156	-49	71	-85
HEL	268	306	313	299	102	141	100	0	414	379	70	25
158	327	381	354	336	182	156	147	0	432	413	112	81
36	98	47	111	-128	69	126	-68	156	0	150	122	-77
STR	168	84	61	122	74	518	522	414	0	40	491	470
38	190	137	109	142	256	536	556	432	0	44	520	505
72	-63	-161	-45	79	-81	-67	131	-49	150	0	-29	133
STR	122	116	88	88	29	495	485	379	40	0	474	446
74	158	147	124	121	243	526	538	413	44	0	506	488
112	40	157	18	-109	54	-75	-150	71	122	-29	0	154
STR	391	417	378	345	68	20	105	70	491	474	0	45
114	432	463	429	424	266	93	130	112	520	506	0	49
148	-127	-30	-145	59	-145	79	49	-85	-77	133	154	0
STR	361	396	364	325	99	68	81	25	470	446	45	0
150	408	455	421	403	253	127	126	81	505	488	49	0

Figure 2. Labelled-graph representation of the 2TNC target protein. The *IJ*-th element of the matrix contains the geometric relationships between the *I*-th and *J*-th secondary structure elements. Each such element contains three components, these being the inter-element angle (°), the distance between the mid-points and the distance of closest approach (Å × 10).

The information that is needed to create these labelled graphs is obtained from the coordinate data in the Protein Data Bank. The α-helices and β-strands are identified using the Kabsch-Sander algorithm [29]; these assignments are checked using the FRODO molecular graphics system [30] and modified if necessary. The residues comprising a secondary-structure element are then input to a routine that calculates the major axis of the secondary-structure element: this is done by mapping idealized α-helices or β-strands to the residue coordinates, using a least-squares fitting procedure. The resulting axes are vectors in 3-D space and the angles and distances between each pair of vectors are calculated: these data then represent the geometry of a protein. The procedure results in a compact database of labelled graphs that can be searched using a subgraph-isomorphism algorithm [14–17] or a maximal common subgraph isomorphism algorithm [18–21]. Both of these search options are available in PROTEP, which outputs matches to a query protein in a format compatible with FRODO, and thus allows the immediate inspection of the highlighted secondary structures from database proteins that match the query.

2.2. Use of the Bron-Kerbosch Clique-Detection Algorithm

The maximal common subgraph procedure that we have used in this work is based on the clique-detection algorithm of Bron and Kerbosch [31], where a clique is a subgraph of a graph in which every node is connected to every other node and which is not contained in any larger subgraph with this property.

The input to the clique-detection procedure is a *correspondence graph*. Given a pair of graphs A and B, the correspondence graph, C, is formed in two stages. In the first stage, the set of all pairs of nodes is created, one from A and one from B, such that the nodes of each pair are of the same type. C is then the graph whose nodes are the pairs from the first step. Two nodes $(A(I), B(X))$, $(A(J), B(Y))$ are marked as being connected in C if the values of the edges from $A(I)$ to $A(J)$ and $B(X)$ to $B(Y)$ are the same. Maximal common subgraphs then correspond to the cliques in the correspondence graph [32], so that the identification of the maximal common substructure for a pair of 3-D chemical molecules is equivalent to the identification of the largest clique in the correspondence graph linking the two molecules. This largest clique then represents the largest possible overlap of one structure on the other, subject to any geometric tolerances that have been defined by the user.

Clique detection was first studied in Sheffield in the context of finding the maximal substructure common to pairs of small 3-D molecules [10], where it was shown that the clique-detection algorithm due to Bron and Kerbosch was the most efficient for this application. A subsequent paper discussed the use of this algorithm for the implementation of similarity searching in files of 3-D small molecules [33] and it has also been used in ligand-binding studies by Crippen and co-workers [34, 35]. Willett [36] gives a detailed account of the use of this algorithm when it is used for the matching of 3-D chemical graphs.

PROTEP uses the Bron-Kerbosch algorithm to identify patterns of α-helices and β-strands in 3-D space that are common to a pair of proteins. Specifically, a user-supplied *target* protein is matched against each of the proteins in the Protein Data Bank in turn to identify the structural similarities that are present (subject to the user's search parameters). The algorithm can be used to identify just the largest common pattern (and hence the term 'maximal common subgraph isomorphism'); however, this is identified only as the result of an exhaustive tree search that identifies all structural patterns that are common to the two graphs that are being compared. PROTEP can thus identify all of the common patterns that are larger than some user-defined threshold size, which is referred to subsequently as *MINCLQ*. This restriction is imposed to ensure that the user is not overwhelmed with a very large output consisting primarily of small common substructures that are of little structural significance.

The precise nature of the output from a PROTEP search is determined by the tolerances that are used. The angular tolerance is specified in terms of numbers of degrees, while the distance tolerances are specified either in Å or as a percentage of the distance in the target structure. Three types of distance match are possible: closest-approach distance (CAD), midpoint distance (MD) or both types of distance (BD). In the experiments reported in Section 3, runs were carried out in which the angular tolerance was varied between 10° and 50° in 10° intervals and the distance tolerance was varied between 10% and 50% in 10% intervals and also between 1 Å and 4 Å in 1 Å intervals. It is also possible to specify that the secondary-structure elements in a database protein that match the target protein are in the same sequence order as in the target; alternatively, the sequence order does not need to be the same. The user thus has a very large degree of control over the number and the quality of the matches that are identified by the program.

3. Searches for calcium-binding proteins

3.1. Introduction

In previous papers, we have presented several examples of the use of PROTEP to identify correctly the structural features that are common to sets of proteins that are known to belong to the same family. For example, Artymiuk *et al.* [19] discuss searches for the dinucleotide-binding fold motif using lactate dehydrogenase (4LDH) as the target molecule, while Grindley *et al.* [20] discuss searches using azurin (1AZU) and myoglobin (1MBD) as the target molecules. Here, we discuss series of searches that have been carried out for calcium-binding proteins using troponin-C (2TNC) as the target structure. These searches were carried out on the 371 labelled graphs that resulted from the use of the database-creation programs described in Section 2.1 on the April 1989 release of the Protein Data Bank. This release contained 12 examples of four known calcium-binding proteins: troponin-C (2TNC, 3TNC, 4TNC, 5TNC), calmodulin (1CLN, 2CLN, 3CLN), parvalbumin (1CPV, 2CPV, 3CPV) and intestinal calcium-binding protein (2ICB, 3ICB).

Troponin-C is a regulatory muscle protein that is a member of the family of calcium-binding proteins, which also includes calmodulin, parvalbumin and intestinal calcium-binding protein. This group of proteins has been extensively studied [37–40] and their 3-D structures are well documented. All the proteins have some degree of sequence homology and each contain a number of calcium-binding sites; parvalbumin and intestinal calcium-binding protein contain two each, while troponin-C and calmodulin are larger proteins that contain four each.

The two calcium-binding sites in parvalbumin are both formed by a helix, a loop and a further helix, with the calcium-binding site being in the loop between the two helices. In parvalbumin, helices **E** and **F** form such a binding site and Kretsinger [38] named this structure motif the '**EF** hand', owing to the relative orientations of the two helices involved. There is a similar motif formed by helices **C** and **D** in parvalbumin, to give the second calcium-binding site. The structure of intestinal calcium-binding protein is very similar to that of parvalbumin.

Troponin-C is approximately twice the size of parvalbumin and contains four calcium-binding sites all of which have very similar '**EF** hand' motifs of helix, loop, helix. These four sites are located in two lobes or domains, with two sites in each domain; helix pairs **AB** and **CD** are in the N-terminal domain and helix pairs **EF** and **GH** are in the C-terminal domain. The two domains are connected by a long, 9-turn α-helix link which comprises helices **D** and **E**, thus giving the protein the appearance of a dumb-bell. There is a further 10-residue helix at the start of the polypeptide chain in the N-terminal domain which does not form part of the '**EF** hand' motifs. Figure 1 shows the structure of the protein troponin-C with its two domains.

Calmodulin has a very similar two-domain structure to troponin-C, with a central linking helix. The two main structural differences between calmodulin and troponin-C are that the 10-residue helix extension in the N-terminal domain is missing and that the central linking helix has a three-residue deletion [40], which means that the orientation of the two domains with respect to each other is different to that in troponin-C. In calmodulin the N-terminal domain is oriented approximately 180° with respect to the C-terminal domain, whereas in troponin-C this value is only approximately 120°.

3.2. Specification of the query pattern

The troponin-C protein used as a pattern (2TNC) contained eight α-helices and four β-strands. The eight helices comprised helices **A**, **B**, **C**, **DE**, **F**, **G**, **H** and the 10-residue N-terminal helical extension as described previously. Note that helices **D** and **E** are given as a single helix, this being the long central linking helix. The four β-strands are the two pairs of loops in the '**EF** hand' motifs.

The domains in the proteins described above all contain two calcium-binding sites and hence contain a similar arrangement of structural motifs. It would thus be expected that matches might be found of the whole structure of troponin-C with the whole structure of calmodulin at higher tolerances. In fact, there was no such match found between the two proteins in any of the runs carried out in this particular investigation, despite the structural similarity of the two domains joined by a long linking helix. This omission was due to the different orientations of the N- and C-terminal domains with respect to each other in the two proteins, which as described earlier constituted a difference of 60° in orientation [40]. As the maximum angular tolerance (*ANGTOL*) employed in any set of runs was 50° this is quite understandable. However, it could be expected that single domains would match each other in both these proteins and, in addition, for either domain in troponin-C to match to the single-domain proteins parvalbumin and intestinal calcium-binding protein.

3.2. Results

Two sets of searches were carried out using the query pattern described above, with the minimum clique size, *MINCLQ*, being set to 5 or to 6. In both cases, the sequence order in the matching proteins was required to be the same as in the target protein. In what follows, we shall focus on the *MINCLQ* = 6 results; Grindley [41] provides further details of the *MINCLQ* = 5 searches in her thesis.

The results from a set of runs can be conveniently displayed by means of a 3-D bar chart, such as those shown in Figures 3 and 4, where the total numbers of cliques found for each set of tolerances are indicated by the

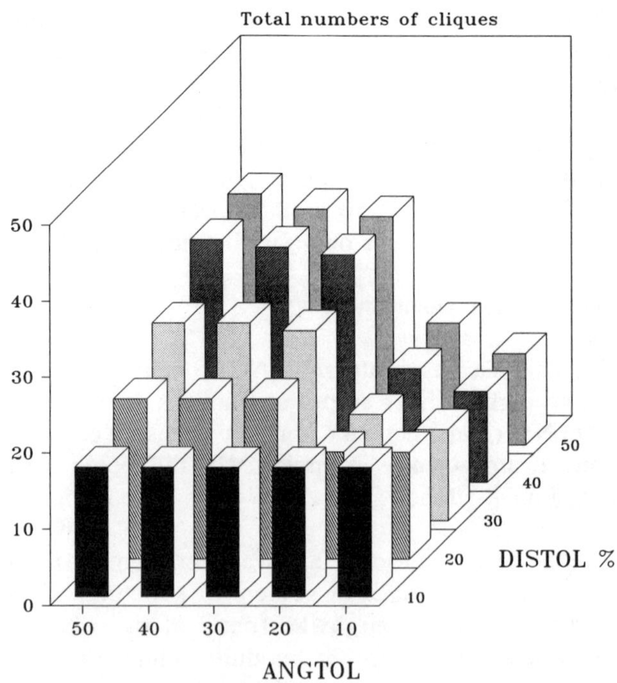

Figure 3. Total numbers of cliques identified in the 2TNC searches with *MINCLQ* set to 6.

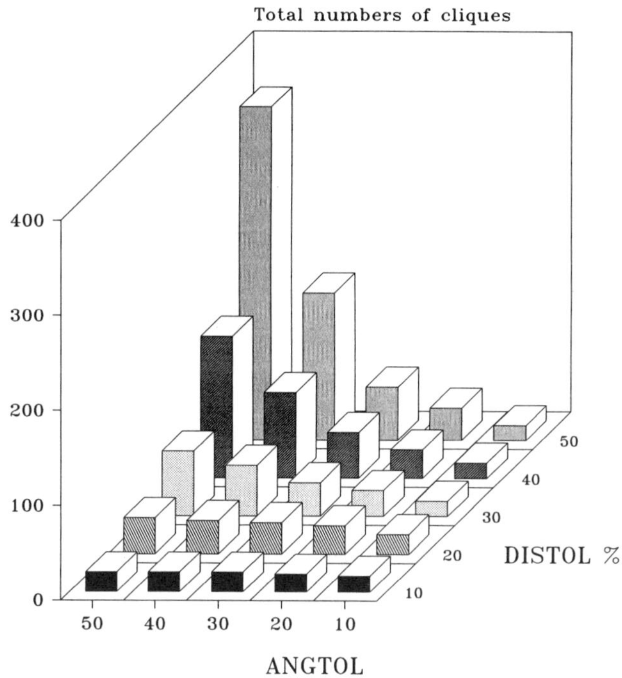

Figure 4. Total numbers of cliques identified in the 2TNC searches with *MINCLQ* set to 5.

height of the relevant bar in the bar chart. These figures summarize the results that were obtained in the CAD searches where the distance tolerances were expressed as a percentage (%) of the distances in the pattern protein.

On examining the *MINCLQ* = 6 set of runs, it was found that more cliques were generally found with % tolerance values than with Å tolerance values in the CAD, MD and BD runs. The vast majority of all of the cliques were found within the known calcium-binding proteins. All those found at lower tolerances (*ANGTOL* ≤ 30° or *DISTOL* ≤ 30%, 4 Å), were purely within calcium-binding proteins. A few non-calcium-binding protein hits were obtained but the cliques from these occurred at higher tolerances, i.e., *ANGTOL* ≥ 40° or *DISTOL* ≥ 30% (there were none found in the Å-tolerance runs carried out). In the remainder of this section, we consider more closely the calcium-binding and non-calcium-binding proteins that were retrieved.

3.3.1. *Calcium-binding proteins*

In the case of both the parvalbumin (1CPV, 2CPV, 3CPV) and the intestinal calcium-binding proteins (2ICB, 3ICB), one clique per protein was found as a match in the set of runs. In all of these cases, this was a size-6 match containing the 'expected' secondary structure element matches, i.e., these cliques were all correct matches of the secondary structure elements in the single domains of parvalbumin and intestinal calcium-binding protein with all the secondary structure elements in the C-terminal domain of the pattern protein 2TNC plus the long linking helix (i.e., helices **EF**, **GH** and the two loop strands connecting the two pairs of helices). These cliques only occured in CAD runs, however, and first appeared at tolerances of 30°/20% or 2 Å in the case of intestinal calcium-binding protein and at 20°/20%–30% or 2 Å–3 Å in the case of the parvalbumins.

Quite large numbers of cliques were found for each troponin-C structure, but in each case the largest cliques found were, in fact, those containing the 'expected' secondary structure element matches. In the case of 2TNC and 3TNC these were cliques of size 12, i.e., a match of all the secondary structure elements present in the pattern protein 2TNC. In the case of the proteins 4TNC and 5TNC, each had two cliques which contained 'expected' secondary structure element matches: these were of sizes 11 and 9 and sizes 11 and 8, respectively. These cliques first appeared at low tolerances of between 10°/10% or 1 Å up to 20°/30% or 3 Å in all three types of distance runs (CAD, MD and BD). However, looking at the lowest tolerance runs (CAD, MD and BD) it could be seen that subsets of these largest 'expected' cliques occured in 3TNC, 4TNC and 5TNC, the sizes of these subsets ranging from 6 upto 10 secondary structure elements. All of the other cliques in the troponin-C proteins were hits which contained some 'unexpected' secondary structure element matches, i.e., 'EF' hand secondary structure elements that were incorrectly matched in the target protein and the database proteins. These 'unexpected' matches first appeared over a wide range of tolerances from 20°/20% up to 50°/50% and at 30°/3 Å, but predominantly at higher tolerances.

In the case of calmodulin, cliques of size 6 were found for 1CLN and 2CLN, but there were no cliques at all for 3CLN throughout the whole set of runs. This was due to what we have termed the 'crossover' problem [41], which arises from the way in which the inter-element torsion angles are calculated for pairs of secondary structure elements that are nearly co-planar. Here, on closer examination, it could be seen that where a hit of the C-terminal domain of 2TNC with the C-terminal domain of 3CLN might have been expected (i.e., matches in both proteins of α-helices **E**, **F**, **G**, **H** and the two inter-connecting β-strands forming the '**EF** hands') this did not happen. This was because the inter-element torsion angle between α-helix **G** and its inter-connecting β-strand was 140° in 3CLN and –150° in 2TNC. Thus, whilst the absolute values of the two angles were very similar (the difference being only 10°), and were also close to the value of 180° denoting a co-planar pair of elements, the signs were different, meaning that one of the elements in the pattern protein was (just) on the opposite side of the plane from that which it occupied in the database structure. When a search is carried out for the pattern protein, PROTEP will attempt to match secondary structure elements in the target protein and in the database structure by checking to see whether the associated angles and distances are equal, to within the search tolerances. Angles in PROTEP are measured in the range –180° to +180°; the difference between the two angles is thus 360°–(140°–(–150°)), i.e., 70°

which is greater than the maximum angular tolerance of 50° that was used in these experiments. Similarly, no size-6 cliques were found between the C-terminal domain of 2TNC and the N-terminal domain of 3CLN as the inter-element angle between α-helix **G** and its interconnecting β-strand was –150° in 2TNC and the corresponding angle between α-helix **C** and its interconnecting β-strand was +139° in 3CLN. Matches of the N-terminal domain of 2TNC with either domain in 3CLN encountered similar problems, with the addition that some helices did not match each other in both proteins to within 50°, the highest specified angular tolerance. Checks have now been incorporated into PROTEP to identify occurrences of this problem and hence to ensure the retrieval of such matches.

With regard to those size 6 cliques found with calmodulins 1CLN and 2CLN, an 'expected' match of the C-terminal domains was found in both 1CLN and 2CLN with 2TNC. These first occurred in CAD, MD and BD runs at 10°/20–30%$ or 2–3 Å for 2CLN and 30°/30% or 2 Å (CAD) and 30°/50% (MD, BD) for 1CLN. A further 'expected' match of the N-terminal domain with the C-terminal domain was found in CAD runs only at 20°/20% or 3 Å for 2CLN and for 1CLN at 40°/30% or 2 Å (CAD) and 40°/50% (MD, BD). A third clique in 1CLN containing 'unexpected' secondary structure element matches with 2TNC was found in the MD runs at high tolerances of ≥ 40°/50%.

3.3.2. *Non-calcium-binding proteins*

Only a very few cliques were found in non-calcium-binding proteins, and none of these occured in the BD runs. In the CAD runs, only one clique in photosynthetic reaction centre protein (1PRC) was found, first appearing at the highest tolerances of 50°/50%. The secondary-structure elements in 1PRC were taken from three different domains, and one could thus question the relevance of such a match in terms of functional significance. The remaining non-calcium-binding type proteins were all located in MD% runs. There were seven such proteins and the cliques with them first occurred at tolerances of 50°/50% (except for those in acid proteinase (2APP) which first occurred at 40°/40%). Two of these protein cliques were examined on an Evans & Sutherland PS390 graphics workstation.

The 2APP clique contained matches with both pairs of strands in 2TNC and with an N-terminal domain helix and the long linking helix. The strand matches were not too bad in terms of distance, but the helix matches were not at all good in terms of position or length. Because of the dispersed nature of the matched secondary structure elements this was not really a meaningful hit. The clique with the immunoglobulin G1 FAB fragment complexed with lysozyme (2HFL) was also examined; this again contained matches with both pairs of strands in 2TNC plus two helices from the C-terminal domain. All of the secondary structure element matches in 2HFL were taken from the lysozyme molecule. If the two matched strands in the N-terminal domain were ignored, it could be seen that the two helices and two strands in the C-terminal domain of the pattern protein, 2TNC, actually looked like quite a feasible match in terms of position, orientation and length with the relevant secondary-structure elements in 2HFL.

3.3.3. *Other Runs*

A further set of runs was carried out with *MINCLQ* set to five. As would be expected, there was a very substantial increase in the number of matches obtained. This finding is exemplified by the barchart results shown in Figure 4, which are far higher than the corresponding runs in Figure 3. Inspection of the hits showed that there was very little change in the numbers of hits in known calcium-binding proteins, except at very high tolerances; the large increase was almost entirely due to the retrieval of very many size-5 cliques in non-calcium-binding proteins. Inspection of the matches suggested that few of them were particularly meaningful in terms of matching secondary structure elements giving a compact folding unit. The primary exceptions were: a size-5 clique in 2HFL at 30°/40%, which was part of the size-6 match mentioned previously; a fair match with phage source lysozymes in the MD runs at high tolerances (50°/50%); and cliques in some globin proteins that first appeared at tolerances of 40°/40% in MD runs. Both of the last two sets of matches represented a five-membered 'helix bundle' motif [42].

In general, then, the experiments showed that good matches were obtained with the known calcium-binding proteins at low tolerances, and that matches with non-binding proteins were obtained only when much less-specific search parameters were used.

4. Structural resemblance amongst prealbumin, protocatechuate 3,4-dioxygenase and thaumatin

The previous section has illustrated the use of PROTEP to identify correctly the structural features that are common to sets of proteins that are known to belong to the same family. However, the main intended use of PROTEP is to identify previously unrecognised structural resemblances. We now report such a resemblance, this being the structural relationship between the β-sandwich domains of three otherwise-unrelated proteins: prealbumin (2PAB) [43, 44], protocatechuate 3,4-dioxygenase (1PCD) [45] and thaumatin (1THI) [46].

A query pattern was defined for 1PCD that consisted of 8 helices and 32 strands. This pattern was searched against the Protein Data Bank using the following parameters: an angular tolerance of 30°; a distance tolerance that was the lesser of 4 Å and of 40% of the distance in the pattern; *MINCLQ* set to seven; and only sequence-ordered hits were retrieved.

The common fold between the three proteins is a β-sandwich consisting of one three-stranded and one four-stranded β-sheet, these motifs comprising a major part of the mixed β-sandwich found in each protein. The proteins and the common motif are illustrated in a schematic fashion in Figure 5. In

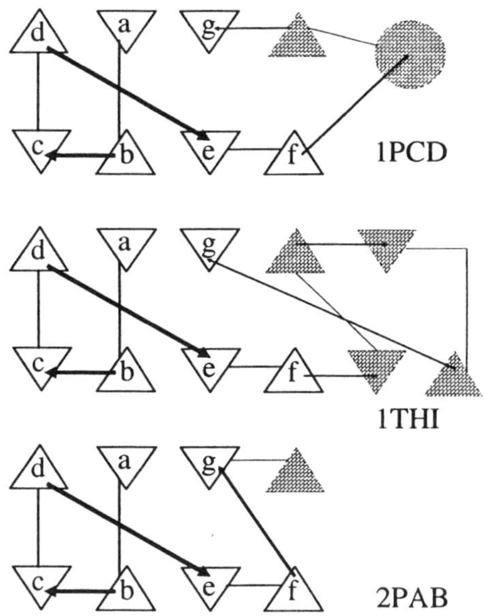

Figure 5. Topological representations of the common β-motif in protocatechuate 3,4-dioxygenase (1PCD), thaumatin (1THI) and prealbumin (2PAB). Circles represent α-helices and triangles represent β-strands (apex down indicates that the strand is viewed from the C-terminus). Open circles and triangles indicate those secondary-structure elements that were found by the PROTEP search to superpose in 3-D, and shaded ones those that do not. The strands in the five-stranded β-sheets are lettered according to their order in the sequence.

this topological diagram, circles represent the α-helices and triangles represent the β-strands (with the apex pointing down indicating that the strand is viewed from the C-terminus). Open triangles represent those β-strands that constitute the common motif, and shaded triangles those that do not. The common strands are labelled *a* to *g* according to their order in the proteins' primary sequences; specifically, the constituent β-strands are labelled *d*, *a* and *g* (for the three-stranded mixed motif) and *c*, *b*, *e* and *f* (for the four-stranded antiparallel motif).

Human prealbumin (which is also known as transthyretin) is a hormone binding protein which has two thyroxine-binding sites and also independent binding sites for retinol binding protein [43], the specific plasma carrier of vitamin A. The crystal structure has been solved to 1.8 Å resolution and detailed binding studies have revealed the nature of the substrate specificity sites. The molecule is a tetramer of identical subunits, each containing 127 amino acid residues and consisting of an 8-stranded β-barrel structure organized into two 4-stranded β-sheets, together with a short stretch of α-helix.

Protocatechuate 3,4-dioxygenase belongs to a family of bacterial dioxygenases that catalyse the cleavage of molecular oxygen with the subsequent incorporation of both oxygen atoms into organic substrates. The crystal structure of 1PCD has recently been solved [45]. The molecule is an oligomer of dissimilar subunits, the α-subunit containing 200 residues and the β-subunit containing 238 residues. Both subunits share a common structural core, which consists of a flattened 9-stranded β-barrel and three short α-helices, but they differ in their terminii and in several surface loops.

Thaumatin (1THI) is a monomeric (207 residues), intensely sweet protein of unknown function, which is extracted from the fruit of an African plant *Thaumatococcus danielli* [47]. The crystal structure has been solved at 3.0 Å resolution and reveals a flattened 11-stranded β-barrel, surrounded by many loops closed by disulphide bonds.

The 3-D resemblance in the common region is good, with the RMS differences over the equivalenced residues ranging from 0.91 Å (for the three-stranded motif in 2PAB and 1PCD) to 2.35 Å (for the four-stranded motif in 1THI and 1PCD). The superpositions show an identical organisation and order of the strands, which also appear to be conserved in position, although the four-stranded motif in 1THI/1PCD superposed less well than the others. A detailed multiple-alignment study failed to identify any statistically-significant sequence homologies between the residues comprising the common structural feature in the three proteins [48] and the question then arises as to why there is such a strong structural resemblance despite the lack of any sequence resemblance. Three possible explanations suggest themselves.

Firstly, it is possible that the proteins may be evolutionarily related, but so remotely that all significant sequence homology has been lost, and only a common substructure remains. A recent example in which familial relationships have been identified in the absence of sequence homology is that between PapD chaperone protein and the immunoglobulin superfamily [50]. An alternative possibility is that some members of the three families may bind to structurally-similar receptors. This explanation could arise by evolutionary processes, and would thus be compatible with the previous suggestion. However, the three proteins are extremely diverse and are not known to bind structurally-similar receptors. Thus, in this case, there seems no reason to believe that there could be any evolutionary similarity; accordingly, the previous explanations are unlikely.

It is more likely that the observed 3-D resemblance may simply reflect a particularly-stable folding motif. An example of this behaviour is provided by the 8-stranded parallel β-barrel proteins, which show strong structural similarity in their β-barrel domains. It has been shown that while some of these proteins are probably evolutionarily related to each other, others are almost certainly not [49]. In fact, the common fold between 2PAB, 1PCD and 1THI resembles the β-sandwich found in another unrelated protein: azurin (1AZA). The only difference is that the outer pair of parallel β-strands in the three-stranded sheet of azurin (i.e., strands *c* and *a*) is N-terminal to the rest of the fold (as shown in Figure 6), whereas these parallel strands (*a* and *g*) are C-terminal to the rest of the fold in the other three proteins.

In addition, five of the β-strands (*d*, *a*, *c*, *b*, *e*) in the common fold between 2PAB, 1PCD and 1THI are structurally very similar to five of the β-strands

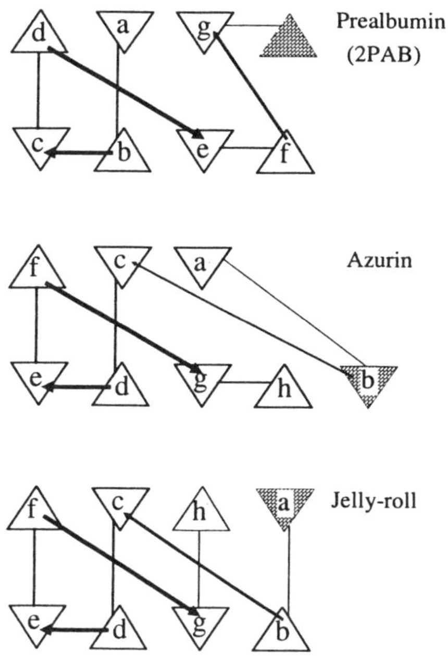

Figure 6. Topological representations of prealbumin (using the structure 2PAB), azurin and a jelly-roll structure.

(*f*, *c*, *e*, *d*, *h*) in a Type 1 jelly roll [51]. The latter motif is also shown in Figure 6, where it will be seen that it comprises two four-stranded β-sheets. Both of these sheets are anti-parallel, with most of the strands that are adjacent in sequence appearing in different sheets. The jelly-roll motif is a stable structural feature that is found in a wide variety of unrelated molecules [52]. Consequently, it is quite possible that the common fold identified in this work is simply a common modification of a particularly stable folding motif, rather than an example of an evolutionary relationship.

5. Conclusions

In this paper, we have described PROTEP, a computer program that allows searches to be carried out on the Protein Data Bank to find those structures that are structurally most-similar to a user-defined target protein. The 2TNC searches for calcium-binding proteins (as described in Section 3) demonstrate the effectiveness of the program since it successfully finds good matches with other known calcium-binding proteins for which structures exist in the Protein Data Bank, and it is possible to rationalise those matches that are not retrieved. The discriminatory power of the program is exemplified by the fact that very few strong matches are found with non-calcium-binding proteins.

The main value of PROTEP, however, is not to confirm known structural similarities, but to identify resemblances that have not previously been identified. Its use for this purpose is exemplified here by the work described in Section 4, which details the discovery of the common β-sandwich domains in prealbumin, protocatechuate 3,4-dioxygenase and thaumatin. This is the third such major 'discovery' made by PROTEP since we started to use it on a routine basis for the analysis of protein structures, and we thus believe that it provides an effective means of exploring the structural information that is contained within the Protein Data Bank (the other discoveries being the similarities between leucine aminopeptidase and carboxypeptidase A [18]; and between HIV-1 reverse transcriptase and hexokinase, heat-shock cognate protein and actin, and within HIV-1 reverse transcriptase itself [21]). The utility of the approach is likely to increase still further over the next few years, owing to the rapid growth in the number of proteins for which 3-D structures are available. In addition to providing an effective means of measuring inter-protein similarity, PROTEP is also highly efficient in operation. A typical search of the entire Protein Data Bank takes a few tens of minutes on a UNIX workstation, thus allowing the user a sufficiently-rapid response to encourage the use of a range of search parameters to maximise the retrieval of relevant material.

The clique-detection procedure we have used permits the identification of all combinations of secondary-structure elements that are geometrically common to a pair of proteins, without requiring that the protein sequences be aligned in any way or that the user provide initial inter-element equivalences. However, it is normally necessary to specify a threshold number of matching elements, the *MINCLQ* parameter, to avoid the retrieval of very large numbers of small cliques that are of little structural significance. Even so, it is often the case that a large amount of output is produced, especially if a broadly-defined, high-tolerance search is carried out, and we are thus currently evaluating the use of clustering procedures that can summarise automatically the main features of the set of hits resulting from a search. We are also extending the scope of the program so that it can process protein graphs that have individual amino-acid residues as the nodes, rather than complete secondary-structure elements as in the work discussed above. The use of these procedures will be reported shortly.

Acknowledgments

We thank Pfizer Central Research (U.K.), the Science and Engineering Research Council, the Wellcome Trust and Tripos Associates for funding. Peter Artymiuk is a Royal Society University Research Fellow and thanks the Medical Research Council for support; David Rice is a Lister Institute Research Fellow. This paper is a contribution from the Krebs Institute for Biomolecular

Research, which is a designated center for the Molecular Recognition Initiative of the Science and Engineering Research Council.

References

1. Thornton, J. M. and Gardner, S. P., *Trends Biochem Sci* **14**, 300 (1989).
2. Islam, S. A. and Sternberg, M. J. E., *Protein Engineering* **2**, 431 (1989).
3. Gray, P. M. D., Paton, N. W., Kemp, G. J. L., and Fothergill, J. E., *Protein Engineering* **3**, 235 (1990).
4. Rawlings, C. J., Taylor, W. R., Nyakairu, J., Fox J., and Sternberg, M. J. E., *J of Mol Graph* **3**, 151 (1985).
5. Ash, J. E., Warr, W. A., and Willett, P., Eds., *Chemical Structure Systems* (Ellis Horwood, Chichester, 1991).
6. Bonchev, D. and Rouvray, D. H., *Chemical Graph Theory: Introduction and Fundamentals* (Abacus Press, New York, 1991).
7. Willett, P., *Modern Approaches to Chemical Reaction Searching* (Gower, Aldershot, 1986).
8. Hagadone, T. R., *J Chem Info Comp Sci* **32**, 515 (1992).
9. Brint, A. T. and Willett, P., *J Mol Graphics* **5**, 49 (1987).
10. Brint, A. T. and Willett, P., *J Chem Info and Comp Sci* **27**, 152 (1987).
11. Martin, Y. C., Bures, M. G., Danaher, E. A., DeLazzer, J., Liko, I., and Pavlik, P. A., *J Comp Aided Mol Design* **7**, 83 (1933).
12. Ho, C. M. W. and Marshall, G. R., *J Comp Aided Mol Design* **7**, 3 (1993).
13. Brint, A. T., Davies, H. M., Mitchell, E. M., and Willett, P., *J Mol Graphics* **7**, 48 (1987).
14. Artymiuk, P. J., Mitchell, E. M., Rice, D. W., and Willett, P., *J Info Sci* **15**, 287 (1989).
15. Artymiuk, P. J., Mitchell, E. M., Rice, D. W., and Willett, P., *J Mol Biol* **212**, 151 (1990).
16. Artymiuk, P. J., Rice, D. W., Mitchell, E. M., and Willett, P., *Protein Enginering* **4**, 39 (1990).
17. Artymiuk, P. J., Grindley, H. M., Rice, D. W., Ujah, E. C., and Willett, P., in *Recent Advances in Chemical Information*, H. Collier, ed. (Royal Society of Chemistry, Cambridge, 1991), p. 91.
18. Artymiuk, P. J., Grindley, H. M., Park, E. J., Rice, D. W., and Willett, P., *FEBS Lett* **303**, 48 (1992).
19. Artymiuk, P. J., Bath, P. A., Grindley, H. M., Pepperrell, C. A., Poirrette, A. R., Rice, D. W., Thorner, D. A., Wild, D. J., Willett, P., Allen, F. H., and Taylor, R., *J Chem Info Comp Sci* **32**, 617 (1992).
20. Grindley, H. M., Artymiuk, P. J., Rice, D. W., and Willett, P., *J Mol Biol* **229**, 707 (1993).
21. Artymiuk, P. J., Grindley, H. M., Kumar, K., Rice, D. W., and Willett, P., *FEBS Lett* **324**, 15 (1993).
22. Bernstein, F. C., Koetzle, T. F., Williams, G. J. B., Meyer Jnr., E. F., Brice, M. D., Rodgers, J. R., Kennard, O., Shimanouchi, M., and Tasumi, M., *J Mol Biol* **112**, 535 (1977).
23. Abola, E. E., Bernstein, F. C., Bryant, S. H., Koetzle, T. F., and Weng, J., in *Crystallographic Databases: Information Content, Software Systems, Scientific Applications*, F. H. Allen, G. Bergeroff, and R. Sievers, eds. (Data Commission of the International Union of Crystallography, Cambridge, 1987), p. 107.
24. Kaden, F., Koch, I., and Selbig, J., *J Theoret Biol* **147**, 85 (1990).
25. Koch, I., Kaden, F., and Selbig, J., *Proteins: Structure, Function, and Genetics* **12**, 314 (1992).
26. Subbarao, N. and Haneef, I., *Protein Engineering* **4**, 877 (1991).
27. Kasinos, N., Lilley, G. A., Subbarao, N., and Haneef, I., *Protein Engineering* **5**, 69 (1992).
28. PROTEP is distributed by Tripos Associates, St. Louis, Missouri, U.S.A.
29. Kabsch, W. and Sander, C., *Biopolymers* **22**, 2577 (1983).

30. Jones, T. A., *Meth Enzymol* **115**, 157 (1985).
31. Bron, C. and Kerbosch, J., *Communications of the ACM* **16**, 575 (1973).
32. Barrow, H. G. and Burstall, R. M., *Information Processing Lett* **4**, 83 (1976).
33. Brint, A. T. and Willett, P., *J Comp Aided Molec Design* **2**, 311 (1988).
34. Kuhl, F. S., Crippen, G. M., and Friesen, D. K., *J Comput Chemistry* **5**, 24 (1984).
35. Smellie, A. S., Crippen, G. M., and Richards, W. G., *J Chem Info Comp Sci* **31**, 386 (1991).
36. Willett, P., *Three-Dimensional Chemical Structure Handling* (Research Studies Press, Taunton, 1991).
37. Holjte, H-D. and Hense, M., *J Comp Aided Molec Design* **3**, 101 (1989).
38. Kretsinger, R. H., *Annu Rev Biochem* **45**, 239 (1976).
39. Kretsinger, R. H. and Barry, C. D., *Biochim Biophys* **405**, 40 (1975).
40. Strynadka, N. C. J. and James, M. N. G., *Proteins: Structure, Function and Genetics* **3**, 1 (1988).
41. Grindley, H. M., *The Use of Graph-Theoretical Techniques for Searching Databases of 3-D Protein Structures* (PhD thesis, University of Sheffield, 1991).
42. Weber, P. C. and Salemme, F. R., *Nature* **287**, 82 (1980).
43. Blake, C. C. F. and Oatley, S. J., *Nature* **268**, 115 (1977).
44. Blake, C. C. F., Geisow, M. J., and Oatley, S. J., *J Mol Biol* **121**, 339 (1978).
45. Ohlendorf, D. H., Lipscomb, J. D., and Weber, P. C., *Nature* **336**, 405, (1988).
46. Kim, S. H., de Vos, A., and Ogata, C., *Trends Biol Sci* **13**, 13 (1988).
47. van der Wel, H. and Loeve, K., *Eur J Biochem* **31**, 221 (1972).
48. Ujah, E. C., *A Study of β-Sheet Motifs at Different Levels of Structural Abstraction Using Graph-Theoretic and Dynamic Programming Techniques* (PhD thesis, University of Sheffield, 1992).
49. Chothia, C., *Nature* **333**, 598 (1988).
50. Holmgren, A. and Branden, C. I., *Nature* **342**, 248 (1989).
51. Stirk, H. J., Woolfson, D. N., Hutchinson, E. G., and Thornton, J. M., *FEBS Lett* **308**, 1 (1992).
52. Gibson, T. J. and Argos, P., *J Mol Biol* **212**, 7 (1990).

3D Molecular Similarity Modelling in Computational Drug Design

DOUGLAS C. ROHRER
Computer Aided Drug Discovery Upjohn Laboratories, Kalamazoo, MI 49007-4940, U.S.A.

Abstract. The ability to compare 3D molecular structures and properties has become an important tool in drug discovery and development. Algorithms for evaluation and comparison of the extended Molecular Electrostatic Potential and Molecular Steric Volume are presented which match the similarity of these field data generated at grid points surrounding the molecules. The results from translation and rotation test comparisons involving polar and non-polar molecules are reported for five similarity functions. These results indicate that this approach provides an excellent means of structurally matching seemingly dissimilar molecules and evaluating their relative similarity.

1. Introduction

The search for suitable pharmaceutical agents in the area of computational drug design has often been based upon the premise that a series of molecules with a similar mechanism of biological action bind at the same site in a similar manner. In order to avoid the need to know the actual structure of the binding site, the concept of a "pharmacophore" [1, 2] has been developed to describe the three dimensional arrangement of functional groups which represent a structural "key" in determining the binding and activity of the series of analogues. This strategy has been used for sometime and has involved a variety of methods and molecular attributes. However, the majority of applications have involved positioning of atoms and/or electrostatic potentials based on visual analysis of the molecular features in the series. One drawback to this method has been selection of similar features when the molecular structures are significantly different or contain numerous possibly important functional groups.

Recently several computational approaches have been described, as reported here and by others [3], which remove the need for visual analysis in matching the features of molecules. These approaches involve applying similarity measures to match the Molecular Electrostatic Potential (MEP) and/or the Molecular Steric Volume (MSV) of the molecules without specific reference to the atomic positions. This strategy is based on the assumption that a group of molecules which have been shown to produce a biological response by interacting with a common subsystem do so by presenting a similar structural and electronic profile to a binding site. Therefore, two molecules which have similar profiles should also have similar biological activity. The nature of the profile presented by the molecules can be modelled by evaluating the electrostatic potential and steric volume at points around the atoms of the

R. Carbó (ed.), Molecular Similarity and Reactivity: From Quantum Chemical to Phenomenological Approaches, 141–161.

molecules. These three-dimensional data are then used as the basis of comparison. The greater the similarity between the profile data of two molecules, the more similar the biological response. Furthermore, once the profile has been evaluated, it can be used to evaluate the similarity and likely activity of new molecular structures.

2. The *mimic* algorithm

A computer algorithm (*MIMIC*) has been developed for matching the electrostatic potential and steric volume of molecules. The first step in the matching process is to evaluate the MEP and MSV at either a rectangular grid of points or concentric spheres of points surrounding the atoms of a prototype molecule. The MEP and MSV of the molecule to be matched to the prototype are then evaluated using the same grid-point coordinates generated for the prototype. Typically each molecule has been positioned so that the center of mass has been placed at the origin to assure molecular overlap for the data being evaluated. The similarities between corresponding data at each grid point are then evaluated and combined into a single similarity index using one of the published similarity functions discribe below. Finally, the position and orientation of the matching molecule can be the changed using a pattern search approach to obtain the best similarity match between the sets of data. Since the starting position and orientation of the second molecule limit the range of the matches that can be obtained during the optimization process, either a systematic search or random choice of starting points are used. In addition, within limits the conformation of rotatable bonds in the second molecule may also be allowed to change during the optimization process to approximate an induced fit of the matching molecule's MEP and MSV to that of the prototype's.

2.1. Calculating the molecular electrostatic potential

The molecular electrostatic potential at each point, MEP_p, is evaluated using a simple point charge model involving the partial atomic charge of the atoms, $Charge_i$, and the distance between the point and the atom, $Distance_{ip}$.

$$MEP_p = \sum_{i=1}^{\#\,Atoms} \left\{ \frac{(1.0 * Charge_i)}{Distance_{ip}} \right\}. \tag{1}$$

The partial atomic charges for each molecule are pre-calculated independent of the *MIMIC* program. Using atomic point charges for evaluating the MEP produces a relatively crude approximation. However, only an estimate of the electrostatic potential is needed for these types of comparisons. To improve the efficiency of the calculation, individual MEP grid points with small magnitudes, usually less than 5.0 kcal/mol, are excluded from the calculation. A fundamental problem with Equation (1) exists when a grid point is very "close" to an atoms position resulting in a very large value for the MEP and the need to exclude the point. This results in no MEP grid points being located close to atom centers. This deficiency has been removed by optionally allowing the contributions from the atoms within 1 Å of a grid point to be to be treated as though the distance were 1 Å. This produces an extended MEP, xMEP, grid for comparison purposes which does not have holes around the atom positions.

2.2. Calculating the molecular steric volume

The MSV data are evaluated by a procedure devised by Marshall *et al.* [4]. A Gaussian function is used to represent the volume contribution from each atom at each grid point, MSV_p

$$MEP_p = \sum_{i=1}^{\#\,Atoms} \left\{ Amp_i * \exp\left(- \frac{Distance_{ip}^{\;2}}{Width_i^2} \right) \right\}. \tag{2}$$

The amplitude factor for each atom, Amp_i, is a function of B_i and the atomic number of the atom, Z_i,

$$Amp_i = Z_i * \sqrt{\left(\frac{4\pi}{B_i} \right)^3}. \tag{3}$$

The atomic width factor, $Width_i$, is solely a function of B_i,

$$Width_i = \frac{\sqrt{B_i}}{(2\pi)}. \tag{4}$$

The value of the atomic B-factor is determined analytically so that the individual atom's contribution to the MSV will be 5.0 when the distance between the atom and a grid point is equal to the van der Waal's radius. Parameter values used in the calculation of the MSV factors are given in Table I.

TABLE I
Molecular Steric Volume parameters

Atom Type	van der Waal's Radius (Å)	Amp_i	$Width_i$
Hydrogen	1.20	10.00	1.441
Carbon	1.65	350.00	0.801
Nitrogen	1.55	584.50	0.710
Oxygen	1.55	771.33	0.677
Fluorine	1.35	1,488.67	0.566
Chlorine	1.80	1,090.50	0.776
Bromine	1.95	2,948.67	1.042
Iodine	2.15	2,464.67	0.863
Phosphorous	1.80	915.50	0.789
Sulfur	1.80	1,002.17	0.782

2.3. Calculating the similarity indices

Any of the functions shown in Equations (5–9) can be used to obtain an estimate of the similarity between a prototype molecule and the matching molecule's MEP, MSV and a weighted combination of the two. The following similarity functions are currently available:

Carbo Similarity Index (C_{AB}) [5]

$$C_{AB} = \frac{\sum_{i=1}^{\#\,Points} (P_A * P_B)}{\sqrt{\sum_{i=1}^{\#\,Points} (P_A^2)} * \sqrt{\sum_{i=1}^{\#\,Points} (P_B^2)}} . \tag{5}$$

Hodgkin/Richards Similarity Index (H_{AB}) [6]

$$H_{AB} = \frac{2 * \sum_{i=1}^{\#\,Points} (P_A * P_B)}{\sum_{i=1}^{\#\,Points} (P_A^2) + \sum_{i=1}^{\#\,Points} (P_B^2)} . \tag{6}$$

Petke Similarity Index (P_{AB}) [7]

$$P_{AB} = \frac{\sum_{i=1}^{\#\,Points} (P_A * P_B)}{\max\left(\sum_{i=1}^{\#\,Points} (P_A^2), \sum_{i=1}^{\#\,Points} (P_B^2) \right)} . \tag{7}$$

Good's Similarity Indices (L_{AB} & E_{AB}) [8]

Linear form

$$L_{AB} = \left(\frac{1}{\#\,Points} \right) * \sum_{i=1}^{\#\,Points} \left(1 - \frac{|P_A - P_B|}{\max(|P_A|, |P_B|)} \right) . \tag{8}$$

Exponential form

$$E_{AB} = \left(\frac{1}{\#\,Points} \right) * \sum_{i=1}^{\#\,Points} \exp^{-\left(\frac{|P_A - P_B|}{\max\,(|P_A|, |P_B|)} \right)} . \tag{9}$$

3. Test calculations

The choice of parameters used in performing these similarity comparisons must be determined in order to assure a successful application of the approach. To help evaluate these parameters, a series of calculations have been performed using molecular test systems involving sets of polar and non-polar planar molecules. The test calculations were designed to evaluate the five similarity functions for the MEP, MSV and weighted average as a function of translation or rotation. The coordinates used for each of the molecules were obtained by molecular mechanics energy minimization employing an MM2 potential [9] and the partial-atomic charges were calculated by a semi-empirical approach using the AM1 method [10]. The ratio of the steric volumes of the molecules evaluated within the van der Waal's radii should provide a measure of the accuracy of the similarity functions ability to evaluate the match between the MSV of the molecules.

3.1. Non-polar molecules

The set of non-polar molecules used consisted of naphthalene as the prototype and benzene as the matching molecule. These were chosen because they have low atomic charges and clearly identifiable positions of high similarity. The largest atomic charge in these molecules was only 0.20. Each of the molecules was positioned prior to any translation or rotation so that the origin was located at the center of mass and the ring was oriented in the XY-plane with the Y-axis aligned with a carbon-carbon bond. The ring fusion carbon-carbon bond was used as the Y-axis bond in the naphthalene molecule. Both translational and rotational test calculations were performed using the two sets of molecules. In order to evaluate both the optimal parameters used in the search and the behavioral characteristics of the various similarity indices, a series of translational scans were performed. In each case, the matching molecule was scanned from –8 Å to +8 Å in the X, Y and Z directions in steps of 0.25 Å relative to the reference prototype molecule. Similarly, rotational scans were performed where the matching molecule was rotated through 360° in steps of 20° without translation. The similarity functions were evaluated at each translational and rotational step location.

3.1.1. *Translation Tests*

The translation tests of these two molecules were performed in all three directions. However, the translation along the X-direction provides the clearest and most interesting set of data and will be used as exemplary of the results obtained. There were three types atom-atom superpostion which occurred as the benzene molecule was translated in the X-direction across the naphthalene as shown in Figure 1; these are when six carbons and four hydrogens of each molecule overlap and two hydrogen of the benzene overlap two naphthalene carbons (±1.25 Å separation), when two carbons of each molecule overlap and two hydrogens from one molecule overlap two carbons of the other (±3.75 Å separation), and when two hydrogens of each molecule overlap (±5.5 Å separation).

MSV results: The volumes of these molecules evaluated within the van der Waal's radius are 84.7 Å^3 and 128.7 Å^3 for the benzene and naphthalene molecules, see Figure 2. The best fit of the volume of the benzene molecule to the naphthalene would then result in only 0.66 of the naphthalene volume

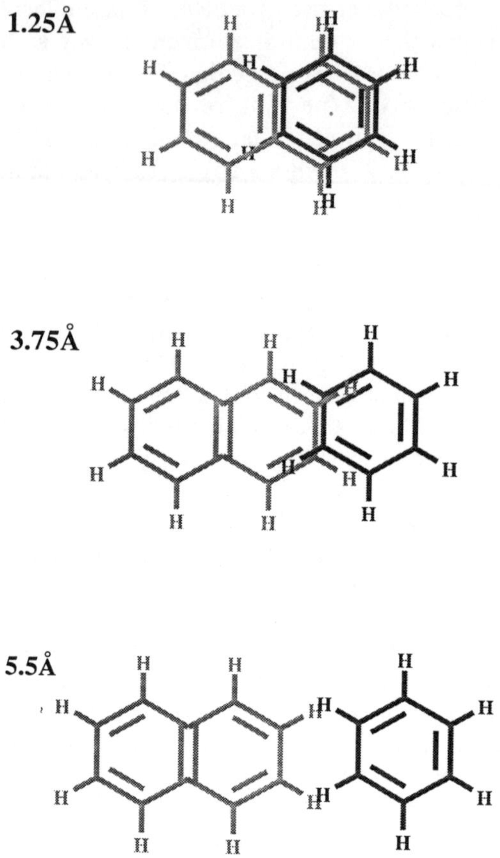

Figure 1. Molecular overlay drawings of benzene (dark lines) and naphthalene (light lines) showing the locations where multiple atom/atom overlaps occur during X-translation of the benzene molecule.

being filled. However, an accurate estimate of the similarity should be less than the ratios of the volumes since the MSV values for the grid points in the region where the two hydrogen atoms of the benzene overlap two Carbons of the naphthalene will be different and similarity functions measure sameness, not volume filled.

Each of the similarity function results for the MSV data, Figure 3a, clearly showed the two locations where the six-member rings overlap at about ±1.2 Å. The Carbo, Hodgkin/Richards and Petke functions all showed maxima at slightly smaller X-values than the sharp peaks produced by Good's functions and had a much shallower minima between the peaks. The magnitude of the maxima for the Carbo (0.82), Hodgkin/Richards (0.80) and Good exponential (0.70) functions were all larger than the ratio of the volumes indicating an over estimate of the similarity, while the Petke function value

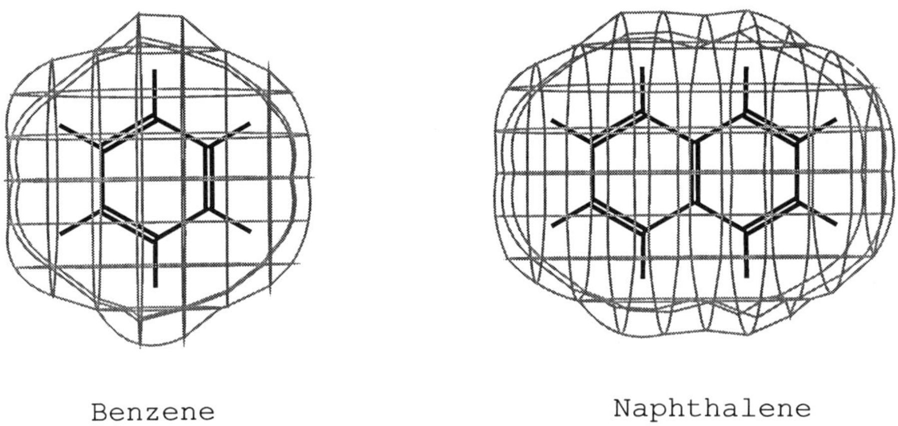

Figure 2. Molecular steric volume enclosures for benzene and naphthalene contoured at the van der Waal's radii for each atom.

(0.63) was very close to the ratio and the Good-linear function (0.55) was below the ratio. However, no clear indication of the other two types of overlapping position could be easily detected from these results. At best, there were only small bumps or shoulders at the expected positions which could be seen in the Good's functions results.

MEP results: The similarity function results for the MEP data, Figure 3b, were more difficult to interpret. The Carbo, Hodgkin/Richards and Petke functions all showed the same shape with the Carbo and Hodgkin/Richards functions producing nearly identical results and the Petke function having smaller magnitudes. However, these functions were not able to distinguish between the two positions where the six-member rings overlap, rather they produced a single maximum at the translation position midway between the two six-member ring overlap positions. Both the functions of Good showed peaks identifying the two overlap positions as maxima with a very shallow minimum between. This lack of sensitivity in the Carbo, Hodgkin/Richards and Petke functions was probably caused by dominance of the contribution of the larger values in determining the similarity indices. This could be a problem when attempting to match molecules which contain regions of relatively small charges or a mixture of large and small charges.

Combined results: The results obtained by combining the MSV and MEP similarity function results are shown in Figure 3c. These values were obtained by taking the weighted average of the two values, weighting each value equally for the test. As expected, the combined results show the same general distribution as observed in the individual results. The MSV results were able to partially overcome the MEP maximum at the zero X-position for Carbo, Hodgkin/Richards and Petke functions producing two maxima, but the a.

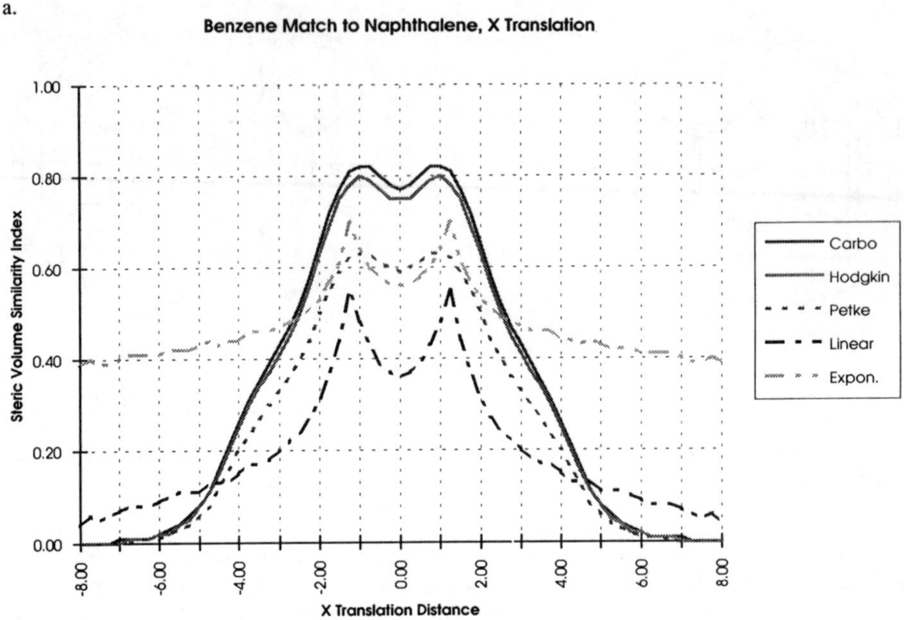

b.

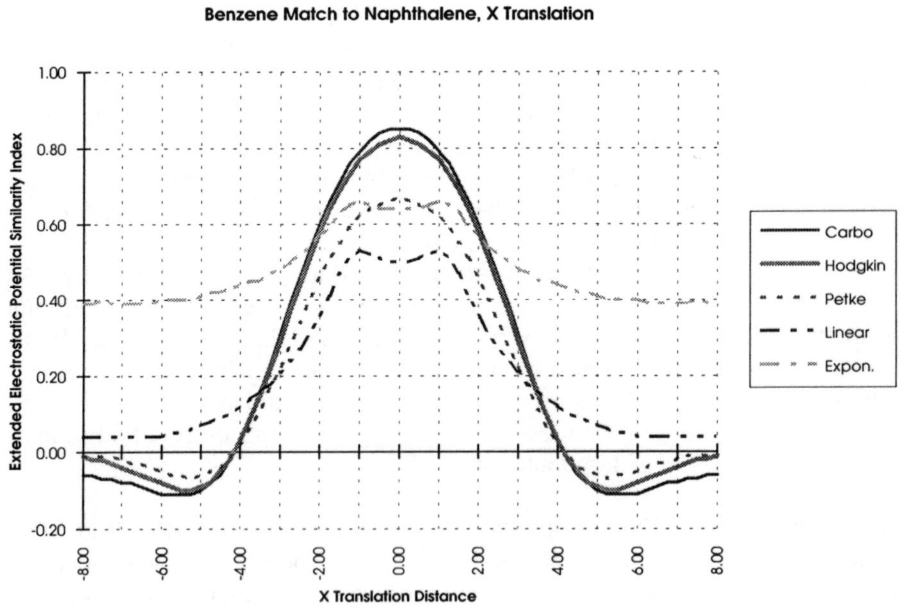

Figure 3. Graph of the similarity indices for benzene matching to naphthalene versus benzene X-translation distance: *a.* Molecular steric volume indices; *b.* Molecular electrostatic potential indices; *c.* Combined indices.

c.

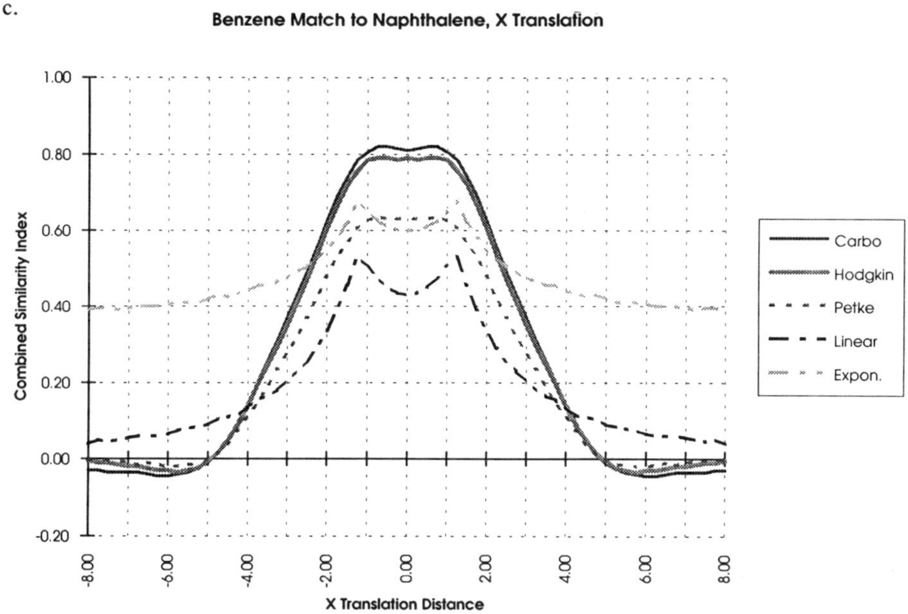

Figure 3 (Continued).

intervening minimum was very shallow. This seemed to be an indication that the weighting for the contribution of the MSV should be increased.

3.1.2. *Rotation Tests*

Like the translation tests of these two molecules, the rotation tests were performed in all three directions. The rotation results did not provide as interesting data because the origins of the molecules resulted in an absence of atom/atom overlap between the benzene and naphthalene molecules. However, the rotation along the X-direction, see Figure 4, provides representative set of data and will be used as exemplary of the results obtained. In this example, the maximum similarity should be obtained only at the 0° and 180° rotation orientations.

MSV results: The same general characteristics described for the X-translation were observed for these results, Figure 4a. The lack of resolution of the Carbo, Hodgkin/Richards and Petke functions was again evident. The width of the peaks were approximately 80° at half-height. The function's of Good show much better resolution having only a 40° width and a much sharper appearing profile.

MEP results: The profiles presented from each of the functions, Figure 4b, were very wide and shallow. One striking feature was the dynamic range of a.

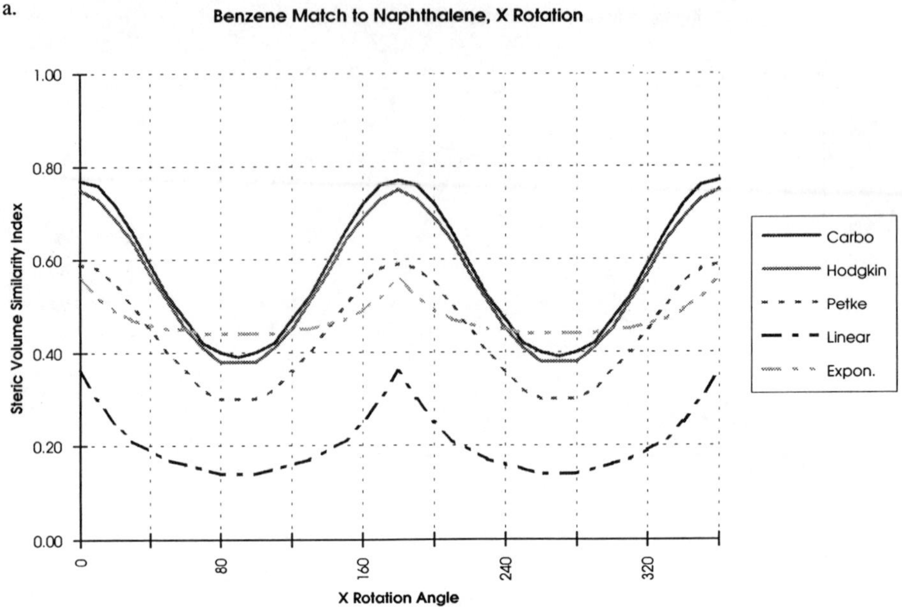

b.

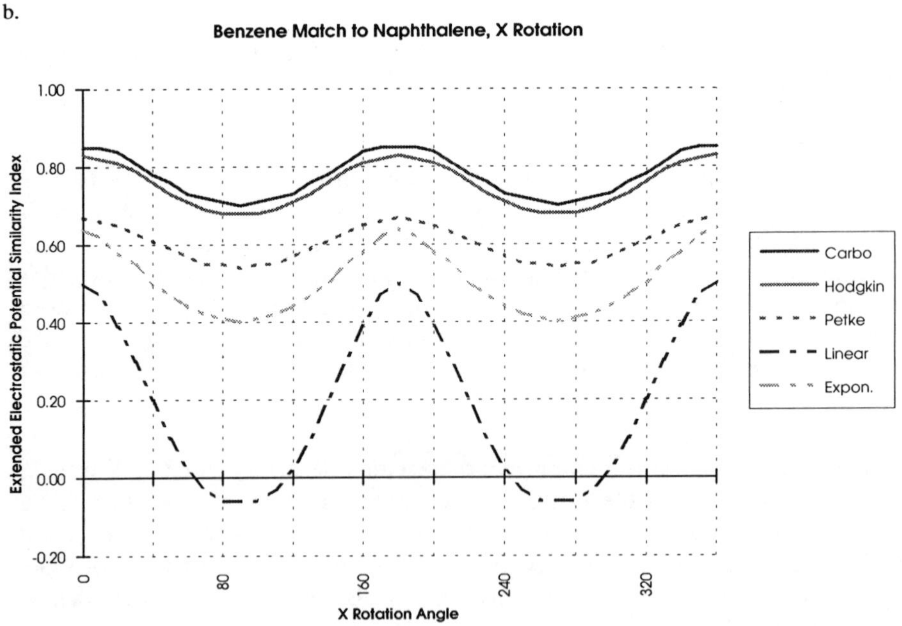

Figure 4. Graph of the similarity indices for benzene matching to naphthalene versus benzene X-rotation angle: a. Molecular steric volume indices; b. Molecular electrostatic potential indices; c. Combined indices.

c.

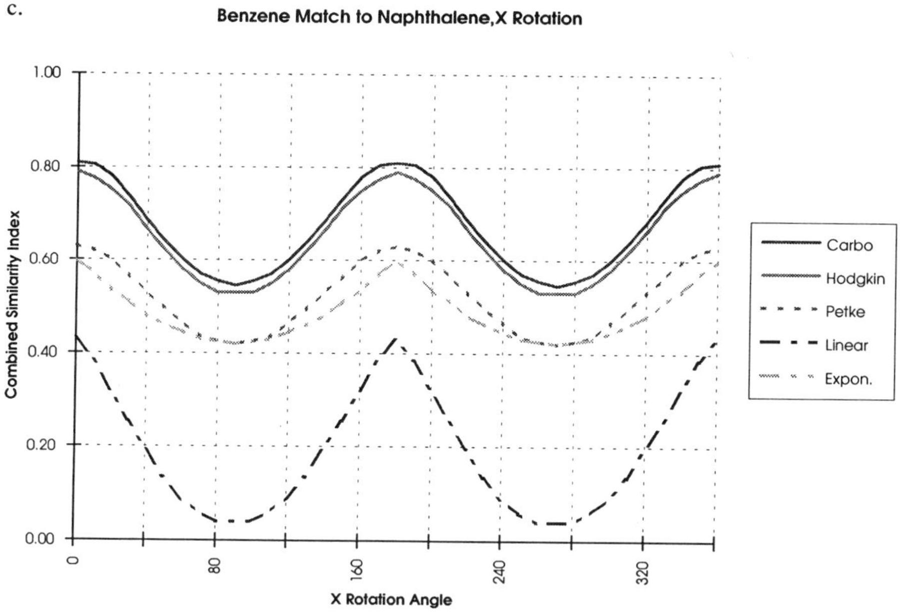

Figure 4. (Continued).

Good's linear function which showed a range 2.5 times that of any of the other functions.

Combined results: With the exception of Good's linear function, the already wide profiles obtained for the MSV were further flattened when combined with the MEP results, Figure 4c.

3.2. POLAR MOLECULES

Guanine for the prototype and cytosine for the matching molecule were chosen as the set of polar molecules because of their larger atomic charges and identifiable similarity location and orientation. The largest atomic charge in these molecules was –0.65. Each of these molecules was oriented prior to any translation or rotation in a manner similar to that described for the non-polar test example above. The origin was located at the center of mass and the ring was positioned in the XY-plane with the Y-axis aligned with a carbon-carbon double bond in the cytosine molecule and the ring fusion carbon-carbon bond in the guanine molecule. The same procedure used for the non-polar molecules was used here, the matching molecule, cytosine, was scanned from –8 Å to +8 Å in the X, Y and Z directions in steps of 0.25 Å relative to the reference prototype molecule, guanine. Similarly, rotational scans were performed where the matching molecule was rotated through 360° in steps of 20°. The MEP, MSV and combined similarity indices for the five functional forms were evaluated at each translational and rotational step location.

3.2.1. *Translation Tests*
As in the test above, the translation tests of these two molecules were performed in all three directions. However, the translation along the X-direction provides the most interesting set of data and will be used as exemplary of the results obtained. Figure 5 shows that the atom types at the six-member ring overlap position (0.8 Å separation) were much different placing an cytosine amine group upon a guanine carbonyl oxygen and a cytosine carbonyl oxygen close to a guanine amine. Figure 5 also shows that the other ring-ring overlap

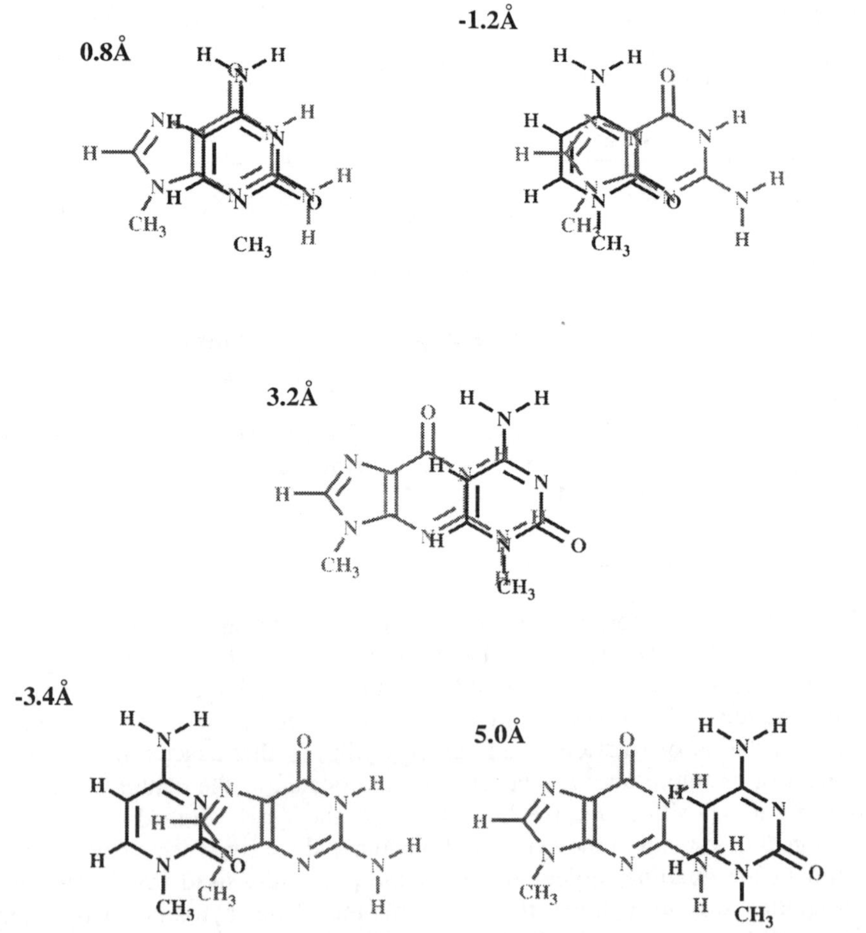

Figure 5. Molecular overlay drawings of cytosine (dark lines) and guanine (light lines) showing the locations where multiple atom/atom overlaps occur during X-translation of the cytosine molecule.

position involved a six-member ring upon a five-member ring (–1.2 Å separation). A third major overlap position involved super imposition of carbon-nitrogen bonds in the six-member rings (3.2 Å separation). Two other positions which produce overlaps were also found (–3.4 Å and 5.0 Å separations).

MSV results: The volumes of these molecules evaluated within the van der Waal's radius were 108.5 $Å^3$ and 130.8 $Å^3$ for the cytosine and guanine molecules, see Figure 6. The best fit of the volume of the cytosine molecule to the guanine would then result in 0.83 of the guanine volume being filled. The expected similarity values at this position should be proportionately smaller than those obtained for the benzene/naphthalene test due to the differences in the atom types involved in the overlay.

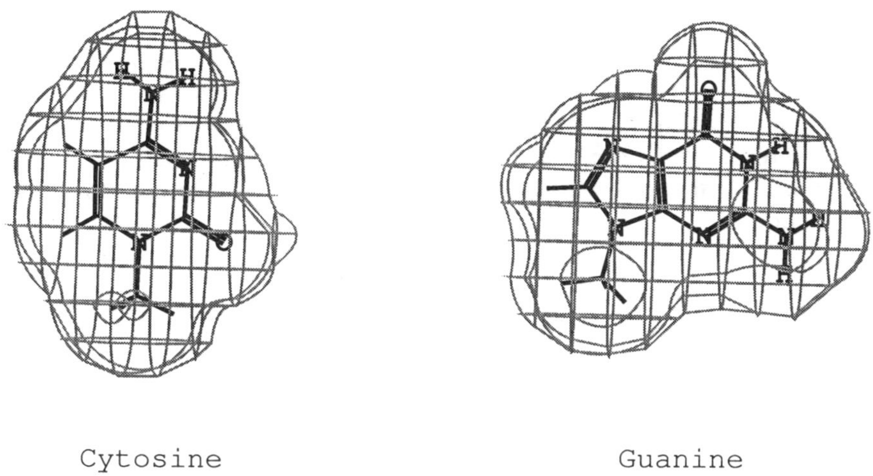

Figure 6. Molecular steric volume enclosures for cytosine and guanine contoured at the van der Waal's radii for each atom.

The Carbo, Hodgkin/Richards and Petke functions all had similar shaped profiles for the X-translation of the cytosine molecule across the guanine with the Carbo and Hodgkin/Richards producing nearly identical results and the Petke function producing smaller magnitudes, see Figure 7a. The two maxima in each functions similarity profile clearly identify the locations were the six-member ring of the cytosine overlays the five at –1.2 Å, Figure 5, and six-member at 0.9 Å, Figure 5, rings of the guanine. The relative intensities of the peaks were consistent with what would be expected, but are larger than would be expected on the bases of the ratio of the steric volumes. The minimum separating the peaks was very shallow. Other overlap positions, see Figure 5, at –3.4 Å, 3.2 Å, and 5.0 Å involving overlap of multiple cytosine atoms with guanine atoms only produce shoulders on the profiles.

The profile results, Figure 7a, from the two functions of Good showed the same dynamic range, magnitude and minimum value characteristics observed for the non-polar molecule test system. However, the sharp peaks have been dulled to rounded maxima. Furthermore, while the two ring/ring overlap positions were represented by maxima, the relative intensities of the peaks are reversed from the expected values. While these results were difficult to explain, the only differences in the MSV compared to the non-polar molecules are that the atom/atom overlaps involve different types of atoms. The largest magnitudes produced by the MSV functions for the different atom types in these molecules were 10, 350, 584 and 771 for Hydrogen, Carbon, Nitrogen and Oxygen respectively. These results seem to indicate that these functions may not be producing accurate measures of the relative similarity when the types matching atoms, the magnitudes of the data and differences between the data are large.

MEP results: The similarity profiles for the MEP matches, Figure 7b, from all of the functions are very broad, approximate 5 Å at half height. Again the Carbo, Hodgkin/Richards and Petke functions produce similar shaped results with the Petke function results having small magnitudes. The profiles from Good's functions are almost flat showing very little definition of a maximum.

a.

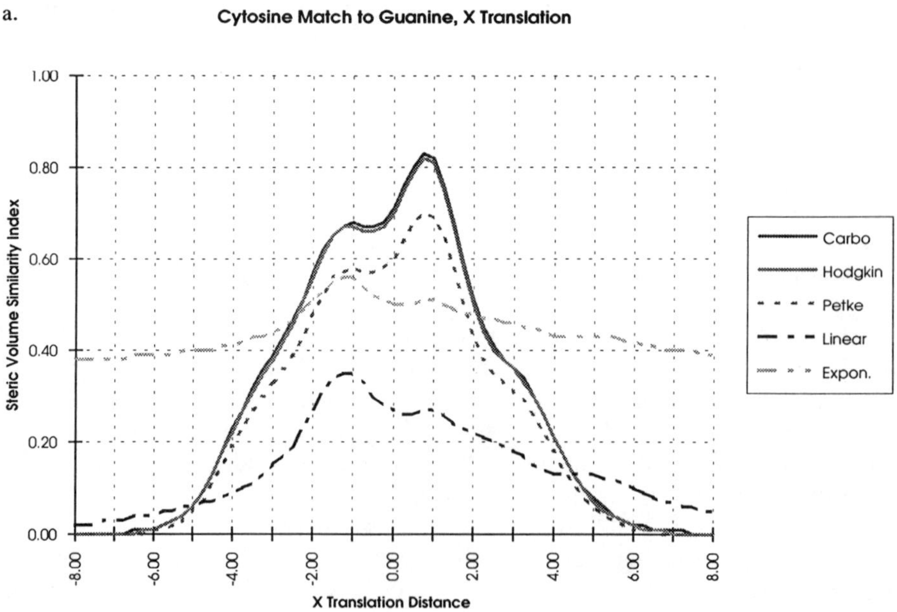

Figure 7. Graph of the similarity indices for cytosine matching to guanine versus cytosine X-translation distance: a. Molecular steric volume indices; b. Molecular electrostatic potential indices; c. Combined indices.

b.

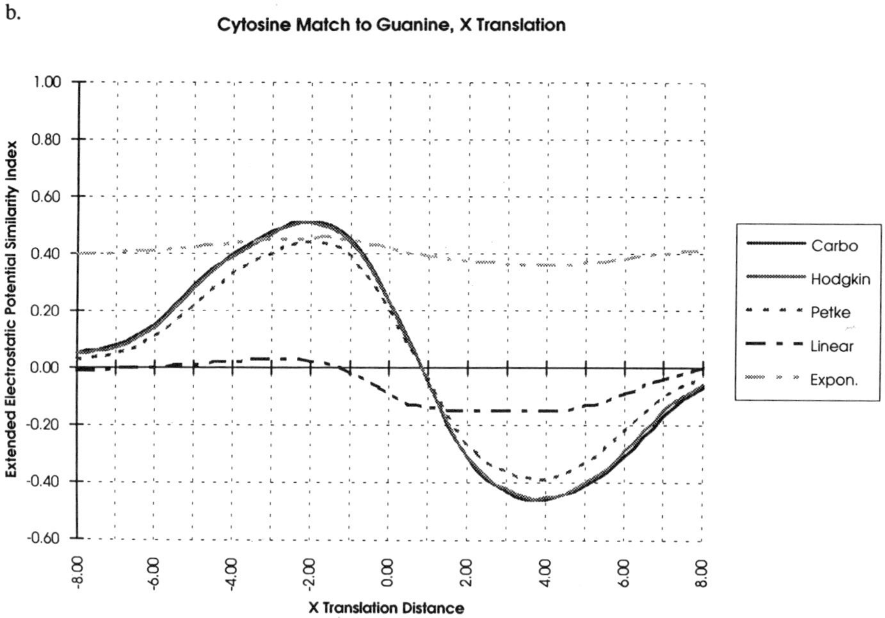

c.

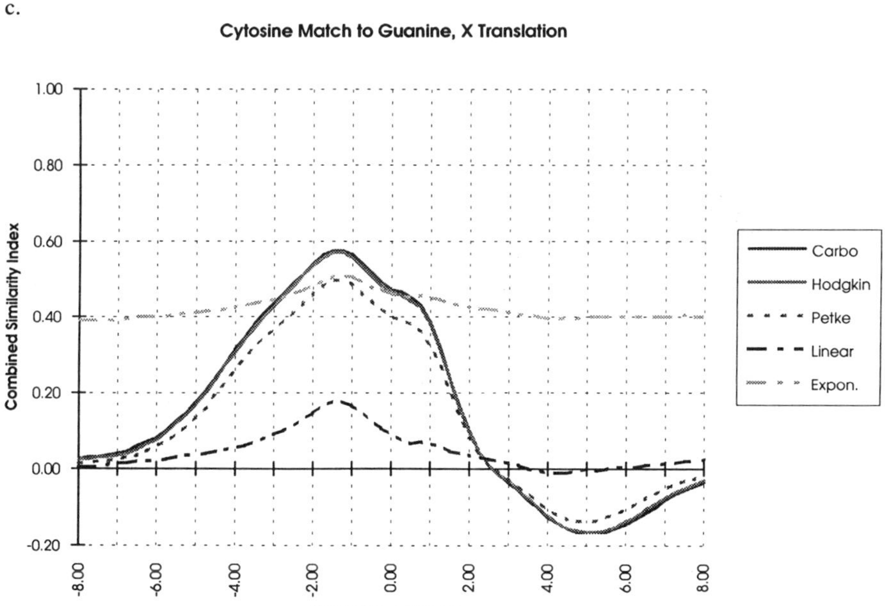

Figure 7. (Continued).

Combined results: The definition of the MSV peaks in the MSV profile, Figure 7c, are dulled by the wide MEP peak, further reducing the resolution of the matches. In addition, the Carbo, Hodgkin/Richards and Petke function relative peak heights for the overlap of the cytosine six-member ring upon the five and six-member ring of the guanine were reversed from those obtained from the MSV alone. The resulting combined shape for all five function were very similar.

3.2.2. *Rotation Tests*

While rotation test calculations for the polar molecules were performed for all three axes, examination of the MEP indicated that Z-rotation would provide the most interesting example for presentation here. Figure 8 shows iso-potential contours for the MEP's of cytosine and guanine. It was clear from these diagrams that the optimal match the MEP involved a rotation the cytosine of –90° about the Z-axis. The MSV results for the Z-rotation were also interesting. However, the placement of the origin at the center of mass in each molecule meant that there were no orientations where atom overlaps would occur between the two molecules. The MSV similarity function results would identify orientations where matches occurred that were not simple atom/atom overlays between the two molecules.

MSV results: In general, the similarity functions showed the same trends identified in the previous analyses, Figure 9a. The Carbo, Hodgkin/Richards and Petke functions were able to identify five orientation were similarity matches

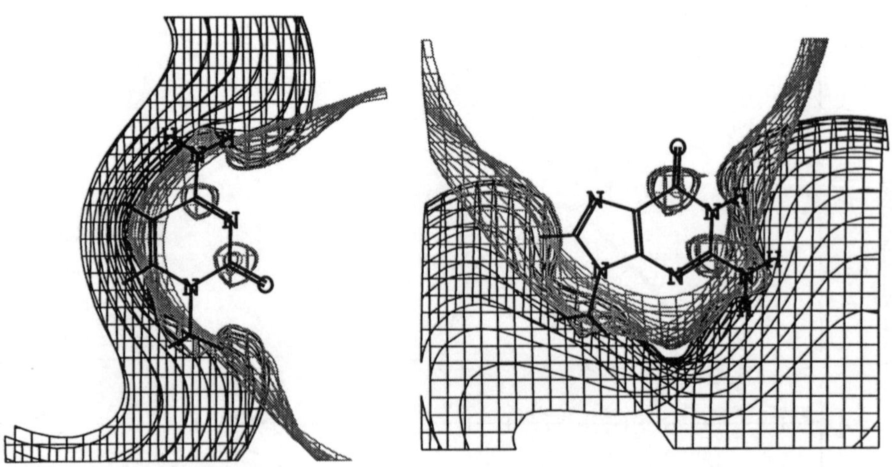

Cytosine Guanine

Figure 8. Molecular electrostatic potential iso-contours at +5.0 kcal (dark lines) and –5.0 kcal (light lines) for cytosine and guanine.

existed. Each of these represented orientations which placed either ring atoms and/or exocyclic atoms of the cytosine close to overlapping atoms on the guanine molecule, Figure 10 shows the five orientations. The greatest similarity occurs at the 110° orientation where the three cytosine nitrogens align with guanine nitrogens and the cytosine carbonly oxygen is close to a guanine ring nitrogen. The two functions of Good fail to show the match which occurred at 185° and only showed a shoulder for the 300° orientation. These orientations seemed to position atoms with the largest type difference in close proximity, for example the cytosine methyl and the guanine carbonyl oxygen. The contribution from these dissimilar atoms which had maximum data values of 350 for the carbon and 771 for the oxygen may explain why the Good function was not able to identify these orientations.

MEP results: The Z-rotation similarity function profiles, Figure 9b, clearly showed what was anticipated from examination of the MEP contours. Each function produce a wide peak with the maxima all located at the –90° orientation.

Combined results: Like the previous examples, combining the results from the MSV and MEP resulted in flattened profile due to the very broad MEP peak Figure 9c.

a.

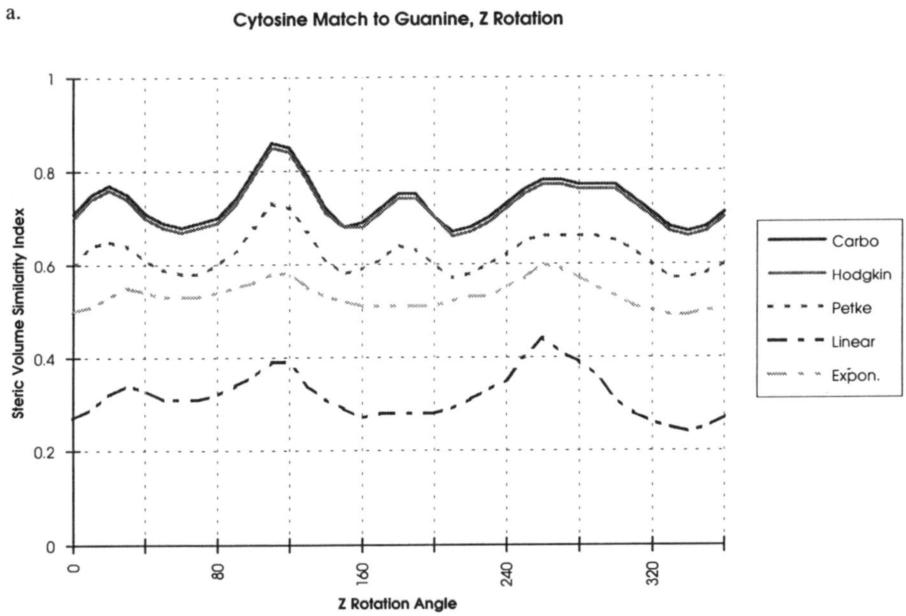

Figure 9. Graph of the similarity indices for cytosine matching to guanine versus cytosine Z-rotation angle: a. Molecular steric volume indices; b. Molecular electrostatic potential indices; c. Combined indices.

b.

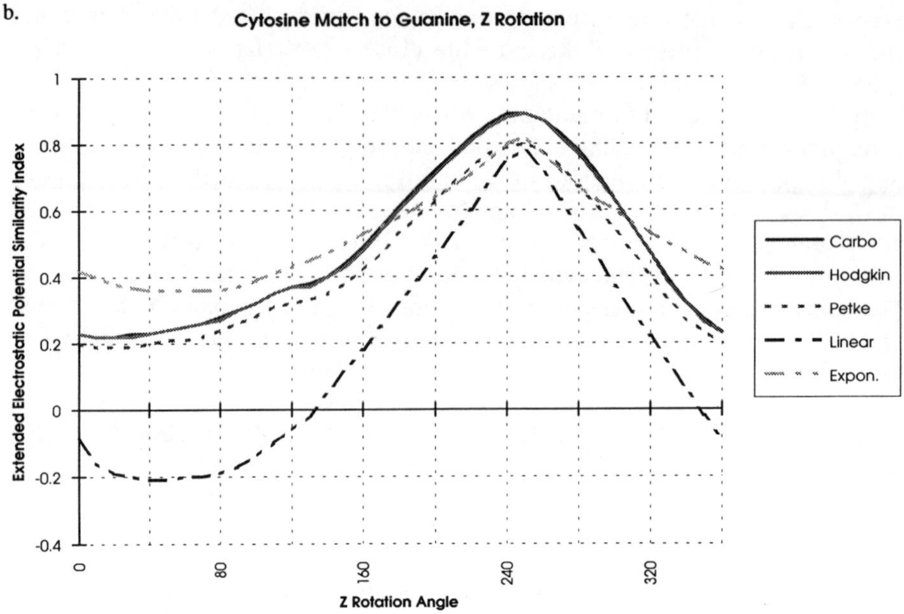

c.

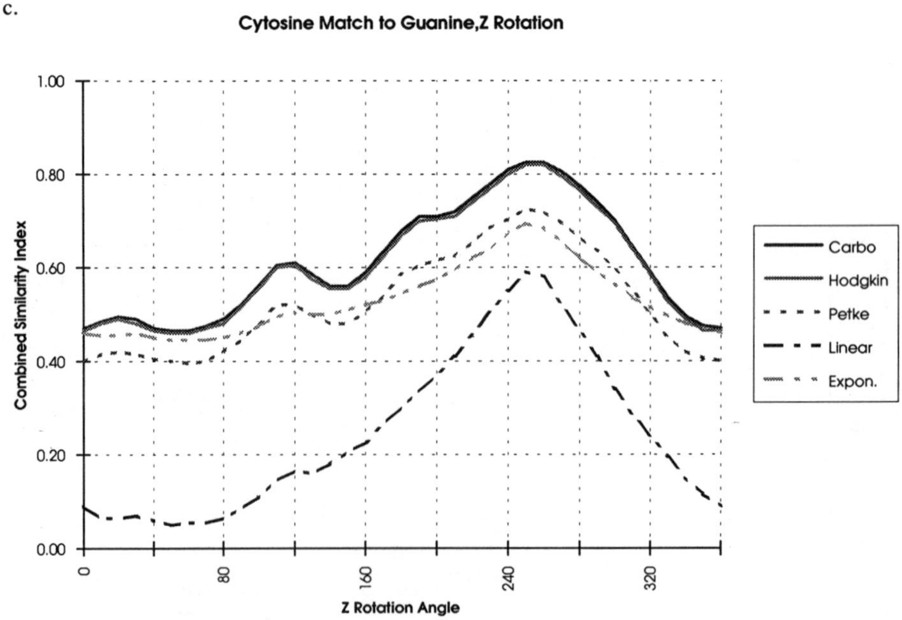

Figure 9. (Continued).

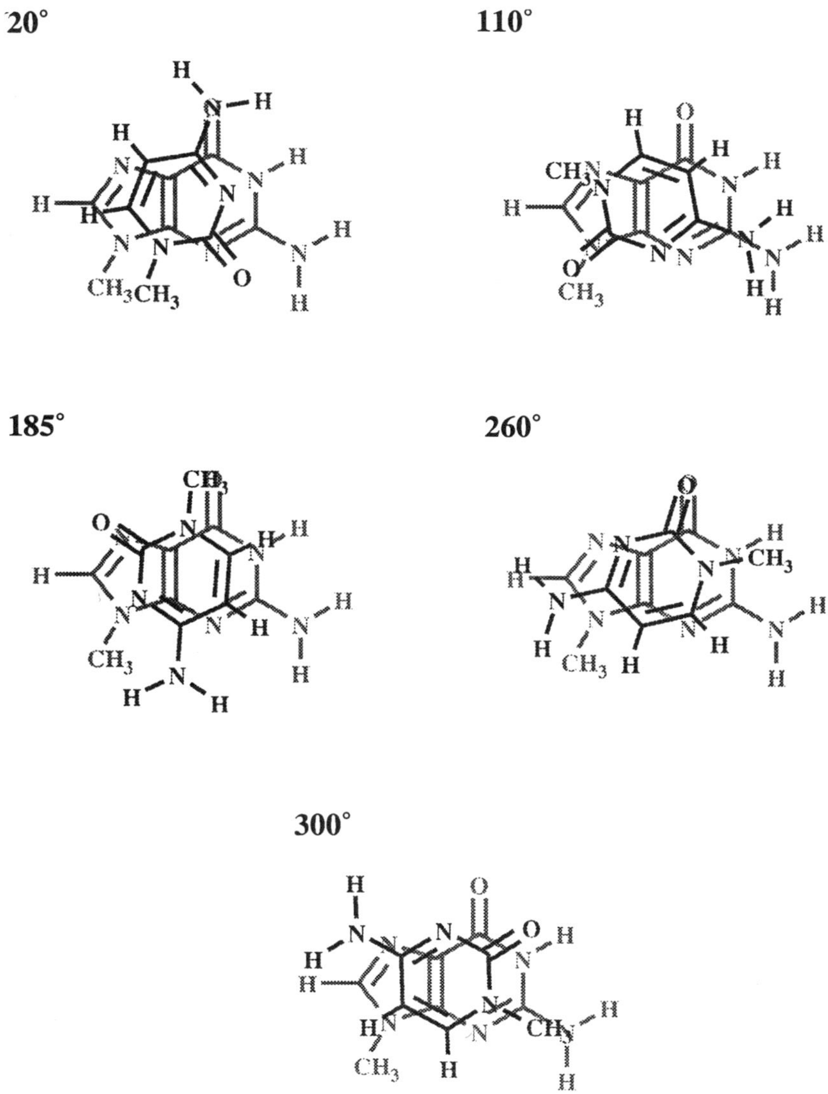

Figure 10. Molecular overlay drawings of cytosine (dark lines) and guanine (light lines) showing the orientations which correspond to maxima in the similarity function profiles.

4. Conclusions

The results obtained from the translation and rotation test of the use of similarity functions to identify matches between non-polar and polar molecules were very encouraging. In each case, the various similarity functions were each able to identify the major matches between the MSV of the molecules. The results for the MEP's were somewhat less accurate, but the general location of the best matches were also identified in each case. The flattening effect of the broad profiles produced by the contribution from the MEP to the combination index seemed to indicate that the MSV contribution should be increased. Considering the expected use of this approach, these functions may servc wcll. However, their inability to resolve the secondary matches could be a handicap under some circumstances. More detailed observations about the similarity functions are as follows:

THE CARBO'S AND HODGKIN/RICHARDS' SIMILARITY FUNCTIONS

- Indicate the best matches without difficulty.
- Relative magnitudes are proportional to expected values.
- Actual similarity values seem to be overestimated.
- Have less resolution than would be desired.

PETKE'S SIMILARITY FUNCTION

- Indicates the best matches without difficulty.
- Relative magnitudes are proportional to expected values.
- Actual similarity values seem to be slightly overestimated.
- Has less resolution than would be desired.

GOOD'S SIMILARITY FUNCTIONS

- Indicate the best matches without difficulty.
- Relative magnitudes are not always proportional to expected values.
- Actual similarity values seem to be close to expected values.
- Have less resolution than would be desired.
- The exponential function has very limited range and does not return to zero.

Acknowledgments

The author would like to thank Jim Petke who has done most of the work on the MEP calculation and the similarity indices and to Gerry Maggiora who has been patient in listening to our ideas and making useful suggestions.

References

1. Gund, P., *Prog. Mol. Subcell. Biol.* **5**, 117 (1977).
2. Marshall, G. R., Barry, C. D., Bosshard, H. E., Dammkoehler, R. A., and Dunn, D. A., in *Computer-Assisted Drug Design, ACS Symposium Series*, No. 112 (American Chemical Society, Washington, D.C., 1979), Chapter 9.

3. Kearsley, S. K. and Smith, G. M., *Tetrahed. Comp. Meth.* **13**, 615 (1990).
4. Humblet, C. and Marshall, G. R., *Ann. Rev. Med. Chem.* **15**, 267 (1980).
5. Carbo, R., Leyde, L., and Arnau, M., *Int. J. Quantum Chem.* **17**, 1185 (1980).
6. Hodgkin, E. E. and Richards, W. G., *Int. J. Quantum Chem. Quantum Biol.* **14**, 105 (1987).
7. Petke, J. D., *J. Comp. Chem.* **14**, 928 (1993).
8. Good, A. C., *J. Mol. Graphics* **10**, 114 (1992).
9. Allinger, N. L., *J. Amer. Chem. Soc.* **99**, 8127 (1977).
10. Dewar, M. J. S., Zoebisch, E. G., Healy, E. F., and Stewart, J. J. P., *J. Amer. Chem. Soc.* **107**, 3902 (1985).

Application of Adaptive Mixture of Local Neural Networks in Chemistry

Prediction and Classification of Resonance and Inductive Effects of Substituents

V. KVASNIČKA

Department of Mathematics, Slovak Technical University, 812 37 Bratislava, Slovakia

Abstract. The theory of adaptive mixture of local neural networks and its application to the classification and prediction of resonance and inductive sigma constants of functional groups is described. The structure of functional groups is described by nonnegative integers – descriptors that correspond to the numbers of appearance of some preselected fragments in the given structural formula. The local neural networks as well as the gating neural network are defined as 2-layer neural networks (perceptrons) without hidden neurons. The adaptation process of this type of neural networks is performed by a straightforward generalization of simple steepest – descent method accelerated by the momentum method. The obtained classification of sigma constants has roughly the same level of precision as the classification performed by the standard feed-forward neural networks composed also of hidden neurons.

1. Introduction

In the series of our recent papers [1–7] we have used different paradigms of neural network theory [8, 9] for prediction and classification of chemical properties (see also refs. [10, 11]). The molecular structure has been always determined by making use of descriptors (nonnegative integers) that correspond to the appearance of some prescribed structural fragments. The purpose of the present communication is to apply the so-called adaptive mixture of local neural networks [12, 13] to the prediction and classification of resonance and inductive effects of functional groups. In this approach a number of neural networks is simultaneously adapted for the same training set of objects (structural formulae). The resulting response of adaptive mixture is determined by the so-called gating neural network specifying which of the local neural networks is used for the classification of the given training object. It means that the whole training set is divided into disjoint subsets so that objects belonging to the same subset are classified by the same local network. In other words, the local networks are understood as expert systems that are specialized for objects from the subsets.

The theory of adaptive mixture of local neural networks is illustrated by the prediction and classification of resonance and inductive sigma constants of functional groups [14]. We have studied the same classification problem by the standard feed-forward neural networks [4] and also by the high-order neural networks [6]. In both these applications it was showed that the given

R. Carbó (ed.), Molecular Similarity and Reactivity: From Quantum Chemical to Phenomenological Approaches, 163–174.

types of neural networks offer valuable classification results. The present application of adaptive mixture of local neural networks to the classification of resonance and inductive effects of functional groups offers an interesting possibility to test simple neural networks without hidden neurons for the classification of objects of relatively complicated structure. For instance, the standard feed-forward neural networks [4] give valuable results only when at least eight hidden neurons are used, or, in the second approach [6], even if hidden neurons are not used, the so-called higher-order terms should be used to achieve a proper classification. The approach of adaptive mixture of local neural networks [12, 13] offers a very flexible neural-network tool capable of simple adaptation process that usually overcomes the often appearing numerical difficulties with local minima and slow convergence. Moreover, these local neural networks without hidden neurons may facilitate simple interpretation of the results based on the concept of perceptron [15].

2. Theory of adaptive mixture of local neural networks

Let $\mathbb{N} = (G, \omega, \vartheta)$ be simple feed-forward neural network [8, 9] represented by an acyclic oriented graph G, and ω and ϑ are the given weight and threshold coefficients, respectively. The activities of input neurons are always kept fixed (input activities) whereas the activities of hidden and output neurons are recurrently determined by

$$x_i = f(\xi_i), \quad \xi_i = \sum_j \omega_{ij} x_j + \vartheta_i, \tag{1a}$$

$$f(\xi) = \frac{1}{1 + \exp(-\xi)}, \tag{1b}$$

where the first derivative of the so-called transfer function $f(\xi)$ is $f'(\xi) = f(\xi)[1 - f(\xi)]$. For fixed weight and threshold coefficients the neural network performs a mapping, which assigns to each input activity vector $\mathbf{x}_I$ an output activity vector $\mathbf{x}_O$, formally,

$$\mathbf{x}_O = F(\mathbf{x}_I; \omega, \vartheta). \tag{2}$$

An evaluation of entries of $\mathbf{x}_O$ is simply carried out recurrently by (1).

Let us have a fixed pair $\mathbf{x}_I/\hat{\mathbf{x}}_O$, where $\hat{\mathbf{x}}_O$ is the so-called desired output activity vector related to the input activity vector $\mathbf{x}_I$. The response of neural network on $\mathbf{x}_I$ is determined by (2). An objective function specifying deviation of the desired output $\hat{\mathbf{x}}_O$ from the calculated output $\mathbf{x}_O$ is determined by

$$E = \frac{1}{2}(\mathbf{x}_O - \hat{\mathbf{x}}_O)^2 = \frac{1}{2}\sum_k g_k^2, \tag{3a}$$

$$g_k = \begin{cases} (x_k - \hat{x}_k) & \text{(output neurons)}, \\ 0 & \text{(otherwise)}. \end{cases} \tag{3b}$$

The first partial derivatives of the objective function E with respect to weight and threshold coefficients are simply determined as follows [7, 16]

$$\frac{\partial \mathrm{E}}{\partial \omega_{ij}} = \frac{\partial \mathrm{E}}{\partial \vartheta_i} x_j, \tag{4a}$$

$$\frac{\partial \mathrm{E}}{\partial \vartheta_i} = \mathrm{f}'(\xi_i)\left(g_i + \sum_l \frac{\partial \mathrm{E}}{\partial \vartheta_l} \omega_{li} \right). \tag{4b}$$

The partial derivatives are useful if the adaptation process of the neural network is performed by a gradient method [8].

After these introductory remarks on standard feedforward neural networks we may turn our attention to a specification of adaptive mixture of local neural networks [12, 13]. Let us consider t local neural networks,

$$\mathbb{N}_i = (G^{(i)}, \omega^{(i)}, \vartheta^{(i)}), \quad (i = 1, 2, \ldots, t), \tag{5a}$$

and one gating neural network

$$\mathbb{N}_g = (G^{(g)}, \omega^{(g)}, \vartheta^{(g)}). \tag{5b}$$

Local networks are assigned to oriented graphs $G_1, G_2, \ldots, G_t$ that are restricted such that all of them contain the same number of input neurons and the same number of output neurons. The oriented graph $G^{(g)}$ assigned to the gating network has also the same number of input neurons as local networks but its number of output neurons is equal to the number of local networks. Let for a given pair $\mathbf{x}_I/\hat{\mathbf{x}}_0$ the output activity vectors of local networks be denoted by $\mathbf{x}_0^{(i)} = (\mathbf{x}_1^{(i)}, \mathbf{x}_2^{(i)}, \ldots)$, for $i = 1, 2, \ldots, t$, and the output activity vector produced by the gating network be denoted by $\mathbf{x}_0^{(g)} = (x_1^{(g)}, x_2^{(g)}, \ldots, x_t^{(g)})$. All entries of the output activity vectors belong to the open interval (0, 1). The so-called proportionality coefficients produced by the gating network are determined as the response on the input vector $\mathbf{x}_I$,

$$p_i = \frac{x_i^{(g)}}{\sum_{j=1}^{p} x_j^{(g)}} . \quad (i = 1, 2, \ldots, t) \tag{6}$$

These coefficients belong to the open interval (0, 1) and their sum is equal to one. In our forthcoming considerations the proportionality coefficients will be interpreted as "probabilities" (see below) that a corresponding local neural network has been involved in the classification of the input pattern $\mathbf{x}_I$.

The objective function (3) is generalized as follows:

$$\mathrm{E} = \frac{1}{2} \sum_{i=1}^{t} p_i(\mathbf{x}_0^{(i)} - \hat{\mathbf{x}}_0)^2 = \frac{1}{2} \sum_{i=1}^{t} \left(\sum_k g_k^{(i)2} \right), \tag{7a}$$

$$g_k^{(i)} = \begin{cases} p_i(x_k^{(i)} - \hat{x}_k) & \text{(output neurons)}, \\ 0 & \text{(otherwise)}, \end{cases} \tag{7b}$$

where $x_k^{(i)}$ is kth output activity of ith local neural network. The partial derivatives of the objective function (7a) with respect to weight and threshold coefficients of local neural networks are determined by (4) with entries g's determined by (7b).

The partial derivatives of E with respect to the coefficients of gating network are determined in a slightly different manner. They are mutually related by (cf. Equation (4a))

$$\frac{\partial E}{\partial \omega_{ij}^{(g)}} = \frac{\partial E}{\partial \vartheta_i^{(g)}} x_j^{(g)} . \tag{8}$$

The partial derivatives $\partial E/\partial \vartheta_i^{(g)}$ are recurrently determined by a formula with the same structure as (4b),

$$\frac{\partial E}{\partial \vartheta_i^{(g)}} = f'(\xi_i^{(g)}) \left(g_i^{(g)} + \sum_l \frac{\partial E}{\partial \vartheta_l^{(g)}} \omega_{li}^{(g)} \right), \tag{9a}$$

where the entities $g^{(g)}$'s are

$$g_i^{(g)} = \begin{cases} \left(\frac{1}{2} (\mathbf{x}_0^{(i)} - \hat{\mathbf{x}}_0)^2 - E\right) \cdot \left(\sum_j x_j^{(g)}\right)^{-1} & \text{(output neurons)} \\ 0 & \text{(otherwise).} \end{cases} \tag{9b}$$

Simple discussion of these expressions for partial derivatives leads to the following conclusions: If both gating network and local networks are simultaneously adapted for a given pair $\mathbf{x}_I/\hat{\mathbf{x}}_0$, then the system tends to use only one local network for classification of the input pattern $\mathbf{x}_I$ so that its output activity is closely related to the desired output activity (classification) $\hat{\mathbf{x}}_0$. Formally, the proportionality coefficients tend to gain binary values,

$$p_j = \begin{cases} 1. & \text{(for } j = i) \\ 0, & \text{(otherwise)} \end{cases} \tag{10}$$

for $j = 1, 2, \ldots, t$, and i is the index of local network with classification nearest to the desired value $\hat{\mathbf{x}}_0$.

An adapted local neural network may be used for classification of patterns from outside the training set (i.e. from the testing set). For an input activity vector $\mathbf{x}_I$ we calculate output activity vectors of all local and gating networks, i.e. we know the vectors $\mathbf{x}_0^{(1)}, \mathbf{x}_0^{(2)}, \ldots, \mathbf{x}_0^{(t)}, \mathbf{x}_0^{(g)}$. Then, using the entries of $\mathbf{x}_0^{(g)}$ the proportionality coefficients p_i's, see (6), are calculated. The input vector pattern $\mathbf{x}_I$ will be now classified by an output activity vector $\mathbf{x}_0^{(i)}$ determined such that the proportionality coefficient p_i is maximal, i.e. $p_i = \max\{p_1, p_2, \ldots, p_t\}$. This property of active process suggests simple schematic interpretation of adapted local neural networks such that a stochastic selector is used [12]. Another interesting possibility how to overcome the ambiguity of which output vector should be used for the classification of the input pattern $\mathbf{x}_I$ is the use of convex combination of output activities formed by local neural networks,

$$\tilde{\mathbf{x}}_0 = p_1 \mathbf{x}_0^{(1)} + p_2 \mathbf{x}_0^{(2)} + \ldots + p_t \mathbf{x}_0^{(t)}. \tag{11}$$

Its entries are bounded from below and above by minimum and maximum values, respectively, of the corresponding entries of output vectors $\mathbf{x}_0^{(j)}$,

$$\min_{1 \le j \le t} \{x_i^{(j)}\} \le \tilde{x}_i \le \max_{1 \le j \le t} \{x_i^{(j)}\}. \tag{12}$$

For an input vector $\mathbf{x}_\mathrm{I}$ taken from the training set the convex combination (11) gives an output vector $\tilde{\mathbf{x}}_0$ equal to the output vector produced by local neural network with unit proportionality coefficient, see Equation (10). For other input vectors taken from outside of the training set the resulting vector $\tilde{\mathbf{x}}_0$ is roughly related to the output vector $\mathbf{x}_0^{(j)}$ with maximum proportionality coefficient.

3. Illustrative example

Basic principles of the theory of adaptive mixture of local neural networks will be illustrated in this section by simple example of nine 2-dimensional patterns that are placed at nods of a square grid, see Figure 1. These patterns obviously form a linearly nonseparable training set. It means, that for correct classification of patterns we have to use a neural network composed of hidden neurons. The hidden neurons create the so-called internal representation of the patterns, which is already linearly separable. A correct classification of

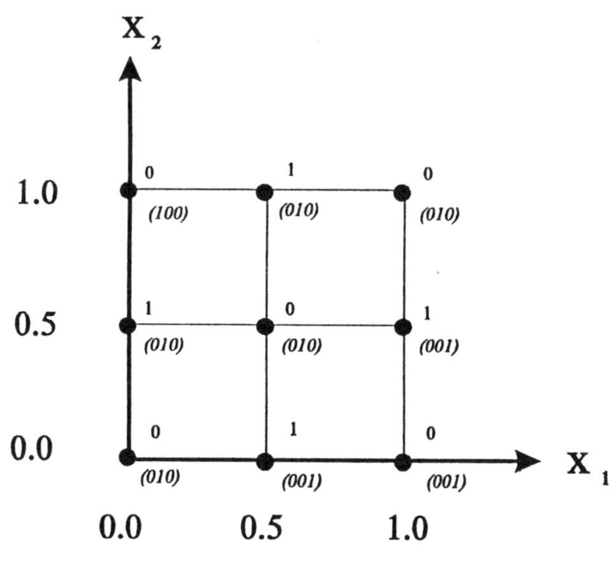

Figure 1. Visualization of training patterns (heavy dots) that are situated at nodes of rectangular grid. As each dot pattern is evaluated either by 0 or 1, it is easy to see that these patterns are not linearly separable. The patterns are divided into disjoint subsets that are evaluated by the same activities (binary integers placed in parentheses) of output neurons of the gating network.

these patterns has been successfully carried out by simple 3-layer neural network with hidden layer composed of (at least) four hidden neurons.

The present theory of adaptive mixture of local neural networks outlined in the previous section is able to classify correctly the above nine patterns so that we employ three local neural networks and one gating network. All local networks are the same 2-layer perceptrons (composed of two input neurons and one output neuron) while the gating network is a 3-layer network (composed of two input neurons, two hidden neurons, and three output neurons), see Figure 2. The activities of output neurons in the gating network separate the training set into three disjoint subsets (see Figure 1), where each of them is correctly linearly classified by the corresponding local network.

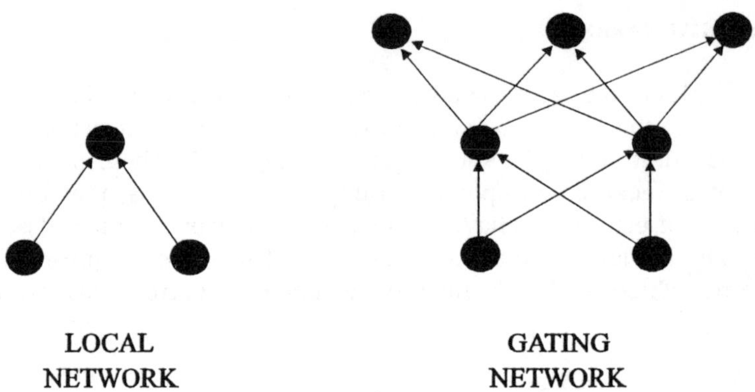

Figure 2. Three local and one gating neural networks are used for the classification of nine patterns, see Figure 1. The local neural networks are perceptrons while the gating network is a 3-layer neural network composed of two hidden neurons and three output neurons. The activities of output neurons of the gating network separate the patterns into three disjoint subsets (see integers in parentheses in Figure 1).

In general, each continuous mapping $\mathbb{R}^m \to \{0, 1\}^n$ may be simulated with the prescribed precision by a 3-layer neural network (unfortunately, this important property is formulated only in a form of an existence theorem [17], it does not specify the number of hidden neurons and the adaptation process). According to this general property, the approach of adaptive mixture of local networks may be, in principle, performed so that the training patterns are divided into disjoint subsets that are already linearly separable and the subset membership is determined by a 3-layer gating network. Then the local networks could be 2-layer perceptrons (their number is equal to the number of disjoint subsets) and the gating network is composed of a number of hidden neurons that is determined experimentally and the number of its output neurons is equal to the number of disjoint training subsets (local neural networks). The above discussed simple example of nine patterns nicely illustrates these general conclusions.

Generalization (or prediction) abilities of adapted neural networks usually depend significantly on the number of hidden neurons. Let us assume that we have two neural networks with different number of hidden neurons and both are trained for the same training set so that each of them classifies the training patterns with the similar precision (or in other words, the optimized adapted values of corresponding objective functions are of the same order of smallness). Then, the neural network with smaller number of hidden neurons usually better classifies patterns outside the training set than the second neural network with the higher number of hidden neurons. According to these general considerations we may expect that *an adaptive mixture of local neural networks is endowed by better generalization abilities than those ones performed by simple neural networks with one or more hidden layers.*

4. Classification of resonance and inductive effects of substituents

The theory of adaptive mixture of local neural networks is used for classification of inductive and resonance sigma constants of functional groups. In our recent publications [4, 6] similar problem has been studied by making use of simple feed-forward 3-layer network composed of eight hidden neurons and by the so-called recurrent neural network. The functional groups have been described by fourteen nonnegative integers (descriptor input activities) that reflect their topology and physical nature of atoms and bonds (cf. [3]). In order to simplify the present study we now left out all descriptors that correspond to phenyl groups, i.e. the number of descriptors was reduced to eleven, see Table I. Examples of descriptors of simple functional groups are given in

TABLE I
Descriptors of functional groups

d_i	Meaning of the descriptor
	First-level descriptors
d_1	Number of lone electron pairs on the first level atoms
d_2	Sum of the main quantum numbers that are decreased by one, of the first level atoms
d_3	Number of hydrogen atoms attached to the first level atoms
	Second-level descriptors
d_4	Number of lone electron pairs on the second level atoms
d_5	Sum of the main quantum numbers that are decreased by one, of the second level atoms
d_6	Number of hydrogen atoms attached to the second level atoms
d_7	Number of pi bonds that connect the first and second level atoms
	Third-level descriptors
d_8	Number of lone electron pairs on the third level atoms
d_9	Sum of the main quantum numbers that are decreased by one, of the third level atoms
d_{10}	Number of hydrogen atoms attached to the second level atoms
d_{11}	Number of pi bonds that connect the second and third atoms

Table II. The training and testing sets are the same as in ref. [4], but now, as has been already mentioned, the functional groups composed of phenyls are omitted, see Table III.

TABLE II
Examples of descriptors of functional groups

—X	d_1	d_2	d_3	d_4	d_5	d_6	d_7	d_8	d_9	d_{10}	d_{11}
—Br	3	3	0	0	0	0	0	0	0	0	0
—SO_2NH_2	0	2	0	5	3	2	2	0	0	0	0
—CH=NOH	0	1	1	1	1	0	1	2	1	1	0
—$NHCOCH_3$	1	1	1	0	1	0	0	2	2	3	1

TABLE III
Experimental and calculated σ_I and σ_R using adaptive mixture of local neural networks

No.	—X	Exp.[a]		Calc.[b] (t = 3)		Calc.[b] (t = 4)	
		σ_I	σ_R	σ_I	σ_R	σ_I	σ_R
		Training set					
1	—Br	0.44	−0.17	0.44	−0.21	0.46	−0.09
2	—F	0.43	−0.34	0.49	−0.41	0.46	−0.43
3	—SO_2F	0.75	0.22	0.68	0.25	0.70	0.21
4	—IO_2	0.63	0.20	0.67	0.27	0.63	0.21
5	—NO_2	0.67	0.16	0.61	0.17	0.66	0.19
6	—N_3	0.30	−0.13	0.36	−0.15	0.34	−0.08
7	—H	0.00	0.00	−0.01	0.02	0.00	0.03
8	—OH	0.29	−0.64	0.19	−0.72	0.27	−0.50
9	—SH	0.28	−0.11	0.20	−0.18	0.27	−0.31
10	—NH_2	0.02	−0.68	0.06	−0.52	−0.01	−0.67
11	—NHOH	0.06	−0.40	0.16	−0.39	0.11	−0.55
12	—SO_2NH_2	0.41	0.19	0.59	0.22	0.56	0.21
13	—N=CCl_2	0.23	−0.08	0.28	−0.09	0.30	−0.09
14	—CF_3	0.38	0.19	0.33	0.19	0.37	0.19
15	—OCF_3	0.38	0.00	0.35	0.00	0.33	0.01
16	—SO_2CF_3	0.73	0.26	0.73	0.26	0.73	0.26
17	—CN	0.51	0.19	0.44	0.19	0.52	0.14
18	—NCS	0.51	−0.09	0.40	−0.06	0.36	−0.07
19	—CHO	0.31	0.13	0.25	0.11	0.29	0.12
20	—CH_2Br	0.10	0.05	0.10	0.05	0.11	0.05
21	—$CONH_2$	0.24	0.14	0.29	0.15	0.26	0.17
22	—CH=NOH	0.25	−0.13	0.26	−0.14	0.26	−0.13
23	—CH_3	−0.04	−0.13	−0.06	−0.15	−0.07	−0.14
24	—$NHCONH_2$	0.04	−0.28	0.18	−0.26	0.16	−0.25
25	—$SOCH_3$	0.52	0.01	0.38	0.01	0.51	0.02
26	—SO_2CH_3	0.54	0.22	0.54	0.20	0.47	0.21
27	—$0SO_2CH_3$	0.39	0.00	0.45	0.00	0.45	−0.04

TABLE III *(Continued)*

No.	—X	Exp.[a]		Calc.[b] (t = 3)		Calc.[b] (t = 4)	
		σ_I	σ_R	σ_I	σ_R	σ_I	σ_R
28	—$NHCH_3$	–0.11	–0.74	0.04	–0.58	0.00	–0.67
29	—$NHSO_2CH_3$	0.25	–0.20	0.20	–0.19	0.26	–0.11
30	—C≡CH	0.19	0.05	0.29	0.09	0.21	0.04
31	—CH_2CN	0.21	–0.18	0.21	–0.18	0.21	–0.18
32	—$COCH_3$	0.32	0.20	0.22	0.09	0.25	0.15
33	—CH_2CH_3	–0.05	–0.10	–0.08	–0.24	–0.06	–0.16
34	—OCH_2CH_3	0.22	–0.44	0.23	–0.36	0.25	–0.44
35	—$NHCH_2CH_3$	–0.11	–0.51	–0.02	–0.50	–0.03	–0.64
36	—$SO_2CH_2CH_3$	0.54	0.22	0.53	0.16	0.50	0.21
37	—$P(CH_3)_2$	–0.08	0.39	0.05	0.07	–0.08	0.22
38	—$PO(OCH_3)_2$	0.37	0.19	0.37	0.16	0.39	0.19
39	—$OCH(CH_3)_2$	0.30	–0.72	0.29	–0.68	0.29	–0.60
40	—$CH_2Si(CH_3)_3$	–0.15	–0.07	–0.14	–0.07	–0.16	–0.06
41	—$B(OH)_2$	–0.07	0.18	0.02	0.17	–0.05	0.15
42	—CH_2OH	0.00	0.00	0.07	–0.04	0.05	0.04
43	—SCH_3	0.20	–0.18	0.19	–0.20	0.20	–0.03
44	—$CH{=}CH_2$	0.07	–0.08	0.11	–0.08	0.14	–0.04
45	—$NHCOCH_3$	0.28	–0.26	0.16	–0.28	0.16	–0.32
46	—$N(CH_3)_2$	0.10	–0.92	0.03	–0.64	0.01	–0.66
47	—$NHNH_2$	0.17	–0.71	0.11	–0.51	0.06	–0.62
48	—$OCOCH_3$	0.41	–0.07	0.41	–0.10	0.37	–0.26
49	—CH_2OCH_3	0.01	0.02	0.01	0.07	0.01	0.09
50	—$CH_2CH_2CH_3$	–0.06	–0.08	–0.14	–0.13	–0.10	–0.11
51	—$C(CH_3)_3$	–0.07	–0.13	–0.11	–0.42	–0.04	–0.19
52	—$Si(CH_3)_3$	–0.04	–0.04	–0.09	0.38	–0.05	–0.02
		Testing set					
53	—Cl	0.41	–0.15	0.46	–0.32	0.46	–0.28
54	—I	0.40	–0.19	0.38	–0.10	0.46	0.00
55	—SCF_3	0.35	0.18	0.41	0.22	0.33	0.19
56	—SCN	0.36	0.19	0.39	–0.03	0.41	0.10
57	—CO_2H	0.33	0.15	0.35	0.17	0.35	0.17
58	—CH_2Cl	0.10	0.03	0.12	0.05	0.10	0.09
59	—CH_2I	0.09	0.03	0.08	0.05	0.11	0.00
60	—NHCHO	0.25	–0.23	0.25	–0.66	0.12	–0.20
61	—CF_2CF_3	0.44	0.11	0.78	–0.01	0.42	0.24
62	—$SCOCH_3$	0.36	0.11	0.47	0.12	0.37	–0.08
63	—CO_2CH_3	0.33	0.15	0.34	0.13	0.39	0.17
64	—SCH_2CH_3	0.23	–0.18	0.19	–0.22	0.26	–0.24
65	—$CH(CH_3)_2$	–0.05	–0.10	–0.09	–0.33	–0.05	–0.17
66	—NHCN	0.26	–0.18	0.32	–0.40	0.21	–0.14
67	—NO	0.50	0.45	0.47	–0.02	0.54	0.19
68	—OCH_3	0.26	–0.51	0.21	–0.48	0.20	–0.26

[a] Experimental sigma constants are from ref. [14].
[b] Estimated sigma constants are calculated according to Equation (11).

The topology of adaptive mixtures of local neural networks is specified so that all local as well as gating networks are 2-layer perceptrons, where local networks always contain eleven input neurons and two output neurons. We examined two different cases; the first mixture contained three ($t = 3$) local networks (then the gating network should be composed of three output neurons), whereas the second mixture contained four ($t = 4$) local networks (and the gating network four output neurons). The gating networks separate the training set of functional groups into disjoint subsets (see Table IV) that are approximately linearly separable with respect to output activities (i.e. resonance and inductive sigma constants of functional groups). Initially, we have expected that this separation of the training set will be done strictly according to the physico-chemical nature of functional groups. Unfortunately, this requirement has not been fully satisfied. The main driving force of the separation is the condition of linear separability of functional groups for each single subset. This condition may be satisfied occasionally for functional groups of different nature and therefore it could not be said that single local networks are "expert systems" in dealing with functional groups that are physico-chemically similar.

The results of both calculations are listed in Table III. The adaptation process of adaptive mixture of local neural networks has been finished approximately

TABLE IV
Training subsets of functional groups determined by gating network for different number of local neural networks

Subset	$t = 3$
1.	$—N_3$ $—N{=}CCl_2$ $—CF_3$ $—OCF_3$ $—SO_2CF_3$ $—NCS$ $—NHCONH_2$ $—OSO_2CH_3$ $—NHSO_2CH_3$ $—OCH(CH_3)_2$ $—CH_2Si(CH_3)_3$ $—NHCOCH_3$ $—OCOCH_3$
2.	$—H$ $—OH$ $—SH$ $—NH_2$ $—NHOH$ $—CHO$ $—CH_2Br$ $—CH{=}NOH$ $—CH_3$ $—NHCH_3$ $—C{\equiv}CH$ $—CH_2CN$ $—CH_2CH_3$ $—NHCH_2CH_3$ $—P(CH_3)_2$ $—CH_2OH$ $—CH{=}CH_2$ $—N(CH_3)_2$ $—NHNH_2$ $—CH_2OCH_3$ $—CH_2CH_2CH_3$ $—C(CH_3)_3$ $—Si(CH_3)_3$
3	$—Br$ $—F$ $—SO_2F$ $—IO_2$ $—NO_2$ $—SO_2NH_2$ $—CN$ $—CONH_2$ $—SOCH_3$ $—SO_2CH_3$ $—COCH_3$ $—OCH_2CH_3$ $—SO_2CH_2CH_3$ $—PO(OCH_3)_2$ $—B(OH)_2$ $—SCH_3$
	$t = 4$
1	$—H$ $—NH_2$ $—NHOH$ $—CH_2Br$ $—CH{=}NOH$ $—CH_3$ $—NHCH_3$ $—CH_2CN$ $—COCH_3$ $—CH_2CH_3$ $—NHCH_2CH_3$ $—CH_2Si(CH_3)_3$ $—CH_2OH$ $—CH{=}CH_2$ $—N(CH_3)_2$ $—Si(CH_3)_3$ $—C(CH_3)_3$ $—CH_2CH_2CH_3$ $—CH_2OCH_3$ $—NHNH_2$
2	$—OCF_3$ $—OCOCH_3$ $—NHCOCH_3$ $—NHSO_2CH_3$
3	$—SO_2F$ $—IO_2$ $—NO_2$ $—SO_2NH_2$ $—CF_3$ $—SO_2CF_3$ $—CN$ $—CHO$ $—CONH_2$ $—SO_2CH_3$ $—SO_2CH_2CH_3$ $—PO(OCH_3)_2$ $—B(OH)_2$
4	$—Br$ $—F$ $—N_3$ $—OH$ $—SH$ $—N{=}CCl_2$ $—NCS$ $—NHCONH_2$ $—SOCH_3$ $—C{\equiv}CH$ $—OSO_2CH_3$ $—OCH_2CH_3$ $—P(CH_3)_2$ $—OCH(CH_3)_2$ $—SCH_3$

after 5000 cycles, the values of objective function (7) are E = 0.4 (for $t = 3$) and E = 0.2 (for $t = 4$), respectively. These results have been statistically evaluated by simple linear regression (least square method), see Table V. The calculated and experimental values are related by $y_{exp} = b_0 + b_1 y_{cal}$. Ideally, if the calculated and experimental values are well correlated, then the coefficient $b_0 = 0$, the slope $b_1 = 1$, and the value of standard error estimate is vanishing. We see that the inductive sigma constants are slightly better predicted than the resonance sigma constants, and that the adaptive mixture of four local networks offers better prediction than the adaptive mixture composed of three local networks (naturally, this last observation could be expected).

TABLE V
Statistical least-square-method evaluation of results

Sigma constants	E[a]	r^b	b_0^c	b_1^c
	$t = 3$			
	Training set			
σ_I	0.072	0.952	0.023	0.902
σ_R	0.107	0.931	–0.012	0.849
	Testing set			
σ_I	0.096	0.903	–0.041	1.254
σ_R	0.197	0.761	–0.108	0.858
	$t = 4$			
	Training set			
σ_I	0.055	0.972	0.014	0.939
σ_R	0.079	0.964	–0.009	0.898
	Testing set			
σ_I	0.050	0.951	–0.023	1.088
σ_R	0.144	0.762	–0.002	0.623

[a] Standard error of estimate, $E = (\Sigma(y_{exp} - y_{cal})^2/(N-2))^{1/2}$.
[b] Correlation coefficient.
[c] Parameters of straightline $y_{exp} = b_0 + b_1 y_{cal}$.

5. Discussion

The main difference between the neural network approach [8, 9] and the perceptron approach [15] lies in the fact that neural networks overcome the principal restriction of perceptrons, which are only applicable to linearly separable patterns by using the hidden neurons. The hidden neurons create the so-called internal representation of classified patterns, which is already linearly separable with respect to the output neurons. The theory of adaptive mixture of local neural networks allows to surmount very naturally this shortcoming of perceptrons so that the local networks are applied only to linearly separable patterns, which are determined by the gating network that may contain or not the hidden neurons. That is, the gating network has a role of an arbiter of separation of the whole training set into disjoint subsets that are itself linearly separable. Then each subset of training patterns is classified by single local neural network (a local expert system) with very simple structure (topology), which potentially offers new unexpected possibilities of how to interpret the resulting classification of the whole adaptive mixture of local networks. The problem of interpretation of results of feed-forward neural networks composed of hidden neurons belongs to very serious and actual questions of general theory of neural networks [8, 9]. Neural networks are often critically classified as *black-box* methods which are unable to explain (or interpret) their results. The possibility to use in the present theory of adaptive mixture of local neural networks only 2-layer perceptrons might be of great value for better understanding and interpretation of the classification produced by them. In particular, since the training patterns are divided into subsets that are itself linearly separable, there may be simply constructed systems of linear inequalities for descriptors according to which the resulting classification is performed. That is, the adapted mixture of local networks may be simply transformed to a system of linear inequalities that will give the same classification as the given mixture of local networks.

References

1. Kvasnička, V., *Chem. Papers* **44**, 19 (1990).
2. Kvasnička, V., *J. Math. Chem.* **6**, 63 (1991).
3. Kvasnička, V. and Pospíchal, J., *J. Mol. Struct. (THEOCHEM)* **235**, 227 (1991).
4. Kvasnička, V., Sklenák, Š., and Pospíchal, J., *J. Am. Chem. Soc.* **115**, 1495 (1993).
5. Kvasnička, V., Sklenák, Š., and Pospíchal, J., *J. Chem. Inf. Comput. Sci.* **32**, 742 (1992).
6. Kvasnička, V., Sklenák, Š., and Pospíchal, J., *Theor. Chim. Acta* **86**, 257 (1993).
7. Kvasnička, V., Sklenák, Š., and Pospíchal, J., *J. Mol. Struct. (THEOCHEM)* **277**, 87 (1992).
8. Rumelhart D. E. and McClelland, J. L. (and the PDP Research Group), *Parallel Distributed Processes* (MIT Press, Cambridge, MA, 1986), Vols. I and II.
9. Simpson, P. K., *Artificial Neural Networks* (Pergamon Press, New York, 1990).
10. Župan, J. and Gasteiger, J., *Anal. Chem. Acta* **248**, 1 (1991).
11. *Tetrahedron Comput. Methodol.* **3**(1) (1990).
12. Jacobs, R. A., Jordan, M. I., Nowlan, S. J., and Hinton, G. E., *Neural Comput.* **3**, 79 (1991).
13. Kvasnička, V., *Neural Net. World* **3** (1993).
14. Hansch, C. and Leo, A., *Substituents Constants for Correlation Analysis in Chemistry and Biology* (John Wiley, New York, 1979).
15. Minsky, M. and Papert, S., *Perception, An Introduction to Computational Geometry* (MIT Press, Cambridge, MA, 1969).
16. Pimeda, F. J., *Phys. Rev. Letters* **59**, 2229 (1987).
17. Hornik, K., Stinchcombe, M., and White, H., *Neural Net.* **2**, 359 (1989).

Section C

Phenomenological Approaches

A Proven Performer in the Similarity Stakes

DENNIS H. ROUVRAY
Department of Chemistry, University of Georgia, Athens, Georgia, 30602-2556, U.S.A.

1. Introductory similarity

At first sight it would seem that the study of similarity has little in common with the noble sport of horse-racing. A moment's reflection, however, is enough to convince us that some striking parallels do indeed exist between the two. In the first place there are many different types of race that can be run, each with its own prize, just as there are many different objectives that we can pursue and attain in our efforts to exploit the concept of similarity. And, just as a variety of horses participate in a race, so there are several kinds of similarity that are available for application. Other parallels also spring to mind. As connoisseurs of racing recognize, there are horses for courses. In an analogous way, modern research has taught us that differing kinds of similarity are to be employed in different ways, that is to say specific similarity types are appropriate for specific applications [1]. Any contender in the similarity stakes with a proven track record will always excite special interest. Such is the case with the breed of similarity known as analogy or metaphor. This particular entry has a long and enviable record and possesses all the hallmarks of a winner [2]. But, before we get carried away with the virtues of analogy, let us briefly pause to take stock of the entire stable. Examination of both the past performance and the present form of each contender will give us some feel for the kind of race being run. After completion of our survey of the field, we shall conclude our chapter by zooming in on our choice for favorite (see Figure 1).

Major interest in similarity is a comparatively new phenomenon and so we start by asking why there has been an upsurge of interest in recent years. The answer is not difficult to find: similarity studies have been burgeoning because of their great potential for commercial exploitation. In particular, a large and growing number of professionals are now involved with the development of novel similarity-based techniques for the design of new molecules and, in particular, the design of new pharmaceutical drugs [3]. The underlying idea here is to discover molecular structures that are similar to those known to be pharmacologically active. This idea rests on the assumption that

R. Carbó (ed.), Molecular Similarity and Reactivity: From Quantum Chemical to Phenomenological Approaches, 177–202.

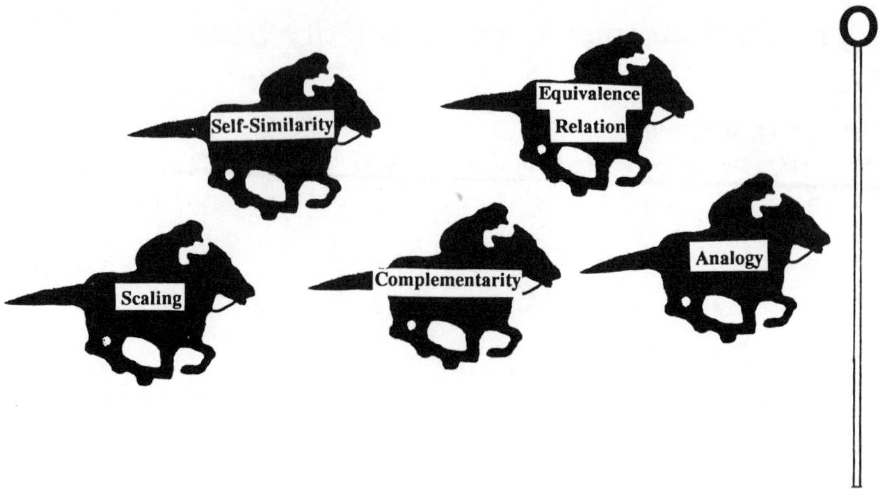

Figure 1. The differing varieties of similarity are viewed as analogous to different contenders participating in a horse race. Our figure shows the contender 'Analogy' winning the race.

molecules with similar structures will exhibit similar bioactivity, a point we shall explore in greater detail later on in our narrative. The elaboration of similarity methods in molecular design, in spite of all the current excitement, represents no more than a small fraction of similarity studies in general [4]. Indeed, similarity in some shape or form is now seen to play an indispensable role in chemistry and in all sciences. It has necessarily been present since our earliest musings on science and continues to be a vital ingredient in twentieth century science. To understand the key importance of similarity, we turn at this juncture to a consideration of what is presently known about the concept itself.

1.1. Similarity is elusive

It has proven to be surprisingly difficult to get to grips with the concept of similarity. There are several reasons for this. We all feel comfortable with the notion of similarity and believe we can discern similarities in objects or events in a direct and instinctive way. It is precisely this aspect of similarity that makes it so intractable when we attempt a rigorous mathematical definition. Psychologists, who have investigated this matter over a period of several decades, have reached the conclusion [5, 6] that we make similarity judgments on the basis of two quite different criteria. These are (i) intensive criteria that relate to such things as the size of the objects being compared, the extent of their illuminance, and signal to noise ratio; and (ii) diagnostic criteria that relate to subjective feelings about the objects under consideration such as their salience for the observer. It is the latter subjective factors that make similarity so hard to quantify, but, since these cannot be completely ignored, our definitions will in general be fuzzy and elusive to some extent. In fact, the best that can be done in practice is to define similarity in relative terms, that is to define it for a set of related objects or events within a given context or set of circumstances. All our measures of similarity will thus retain some degree of arbitrariness and cannot ever be regarded as absolute.

1.2. Similarity in concept formation

In spite of the somewhat capricious nature of similarity, the concept plays a decisive role in many of our thought processes. For instance, it is of key importance in both the formation of mental concepts and in our classification of the multifarious entities and situations in the world about us. Such processes begin when we decide that a given entity is a typical representative – a so-called *prototype* – of a species we wish to define. We then seek for entities that are similar to the prototype and add these in successive layers to form a structure that becomes essentially onion-like [7]. The cut-off point is arbitrarily decided by ourselves. As ever new layers are added to the structure, the entities that we accommodate will be less and less similar to the prototype. This expansion process thus forces us to be ever less discriminating in what we allow to be included. Our demands concerning the necessary attributes of entities for inclusion thus become progressively less rigorous, with the end results that the concept or classification we create is unsharp and ill-defined. It might then be asked: what holds the structure together? In the words of Oden and Lopes [8], the answer seems to be that "although similarity must function at some level in the induction of concepts, the induced categories are not 'held together' subjectively by the undifferentiated 'force' of similarity, but rather by structural principles."

The structural principles referred to are not particularly well understood at present. However, some progress has been made on this front, and one leading worker has attempted to characterize the principles underlying concept formation [9]. These are that (i) concepts are far more than compilations of entities; (ii) entity properties exist at multiple levels of abstraction; (iii) entity properties are in general not independent but are linked by a variety of interproperty relationships; and (iv) similarity needs to embrace attributes, relations, and even higher-order relations. There is currently much research activity in this area and what the final outcome will be only time can tell. We do know however that the effective performance of the mental processes outlined here – vague though they may seem – is essential for our very survival. If they could not be performed adequately for any reason, we should rapidly become overwhelmed by the massive onslaught of unprocessed sense data impinging on our sensory organs. This would have the effect of immobilizing us and so would make life impossible. It is therefore vital that each individual develop a comprehensive armory of concepts and classifications. Rather surprisingly, in view of the colossal number of ways in which sense data from our world might be partitioned, there is widespread agreement on the way it should be done. What constitutes a meaningful category or classification appears to transcend linguistic and cultural barriers [10], to apply also to young children [11], and even to hold for vertebrate members of the animal kingdom [12].

2. Ubiquitous similarity

Because the notion of similarity occupies so central a position in our thought processes, it is to be anticipated that similarity in some shape or form will find expression in much of our human activity. Indeed, a key component in human experience is that of classifying objects, events, systems or biota into certain basic categories and then deciding on what an appropriate fate for the members of such categories should be! All of the arts and sciences make extensive use of classification processes and often a considerable amount of time is spent in coming to agreement on the classification of the entities under study. Thus we find archaeologists arguing over artifacts, botanists puzzling over plants, geologists wrangling over rocks, and zoologists feuding over fauna, all with the same ultimate objective of arriving at some generally acceptable classification scheme. Examples of the results of such endeavors are collected together in Table I, which presents instances of the use of similarity for classification purposes in a wide variety of disciplines. The

TABLE I
A compilation of similarity-based classifications that are employed in a wide variety of arts and sciences

Discipline	Similarity-based classification
Archaeology	Similarity of ancient artifacts
Anthropology	Similarity of peoples and nations
Art	Classification by stylistic similarity
Bacteriology	Classification of bacteria
Botany	Taxonomic classification of plants
Computer Science	Data clustering; use of prototypes
Ecology	Classification by allometry
Engineering	Simulation model classification
Genetics	Classification of genotypes; homology
Linguistics	Speech spectra classification; language groups
Mathematics	Congruity analysis; similarity transforms
Medicine	Classification of electrocardiographs
Music	Similarity of different compositions
Pharmacology	Data clustering in drug design
Physics	Classification of particles; self-similarity
Psychology	Classification of behavior patterns
Sociology	Classification of social attitudes
Statistics	Data clustering; consensus methods
Virology	Classification of viruses
Zoology	Classification of fauna; mammalian studies

classification process itself has come under intense scrutiny in recent years. We mention in passing here that methods have now been developed for (i) the comparison of different classifications based on the same data [13]; (ii) the comparison of classifications deriving from different data sets [14]; (iii) the classification of data sets with ill-defined entities [15]; and (iv) the construction of classification schemes via the use of neural networks [16].

2.1. Similarity in chemistry

The advancement of chemistry has always depended very heavily on the application of similarity notions. Since the emergence of chemistry as a major scientific discipline some two centuries ago, numerous classifications have proven their fundamental importance for the continuing growth of chemistry. We consider now three of the most important classifications made during the past 200 years. One of the earliest examples was that of Dalton [17], who in 1808 classified atoms. His atomic law stated that all atoms of a given kind were identical, had the same mass, and therefore could be expected to behave in identical fashion. Some sixty years later, in one of the greatest classifications of all time, atoms were classified into groups based on analogies in their chemical properties. We refer to the Periodic Law put forward by the Russian chemist Mendeleev [18] in 1869, a theme we return to later on in our discourse. A third significant classification, which usually goes under the name of the Principle of Analogy [2], concerns mainly molecules. This principle is less cut and dried than the above laws and is manifested in a variety of different ways. Examples include (i) the homology principle [19], which states that successive members of homologous series will exhibit similarities in their behavior; (ii) the more general principle that like structures will display similar behavior [2], an observation that has been widely exploited in molecular design studies; and (iii) the manifold variants of the Principle of Least Action [21], such as Le Chatelier's Principle [22], all of which state that, whenever a chemical system undergoes some change, it does so in a way that minimizes the energy expended. This implies that products will be as similar to reactants as possible, consistent with the change that is effected. Table II lists these principles along with other instances of the use of similarity in chemistry over the past two centuries.

2.2. Similarity in scientific laws

The principles included in Table II are but specific examples of laws based on the notion of similarity. What of other scientific laws: are these also similarity-based? The various kinds of laws encountered in chemical contexts have been discussed by Caldin [23], and the laws of physics, such as Newton's laws or the laws of quantum mechanics, have been characterized in inimitable style by Feynman [24]. It is now realized that all scientific laws, whatever their provenance, may be interpreted as statements of similarity. To see why

TABLE II
A compilation of some of the manifold uses of similarity concepts in the chemical domain over the past 200 years

Author(s)	Year	Application of similarity
Richter	1793	Chemical equivalence of species
Proust	1804	Law of constant proportions
Dalton	1808	Classification of atoms
Gay-Lussac	1808	Law of combining volumes of gases
Berzelius	1810	Law of multiple proportions
Avogadro	1811	Avogadro's Law
Mitscherlich	1811	Isomorphism in crystals
Petit & Dulong	1819	Law of specific heats
Wöhler	1828	Behavior of bioactive molecules
Döbereiner	1829	Analogies in behavior of elements
Berzelius	1830	Concept of isomerism
Dumas	1839	Similarity of organic and inorganic species
Berzelius	1845	Concept of allotropy
Gerhardt	1845	Concept of homology
Hofmann	1849	Theory of molecular types
Laurent	1854	Principle of Analogy
Mendeleev	1869	Periodic classification of the elements
Le Bel; van't Hoff	1874	Concept of stereoisomerism
Le Chatelier	1888	Le Chatelier's Principle
Fischer	1894	Complementarity of drug and receptor
Soddy	1813	Concept of isotopy
Ruzicka	1921	Isoprene rule for polyterpenes
Franck; Condon	1925	Franck-Condon Principle
Hückel	1934	Principle of minimum structural change
Rice & Teller	1938	Rice-Teller Principle
Fox	1953	Molecular sequence comparisons
Hammond	1955	Transition state postulate
Hansch & Fujita	1964	Structure-activity correlations
Woodward & Hoffmann	1964	Woodward-Hoffmann Rules
Hine	1977	Principle of least nuclear motion
Ugi *et al.*	1980	Principle of minimum chemical distance
Carbó *et al.*	1980	Molecular charge similarity measure
Mezey	1985	Molecular shape group measure
Cioslowski	1993	Similarity of atoms in molecules

this is so, let up first examine what is meant by a scientific law. Broadly speaking, we may say that a scientific law is a statement about the natural world that satisfies several criteria [25], viz. a law (i) is universally applicable in all regions of spacetime; (ii) can be confirmed by numerous specific instances; (iii) possesses both explanatory and predictive power; (iv) has systematic connections with broader-based theories; and (v) is subject to conditions that may need to be updated from time to time in the light of new knowledge. The kinds of statements made by scientific laws can be illustrated by reference to the laws cited in Table II. Dalton's Law states that all atoms of a given type will display identical properties. This is clearly a statement of similarity, for objects that are behaving identically are certainly similar. The Periodic Law states that elements within specified groups in the Periodic Table will behave similarly, another statement of similarity. The Principle of Analogy is self-evidently a statement of similarity. A little reflection reveals the general pattern: scientific laws tell us that related systems always behave in a similar way, provided that certain prescribed conditions are fulfilled.

2.3. The role of fuzzy logic

Scientific laws make statements about specific categories of objects, and so, before proceeding to the next section, it will be necessary to say something about categories. Having already discussed concepts, we mention here that categories do not differ from concepts in any fundamental way. Indeed, in the psychological community, the two terms are often used interchangeably [26]. As indicated earlier, the construction of categories, i.e. the act of classification, is essential if we are to make sense of the natural world. But, as Bateson [27] has put is, "the division of the perceived universe into parts and wholes is convenient and may be necessary, but no necessity determines how it shall be done." Categories that are quasi-stable are those for which it is possible to specify some defining rule that exclusively identifies all their members; such categories are said to be well-defined [28]. Examples of such categories in chemistry include most groups of compounds that have been designated with a special name, e.g. the alkanes or the organometallics. Categories for which the above does not apply are termed fuzzy categories [29]; these also occur in chemistry – for instance, the category of aromatic compounds. The use of fuzzy set theory [30] or fuzzy logic [31] is appropriate for those situations in which category membership is a matter of degree rather than a certainty or where the boundaries between different categories overlap. Fuzzy sets were introduced by Zadeh [3] in 1965 for the purpose of characterizing systems with such inherent uncertainties. The underlying idea in fuzzy set theory is that each member of the set is neither definitely in nor definitely out but that it belongs to some extent. The assignment of this extent is achieved by associating with each member a function that is a real number on the line segment [0, 1]. This number should not be regarded as the probability that the member belongs but rather as a measure of how true it is that the member belongs to the set.

3. Multifarious similarity

As indicated above, similarity plays a key role in the process of concept formation and in the construction of all kinds of classifications. In fact, similarity manifests itself in the most multifarious of contexts. Because of its interdisciplinary nature – it would be no exaggeration to speak of its universal nature – similarity has been called upon to fulfill many different functions.

In the process of accommodating the concept of similarity to the manifold roles it has had to play, the concept itself has become ever more elaborated. Currently we have not one but numerous different varieties of similarity. This process of differentiation began long ago and was certainly known to the ancient Greeks. Thus, in Aristotle's *Rhetorics* [32] the claim is made that "metaphors are of four kinds; of these the best-liked are the proportional." At the present time, similarity is not only classified into different varieties, but most of the varieties themselves have been further subdivided into subvarieties [2]. We present in Table III a listing of some of the major varieties of similarity. Of the several varieties listed, we shall discuss here analogy, complementarity, equivalence relationships, and scaling. For each we offer a brief description followed by a mention of certain of its more significant chemical applications. After shedding some light on these basic varieties, we shall turn our spotlight with full intensity on analogy. This will enable us not only to examine the various subvarieties of analogy that have been defined [2], but also to explore in some depth the remarkable role that analogy plays in chemistry.

TABLE III

A compilation of some of the many varieties of similarity that exist with an indication for each of the number of attributes in common and the extent of the mapping between the original system and its transform

Type of similarity	Common attributes	Relations mapped
Abstraction	Several	Several
Allegory	Few	Few
Allometry	Several	Several
Analogy	Few	Many
Anomaly	Few	Few
Chirality	Many	All
Complementarity	Few	Many
Identity	All	All
Isomerism	Several	Several
Isomorphism	Several	Several
Matching	Few	Several
Metaphor	Few	Few
Modeling	Many	Many
Non-Identity	None	None
Scaling	Several	Many
Self-Similarity	Several	Many
Simile	Few	Few
Similitude	Several	Many
Simulation	Many	Many
Topological Transformation	Few	Many

3.1. ANALOGY

Analogy is a kind of similarity that involves the description and interpretation of some complex system in terms of one or more simpler systems which are supposedly well understood. An example that immediately springs to mind is that of the atom which has been likened to a tiny solar system. Analogy operates by altering the context of our perception and thus makes it possible for us to comprehend aspects of the natural world that seem obscure or puzzling. The same role is played by metaphor which is essentially a more poetic expression of analogy. The interpretation of one system in terms of some other system clearly implies that there must exist correspondences between the systems, i.e. that parts of the systems can be mapped on to one another. The various kinds of mappings that can be made are investigated in detail in the following section. Suffice it to point out here the great fruitfulness of the many analogies that have been drawn in the past; without them, chemistry and science in general could not have grown and flourished [2, 33]. Some instances of the more successful analogies that have been made in the domain of chemistry are collected together in Table IV. More recent instances of the use of analogy in chemistry include a study by Wold and Sjöström [34] of ^{13}C nuclear magnetic resonance data to determine whether patterns were present that could be exploited in the assignment of NMR spectra to both exo and endo molecular structures, the development by Wipke and Hahn [35] of a computer program capable of automatically invoking analogy in the construction of

TABLE IV
A compilation of some of the analogies that have been adopted to interpret the behavior of physical and chemical systems

Proponent	Year	Description of proposed analogy
Democritus	ca. 420 B.C.	Atoms are like tiny particles that differ in both shape and size
Boyle	1660	Gases consist of particles that behave like miniature springs
Newton	1718	Atoms are like hard, impenetrable movable spheres
Bernoulli	1738	Gases behave like billiard balls in a box
Black	1803	The flow of heat through a system resembles the flow of water
Dalton	1804	Molecules are analogous to hard spheres connected together by sticks
Davy	1806	Electrolytes behave like hard charged spheres
Kekulé	1865	The structure of benzene is like that of a snake biting its tail
Baeyer	1885	Molecules are analogous to hard spheres connected together by springs which may contain strain energy
van't Hoff	1888	Dilute solutions behave like gases
Fischer	1894	A drug at a biological receptor behaves like a key turning a lock
Thomson	1899	Electrons collectively behave like clouds
Nagaoka	1904	Atoms resemble planar solar systems
Rutherfold	1911	Atoms resemble three-dimensional solar systems
Bohr	1939	The atomic nucleus behaves like a drop of liquid
Rouse	1953	Long polymer molecules behave like beads on a spring
Lutz & Schmidt	1992	Chemical cluster molecules behave like the atomic nucleus

minimum energy molecular models, and a method of estimating the properties of molecules from their structure and of designating which component of the molecule could be said to be responsible for the manifestation of a specify property [36, 37].

3.2. Complementarity

Complementarity is a kind of similarity that involves a certain kind of matching up of two distinct entities: the two entities fit or mesh together in such a way that they form a new composite whole. An example is provided by the fitting together of building blocks: when the fit is good the blocks are said to possess a high degree of complementarity (see Figure 2). Knowledge of this kind of similarity is very old, and once again it was well known to the ancient Greeks. For instance, Democritus [38] put forward the idea that atoms were possessed of differing shapes and that the various ways in which they could be fitted together explained the observed properties of bulk matter. In the last century this idea was updated by Fischer [39] who advanced the lock and key model to explain the physiological action of drugs. His idea was that the drug molecule slots into an appropriate biological receptor just as a key fits into and actuates a lock. The better the fit, the greater the efficacy of the drug would be. This idea has been progressively honed in recent years and it is now thought that optimal bioactivity in a drug occurs when there is complementarity in size, shape, functional groups and electrostatic charges [40]. The new field that has emerged from such studies is commonly referred to as molecular recognition research [41–43], and such research is now playing an increasingly important part in the design of new drugs [44, 45]. Other workers have investigated molecular recognition at crystal interfaces [46] and the matching up of different polymeric species in solution [47]. The notion of complementary, though in a more figurative sense, is also used widely in quantum chemistry. Here the term refers to the coexistence of both corpuscular

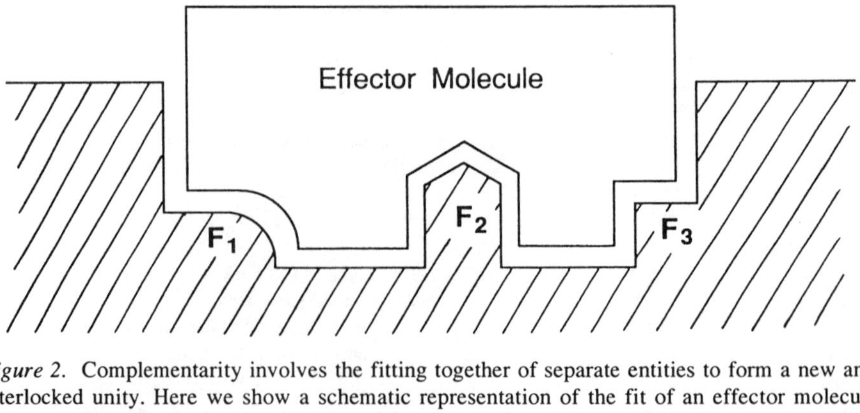

Figure 2. Complementarity involves the fitting together of separate entities to form a new and interlocked unity. Here we show a schematic representation of the fit of an effector molecule at a biological receptor.

and wavelike aspects of fundamental particles rather than to the fitting together of components [48]. This latter notion of complementarity, advocated powerfully by Bohr [49], can also be given a rigorous mathematical formulation [48].

3.3. Equivalence relations

The equivalence relation is essentially a rule that enables us to partition sets of objects into mutually disjoint subsets. It provides the basis for all rigorous classifications and in many ways can be viewed as a prototype for the various manifestations of similarity in the world. Certain criteria need to be satisfied if a relationship is to meet the requirements of an equivalence relationship. Suppose that we have a set of objects $\mathbf{C} = \{c_1, c_2, \ldots, c_n\}$ that we wish to partition into equivalence classes. To do this we need to apply a relation $\mathbf{R}$ that satisfies the criteria of being (i) reflexive, i.e. the relationship $c_x\mathbf{R}c_y$ holds for all $c_n \in \mathbf{C}$; (ii) symmetric, i.e. if the relationship $c_x\mathbf{R}c_y$ holds then $c_y\mathbf{R}c_x$ must also hold for all $c_n \in \mathbf{C}$; and (iii) transitive, i.e. if the relationships $c_x\mathbf{R}c_y$ and $c_y\mathbf{R}c_z$ hold, then the relationship $c_x\mathbf{R}c_z$ must also hold for all $c_n \in \mathbf{C}$. In the realm of chemistry, examples of equivalence relationships abound because any classification of chemical entities into exclusive sets affords an instance of the operation of an equivalence relation. In the case of atoms, equivalence relations partition all naturally occurring atoms as well as artificially produced isotopes in element classes based on the atomic number. The Periodic Table, probably the greatest classification ever made in chemistry, uses equivalence relations to assign each element unambiguously to a group within the table. As we discuss later, the ramifications of this particular classification have been enormous. In the case of molecules, the equivalence relation has been relied upon in numerous instances to classify molecules. For example, molecules may be classified according to some specific constituent that they contain and this gives rise to a very large number of molecular sets, such as the sets comprising the members of various homologous series, the set of radioactively labeled molecules, or the set of chiral molecules. The history of the use of equivalence relationships has been explored more fully elsewhere [50].

3.4. Scaling

Scaling is a kind of similarity that can be expressed in at least two quite different ways. The first of these, which we might more accurately refer to as scale modeling, involves a substantially magnified or reduced reproduction of some structure or system of interest. Scale models need not reproduce all the features of the system but may instead reproduce only key parts of the system. The basic idea behind scale modeling is that systems of dimensions such that they are inaccessible or inconvenient for human investigation can be made accessible by appropriate scaling up or down. The philosophy behind scaling was put forward by Spinoza [51] who declared that "if the parts composing an individual become greater or less, but in such proportion that they preserve the same mutual relations . . . , the individual will still preserve its original nature." Early in the present century this reasoning was formalized into the so-called Principle of Similitude [52], which states that "the fundamental entities out of which the physical universe is constructed are of such a nature that from them a miniature universe could be constructed exactly similar in every respect to the present universe." We now know however that this principle is of limited validity in that it cannot be reliably extended over several orders of size magnitude. For instance, it is known that tiny quantum systems cannot be described in terms of miniaturized macrosystems; the laws for the two systems are quite different [53]. This observation throws into doubt the reliability of the ball-and-spring models that are widely used to model molecular species. Even scaling up has its problems, for we know that a small-scale laboratory preparation of some chemical can differ significantly from full-scale industrial production of the same chemical [54]. The principle has proven its value in one specific area however. Because of the existence of dimensionally invariant laws in the physical sciences [55], it has often been used to determine the form assumed by functional relations that connect physical magnitudes [56, 57].

3.5. Self-Similarity

The remaining kind of scaling similarity that we discuss is known as self-similarity. This particular kind of similarity came into vogue during the past decade following the appearance of a popular book on chaos and fractals by Gleick [58] and a now classic textbook on fractals by Mandelbrot [59]. The latter work revealed that most of the objects in the natural world could be described as fractal objects, that is to say, although they may appear to be irregular, fragmented or even tangled, they nevertheless exhibit a hidden invariance described as self-similarity. This invariance becomes apparent whenever fractal objects are magnified in size, for upon magnification it is found that the overall geometric features of such objects remain the same (see Figure 3). This scale invariance typically holds for progressive dilation of the structures until roughly atomic features are reached. A common way to characterize this kind of invariance is via the fractal dimensionality of the object in question. The fractal dimensionality, d_f, is an exponent that relates the mass of a fractal object, M, to its size, Q, through the equation:

$$M \sim Q^{d_f}$$

Fractal dimensionalities, which typically assume values in the range $1 \leq d_f \leq 3$, should not be interpreted as geometrical or euclidean dimensions. They are to be understood rather as a measure of the kinkiness or tortuousness of an object. It is for this reason that a one-dimensional object such as a highly twisted wire can have a fractal dimensionality that exceeds two [60]. Chemists have investigated the fractal dimensionalities of many large

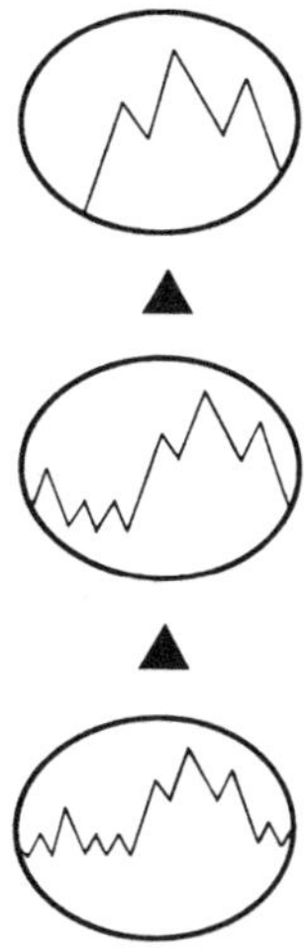

Figure 3. Illustration of imaginary surface contours of some solid shown at different magnifications. Note that the overall geometric features remain essentially the same because the surface is self-similar.

molecules, including protein molecules [61] and long-chain alkanes [62]. Also studied have been a variety of molecular surfaces, including those of complex morphology [63] and reactive surfaces [64]. Other applications have embraced the study of fractal reaction kinetics [64], fractal powders and clusters [65], and even fractal foodstuffs [66].

4. Analogical similarity

Having portrayed several of the major varieties of similarity along with their chemical applications, we now shift the focus of our attention to what is arguably the most important variety of all: analogy. We are convinced of the special importance of analogy for a number of reasons. In the first place we would mention that analogy and the various roles it plays have been appreciated for thousands of years; as we pointed out earlier, it was certainly well known to the ancient Greeks [32]. In much more recent times, psychological researchers have become increasingly aware of just how fundamental analogical thinking is in our cognitive processes [68]. The current involvement with analogy would appear to derive from the fact that analogical thought processes are typical of the kind of reasoning that is employed when we formulate concepts or construct classifications. Since the whole of science rests upon meaningful concepts and classifications, it is evident that without analogy there would be no science. In the specific instance of chemistry, there is wide acknowledgement that analogical reasoning has been indispensable for the advancement of the discipline [33, 69]. As is apparent from Table IV among others, many of the great breakthroughs in chemistry have depended on an appropriate use of analogy. This has been so because analogy has the effect of altering the context of our perception of reality. This change in our perspective is often enough to afford us with new insights into the workings of nature and this can in turn render seemingly intractable problems soluble. We now consider what some of the more famous chemists of the past had to say on the subject of analogy.

Joseph Priestley, the discoverer of oxygen, was of the opinion [70] that "if we could content ourselves with the bare knowledge of new facts, and suspend our judgment with respect to their causes, till, by their analogy, we were led to the discovery of more facts, of a similar nature, we should be in a much surer way to the attainment of real knowledge." The Principle of Analogy was put forward by the French chemist Auguste Laurent (who in 1854 drew the first formula for benzene) in the following terms [20]: when two analogous bodies are submitted to the same metamorphoses, the new products will still be analogous. This principle was drawn up after Laurent's extensive work on homologous series of compounds. The physical chemist Jacobus van't Hoff acknowledged his reliance on analogy in developing his kinetic theory of osmotic pressure in dilute solutions when he stated [71] that he was "not dealing with a fanciful analogy, but with one that is fundamental; for the mechanism which, according to our present conceptions, produces gaseous pressure, and in solutions osmotic pressure, is essentially the same." In publishing his theory of ring strain in organic molecules, Adolf Baeyer did not specifically mention the mechanical analogy he was using, but hoped [72] that "the simple relationship established between the explosivity of acetylene compounds and the laws of ring closure will help experts overcome their rejection of such speculations and recognize the necessity for the further development of our discipline in this direction." Not everyone agreed however. Walther Hückel responded [73] by proclaiming that "chemical events have no mechanical analogies; the laws that are valid for atomic and molecular structures . . . are quite different from those that exist in the dimension of the action of a great number of atoms and molecules."

4.1. The types of analogy

Although analogy has clearly had a massively important part to play in the past, as Hückel [73] has warned, analogies certainly have limitations and may not always be valid. We now come to a consideration of the modern usage of analogy, which has been described by one worker [74] as "essential in stimulating research and in scientific creativity". In our discussion we shall not only explore ways in which analogies can be profitably exploited but also attempt to understand the constraints to which they must be subject if they are to be used legitimately. We start by enumerating the various kinds of analogy that might be used. As there is no general agreement at present on the way in which the differing kinds of analogy should be defined, we shall follow the lead of Sarlemijn and Kroes [75], who argued that analogy could be divided up into five different varieties, at least in a scientific or technological setting. The five varieties (in alphabetical order) are: (i) attributive analogy; (ii) functional analogy; (iii) inductive analogy; (iv) proportional analogy; and (v) structural analogy. After giving a working definition for each of these varieties, we go on to consider the applications of each in the chemical domain. Our emphasis will fall on the first three of the varieties as these will be seen to be of special relevance to chemists. We conclude on a somewhat philosophical note by speculating on the possible future role that analogy might have in chemical research.

4.2. Attributive analogy

To define this type of analogy, we start by imagining two objects, A and B, that have the respective attributes or properties α and β. When there is some transference between the property α and the property β, or when α represents β, the two properties are said to be attributively analogous. An example is provided by a thermometer that is used to determine the boiling temperature of a liquid. There exists an attributive analogy between the length of the mercury column of the thermometer and the boiling point of the liquid. An important practical application of this kind of analogy occurs in the study of quantum-mechanical systems. Such systems often exhibit analogies to classical systems, e.g. the canonical momentum of a classical particle is represented by the quantum-mechanical operator $(h/2\pi i)\partial/\partial x$. Two systems, each with a distinct set of properties and some transference between the properties, is clearly a situation in which the attributive analogy might be applied. Because the analysis of quantum-mechanical systems with traditional statistical procedures leads to a number of paradoxes, such as collapsing wave functions [76], Schrödinger's cat problem [77], and the Einstein-Podolsky-Rosen paradox [78], many attempts have been made to improve matters by changing the underlying analogies that are used for these systems [79]. One such attempt by Youssef [80] maintains that quantum-mechanical systems are analogous to Bayesian systems, i.e. they operate on the principles of Bayesian statistics [81] rather than traditional statistics [82]. Using this analogy, he reformulated quantum mechanics in such a way that all the predictions of standard quantum theory were preserved while at the same time the paradoxes were satisfactorily explained. He showed that the paradoxes arise because the wave function is mistaken for the state of the system. In Bayesian analyses amplitudes do not represent system states but only the best estimate of certain truths about the system [80]. Bayesian approaches also appear to offer advantages when applied to classical systems, and have included the derivation of a theoretical formula for the diffusion coefficient in gases, liquids and solids [83], improved analysis of nuclear magnetic resonance signals [84], and a determination of the thermodynamic efficiency of muscular systems in various animal species [83].

4.3. Functional analogy

The functional analogy relates either to different objects A and B that have the same function or to different systems $\mathfrak{A}$ and $\mathfrak{B}$ in which one or more of the component parts play the same role. An example of the first kind is provided by the many different types of molecule that can be used as explosives under appropriate conditions, e.g. mercuric fulminate and trinitrotoluene. The second kind of functional analogy is encountered when different molecules have similar surface protuberances or other features that make it possible for them to fit into the same biological receptor. This theme has already been addressed in our Section 3.2 on complementarity and the methods of modeling this type of interaction have recently been reviewed [85]. Both the functional analogy, and the inductive analogy we discuss below, play important roles in the field of molecular design. This field now embraces not only the design of pharmaceutical drugs but also agrochemicals, new fuels, and replacement body fluids such as artificial blood [3]. As such, this field is of considerable import in the commercial and industrial life of many nations.

Because the objective in molecular design studies is to identify molecular structures that mimic or supersede molecules already known to be effective in some specific application, many different characteristic quantities have been pressed into service in the study of molecular similarity. These have included (i) measured physicochemical properties of the bulk material, e.g. refractive index or octanol-water partition coefficient; (ii) theoretical parameters derived from the chemical graph of the molecule, e.g. topological indices or complexity measures; (iii) quantum-chemical parameters calculated for parts of the molecule, e.g. charge densities or HOMO-LUMO separations; (iv) thermodynamic parameters, e.g. heats of formation or entropies, and (v) theoretical measures of structural similarity, e.g. the Tanimoto coefficient [86] or the charge similarity index of Carbó *et al.* [87]. To gain an idea of the overall similarity of two molecules, resort is often made to some distance-based similarity coefficient [86]. This is calculated by plotting the various quantities referred to above in some appropriate multidimensional space and then determining their distance apart for each of the spatial dimensions. The coefficient is now readily obtainable, for distance-based coefficients commonly assume the form of a Minkowski metric [86], that is to say, for the pair of points $x = (x_1, x_2, \ldots, x_m)$ and $y = (y_1, y_2, \ldots, y_m)$ the metric has the form:

$$d(x, y) = \sum_{i=1}^{m} [|x_i - y_i|^r]^{1/r},$$

where $d(x, y)$ is the distance between the points and m is the dimension of the space. When the variable $r = 1$, the so-called city-block metric is obtained, whereas when $r = 2$ the more familiar Euclidean metric results [88]. A proximity measure is a metric if it satisfies the conditions: (i) $d(x, y) = 0$ when $x = y$; (ii) $d(x, y) = d(y, x)$; and (iii) $d(x, z) \leq d(x, y) + d(y, z)$. The mathematical spaces used in these and related applications have been reviewed by Johnson [89]. Proximity measures tell us how dissimilar two molecules are, i.e. the larger the value of the measure the lower the similarity. When such measures are normalized to unity, the similarity is given by the expression:

$$\text{Similarity} = 1 - \text{Dissimilarity}.$$

4.4. Inductive analogy

Of even greater importance in molecular design studies and in chemistry in general is the inductive analogy. This analogy is defined for a set of objects A, B, C, D, etc., all of which possess the attributes P and Q. Now, if the objects A, B, and C are found to have the attribute R, by inductive analogy we may reason that object D will probably also exhibit attribute R. An instance of this analogy is provided by the set of penicillin molecules that are known to be bioactive and nontoxic. Inductive analogy leads us to conclude that any new penicillin molecule we may synthesize that is shown to be bioactive will also prove to be nontoxic. This type of analogy takes us one step further than the functional analogy in that new, specific analogies can now be inferred in addition to the broad, general statements furnished by functional analogy. To illustrate this, we present in Figure 4 a schematic representation of inductive analogy. This figure shows two systems (e.g. molecules), 1 and 2, that are comprised of the respective components (e.g. atoms) A_1 through G_1 and A_2 through G_2. Some of the components in system 1 can be mapped in a one-to-one mapping on to components in system 2 as indicated by the arrows. Those components that can be mapped in this way are said to constitute the positive analogy; those that cannot be so mapped form the negative analogy. Because most of the components can be mapped in the case illustrated, the positive analogy between the two systems is clearly large. For such systems, it is commonly assumed that the behavioral characteristics of the two systems will be similar. Thus, if system 1 exhibits some specific kind of behavior α and the equivalent behavior for system 2 is β, then by inductive analogy we may reason that α and β will be similar. This type of reasoning forms the

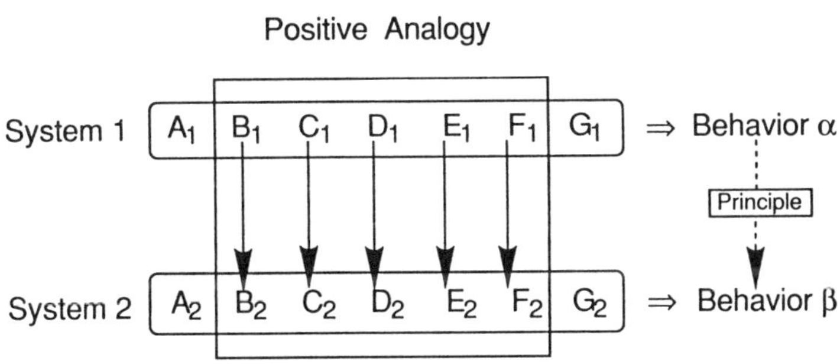

Figure 4. A schematic representation of an inductive analogy that is drawn between a base system and a target system. The behavior of the target is predicted by use of the Principle of Analogy.

basis on which most molecular design research rests. The inductive analogy has been exploited by numerous workers; specific instances of its use were cited earlier in our discussion of analogy [34–37].

The Periodic Table provides a particularly instructive illustration of the use of inductive analogies. Early work by Döbereiner [90] in 1829 revealed that triads of elements, such as {calcium, strontium and barium} or {chlorine, bromine and iodine}, had analogous properties with the middle element exhibiting roughly mean behavior with respect to the two outer elements. In due course, numerous other analogical relationships were established, such as the Law of Tetrads [91] and the Law of Octaves [92] which related together groups of four and eight elements respectively. Upon construction of the Periodic Table in 1869, Mendeleev [18] was able to predict the existence of elements unknown at the time purely on the basis of inductive analogy; most of these elements were subsequently discovered [93]. Such predictions did not stop with the work of Mendeleev and in fact are still being made at the present time, though the predictions nowadays usually concern the nature of compounds formed by the elements. Recent examples of this modern usage of the Period Table include those of Haas [94] who devised a periodic system for functional groups, Godovikov and Jariya [95] who developed a mineralogical-crystallochemical classification of the elements based on their ionization potential and ionic radius, and Allen [96] who introduced the average one-electron valence shell energy of ground-state free atoms as a third factor for the classification of elements in the Periodic Table. The Table has also been widely used to predict the properties of many different kinds of materials, recent instances being the prediction of the behavior of functional glasses formed from oxides and halides [97], the properties of a variety of metal alloys [98], and the thermodynamic functions of inorganic compounds [99]. Novel kinds of Periodic Table have become popular over the last two decades. These classify compounds rather than elements and so the number of atoms per entry in the Table varies according to the size of the molecule being classified. Examples of molecular Periodic Tables are those of Dias [100] for polycyclic aromatic hydrocarbon species (see Figure 5), Kong [101] for diatomic species, and Hefferlin *et al.* [102] for molecular species and clusters containing n atoms with $1 \leq n \leq 4$. According to Dias [103] the three criteria for a set of chemical species to form a Periodic Table are that (i) the set is partially ordered, i.e. the species satisfy a relation R that is reflexive, antisymmetric and transitive (see Section 3.3 for an explanation of these terms); (ii) the Table is two-dimensional and of infinite extent; and (iii) the species satisfy the triad principle, i.e. any central species has a metric property that is the arithmetic mean of the two oppositely adjacent species. The structure of Periodic Tables has also been explored in some detail by Hefferlin [104].

Forbidden Region — $N_c/N_H > 3$ — Least Reactive — Most Reactive — $N_c/N_H < 2$ — $2 < N_c/N_H < 3$ — Intermediate Reactive

$N_c=3N_H+4$	$N_c=3N_H+2$	$N_c=3N_H$	$N_c=3N_H-2$	$N_c=3N_H-4$	$N_c=3N_H-6$	$N_c=3N_H-8$	$N_c=3N_H-10$	$N_c=3N_H-12$	$N_c=3N_H-14$	$N_c=3N_H-16$	$N_c=3N_H-18$	$N_c=3N_H-20$	$(N_c=2N_H-6+N_{Ic})$
$C_{64}H_{20}$	$C_{62}H_{20}$	$C_{00}H_{00}$	$C_{52}H_{18}$	$C_{50}H_{18}$	$C_{42}H_{16}$	$C_{40}H_{16}$	$C_{32}H_{14}$	$C_{24}H_{12}$	$C_{10}H_8$	$C_{14}H_{10}$	$C_{18}H_{12}$	$C_{22}H_{14}$	$N_c=2N_H-6$
$C_{70}H_{22}$	$C_{68}H_{22}$	$C_{54}H_{18}$	$C_{58}H_{20}$	$C_{56}H_{20}$	$C_{48}H_{18}$	$C_{46}H_{18}$	$C_{38}H_{16}$	$C_{30}H_{14}$	$C_{16}H_{10}$	$C_{20}H_{12}$	$C_{24}H_{14}$	$C_{28}H_{16}$	$N_c=2N_H-4$
...	...	$C_{60}H_{20}$	$C_{64}H_{22}$	$C_{62}H_{22}$	$C_{54}H_{20}$	$C_{52}H_{20}$	$C_{44}H_{18}$	$C_{36}H_{16}$	$C_{22}H_{12}$	$C_{26}H_{14}$	$C_{30}H_{16}$	$C_{34}H_{18}$	$N_c=2N_H-2$
		$C_{66}H_{22}$	...	...	$C_{60}H_{22}$	$C_{58}H_{22}$	$C_{50}H_{20}$	$C_{42}H_{18}$	$C_{28}H_{14}$	$C_{32}H_{16}$	$C_{36}H_{18}$	$C_{40}H_{20}$	$N_c=2N_H$
		...			...	...	$C_{56}H_{22}$	$C_{48}H_{20}$	$C_{34}H_{16}$	$C_{38}H_{18}$	$C_{42}H_{20}$	$C_{46}H_{22}$	$N_c=2N_H+2$
							...	$C_{54}H_{22}$	$C_{40}H_{18}$	$C_{44}H_{20}$	$C_{48}H_{22}$	$C_{52}H_{24}$	$N_c=2N_H+4$
								$C_{60}H_{24}$	$C_{46}H_{20}$	$C_{50}H_{22}$	$C_{54}H_{24}$	$C_{58}H_{26}$	$N_c=2N_H+6$
								...	$C_{52}H_{22}$	$C_{56}H_{24}$	$C_{60}H_{26}$	$C_{64}H_{28}$	$N_c=2N_H+8$
									$C_{58}H_{24}$	$C_{62}H_{26}$	$C_{66}H_{28}$	...	$N_c=2N_H+10$
									$C_{64}H_{26}$	$C_{68}H_{28}$	...		$N_c=2N_H+12$
									...	...			$N_c=2N_H+14$
													$N_c=2N_H+16$
													$N_c=2N_H+18$
													$N_c=2N_H+20$
													$N_c=2N_H+22$
													$N_c=2N_H+24$
													$N_c=2N_H+26$
													...

$N_c=3N_H-22$	$N_c=3N_H-24$	$N_c=3N_H-26$	$N_c=3N_H-28$	$N_c=3N_H-30$	$N_c=3N_H-32$	$N_c=3N_H-34$	$N_c=3N_H-36$	$N_c=3N_H-38$	$N_c=3N_H-40$	$N_c=3N_H-42$	$N_c=3N_H-44$	...	
$C_{26}H_{16}$	$C_{30}H_{18}$	$C_{34}H_{20}$	$C_{38}H_{22}$	$C_{42}H_{24}$	$C_{46}H_{26}$	$C_{50}H_{28}$	$C_{54}H_{30}$	$C_{58}H_{32}$	$C_{62}H_{34}$	$C_{66}H_{36}$	$C_{70}H_{38}$	...	$N_c=2N_H-6$
$C_{32}H_{18}$	$C_{36}H_{20}$	$C_{40}H_{22}$	$C_{44}H_{24}$	$C_{48}H_{26}$	$C_{52}H_{28}$	$C_{56}H_{30}$	$C_{60}H_{32}$	$C_{64}H_{34}$	$C_{68}H_{36}$	$C_{72}H_{38}$	...	...	$N_c=2N_H-4$
$C_{38}H_{20}$	$C_{42}H_{22}$	$C_{46}H_{24}$	$C_{50}H_{26}$	$C_{54}H_{28}$	$C_{58}H_{30}$	$C_{62}H_{32}$	$C_{66}H_{34}$	...	...	...		...	$N_c=2N_H-2$
$C_{44}H_{22}$	$C_{48}H_{24}$	$C_{52}H_{26}$	$C_{56}H_{28}$	$C_{60}H_{30}$	$C_{64}H_{32}$	$C_{68}H_{34}$	$C_{72}H_{36}$					...	$N_c=2N_H$
$C_{50}H_{24}$	$C_{54}H_{26}$	$C_{58}H_{28}$	$C_{62}H_{30}$	$C_{66}H_{32}$	...	...	...					...	$N_c=2N_H+2$
$C_{56}H_{26}$	$C_{60}H_{28}$	$C_{64}H_{30}$	...	...								...	$N_c=2N_H+4$
$C_{62}H_{28}$	$C_{66}H_{30}$	...										...	$N_c=2N_H+6$
$C_{68}H_{30}$	...											...	$N_c=2N_H+8$
...												...	$N_c=2N_H+10$
													...

$2 < N_c/N_H < 3$

Figure 5. A Periodic Table developed by Dias that has for its entries benzenoid polycyclic aromatic hydrocarbons. This table dates from 1985.

4.5. Other analogy types

Although the remaining two types of analogy referred to in our list, namely proportional and structural analogies, are not of immediate interest to us here, we mention them briefly for the sake of completeness of our coverage of analogy. A proportional analogy refers to numbers that are in the same ratio or to properties of objects that are commensurate. For instance, if the objects A, B, C and D have the respective attributes α, β, γ and δ, a proportional analogy between the latter is expressed by the notation $\alpha : \beta :: \gamma : \delta$ which is read as "α is to β as γ is to δ". An example from the chemical context is: an atom is to an element as a molecule is to a compound. This type of analogy was used primarily by ancient Greek philosophers and is now largely of historical interest only; a full account of its early use has been offered by Szabó [105]. The structural analogy refers to two objects or systems, one of which is a scale model of the other. The relationship between them is that both can be described in terms of the same set of mathematical equations. This type of analogy has already been discussed in our Section 3.4 on scaling, where it was pointed out that such analogy cannot be reliably extended over several orders of size magnitude. Scaling is legitimate only when a relatively small number of orders of magnitude is involved, that is to say, the Principle of Similitude [52] is always subject to severe restrictions on its domain of validity.

5. Concluding similarity

In this survey on similarity and the role that it plays in chemistry, we have endeavored to discuss the nature of similarity per se, explored the differing varieties of similarity that exist, and have also identified numerous applications of similarity in the realm of chemistry. Several of the manifold varieties of similarity have been listed in our Table III. Whatever variety of similarity we choose to focus on, use of the concept inevitably brings in its wake certain advantages and disadvantages. Some of the advantages include the fact that similarity is a universally familiar and instantly recognizable abstraction experienced by all observers, the knowledge that similarity can be quantified at least in relative terms, and the realization that similarity can have the effect of transforming our perception of the natural world, thereby making it possible for us to interpret aspects of our world in a new light. Some of the disadvantages inherent in the use of the concept are that it is elusive, subjective, and thus hard to define, that similarity is never an absolute magnitude but is always a relative one, and that some degree of arbitrariness is always associated with the concept. The advantages of using similarity concepts, however, outweigh the disadvantages and for this reason, if for no other, we shall likely witness an extensive and ever growing reliance on similarity-based methods in the future. In fact, it is probably no exaggeration to assert that the progress of science as a whole will be closely linked to the multifarious ways in which similarity notions are exploited in the future. The notion that progress in science depends on the continual reformulation of the similarity classes employed to systematize the natural world is not a new one [106]. In recent years this notion has begun to be adopted within the context of neurocomputing [16].

At this point we might ask whether any particular type of similarity has characteristics that distinguish it from all the others and thus make it in some sense preferred over the others. Although there can be no doubt that all of the varieties of similarity have their respective roles to play, we incline to the view that analogy occupies a special place in the similarity pantheon. We have four main reasons for this viewpoint. First, the notion of analogy in one form or another has been with us for a very long time, the word itself deriving from the Greek *analogia*, meaning proportion. Indeed, Plato declared [107] that reasoning by analogy was the "most beautiful form of argument." It seems likely that any concept that has survived for so long must have features that make it intrinsically valuable. A powerful predilection for analogical reasoning appears to be an innate part of human cognition [108]. Second, we would mention the role that analogy plays in the founding of new scientific theories, a topic that has been examined in some depth by Hesse [109]. We have given some indication of how vital analogies can be in the development of scientific theories in our Table IV; many of the analogies exploited in the domain of physics have been more fully discussed by Shive and Weber [110]. Third, analogy plays an important part in solving scientific problems. The basic idea here is that, by the use of an appropriate analogy, a solution may be found to an otherwise intractable problem. Thus, by initially moving away from a problem and viewing it in a new light we may ultimately come to a solution via the use of analogical inference. The ways in which we select analogies in such problem solving have been investigated by Reed *et al.* [111]. Finally, analogy – especially in the form of inductive analogy – can be utilized to make predictions about the behavior of many different kinds of systems. In our Section 4.4, for instance, we showed how analogy could be used to predict the properties of both elements (in the Periodic Table) and compounds (in the design of new molecules such as drugs). In fact, it would not have been possible to construct the Periodic Table at all without the application of inductive analogy.

Having selected analogy as our choice variety of similarity, we come now to the question whether analogical reasoning is valid. In spite of the numerous successes achieved over many centuries, not all observers are convinced of the legitimacy of using analogy in scientific arguments. For instance, one philosopher has gone as far as to claim that "arguments from analogy may be fertile but they are all invalid" [112]. This prompts us to ask if there are circumstances in which such reasoning might be considered legitimate. We have to admit from the outset that there is no logical basis for demonstrating the validity of inductive analogical arguments. The best that can be said in their support is that they are often spectacularly successful in practice. Weitzenfeld [113] carried out a study on the validity of such reasoning and concluded that the patterns of reasoning varied greatly from field to field and so the amount of commonality present was indeterminate. Clement [114] suggested that individuals resort to three different stratagems when generating analogies, namely (i) the use of some general principle; (ii) an association of ideas; and (iii) a transformation of some aspect of the original problem. The latter two were found to occur frequently whereas reasoning from some general principle was observed only rarely. Gick and Holyoak [115] presented experimental proof that analogical thinking does indeed play a role in creative problem solving and the analogies are sometimes sought in domains that are very far removed from the context of a given problem.

Clearly what is needed to make analogical reasoning effective is a set of guidelines or constraints that can be applied to the mappings from the base to the target system. In terms of our Figure 4, that is to say that some of the mappings in the positive analogy are allowed whereas others are not. How are we to decide which mappings are allowed? Although we do not yet have the final answer, a considerable amount of work carried out in the psychological sphere points the way forward. Three major approaches to the resolution of the problem have been advanced, namely those of (i) Holyoak [116], who proposed that mappings be selected on the basis of their relevance to the ultimate goal; (ii) Ortony [117], who suggested mappings be used that display so-called salience imbalance, that is to say highly salient features in the base are mapped on to minimally salient features in the target, and (iii) Clement and Gentner [118], who advocated what they termed a systematicity principle, which means that among the set of common relations only those that participate in common systems of relations, i.e. those relations that are linked to a larger matching system, are mapped. The latter approach would appear to be the most promising at present as there is evidence [118] that judgments about whether an analogy might yield justifiable inferences are positively related to the systematicity of the mapping. This seems to confirm the insight of Rosen [74], who viewed analogy as the invariance of a statement or relation under changes of the elements involved in it.

Arguments based on analogical reasoning can afford powerful new insights, may provide no significant understanding of the systems under study, or may even be quite misleading. The analogy that we drew in our opening paragraph is of the second kind. There we pointed out that the differing kinds of similarity may be liked to horses that are participating in a horse race. This analogy is not particularly enlightening as it does not lead to any profound new understanding of either similarity or our physical world. It did, however, enable us to make the claim that analogy is ahead of the other varieties of similarity in terms of its overall usefulness and ubiquity – a claim that we trust has been duly established in our foregoing discourse. The other two possible outcomes of analogical reasoning, namely that they yield powerful new insights or mislead us are clearly extremes and suggest that a subtle and judicious approach to the use of analogy may be called for. As North [119] would have it, "analogy is the basis for much scientific conjecture, but even conjecture is an art which can be done well . . . even though it might prove in the end to have yielded a false conclusion." Or, as Bunge [112] more graphically put it, "analogy is undoubtedly prolific, but it gives birth to as many monsters as healthy babies." To cut down on the number of monsters generated, some workers [12] have now gone as far as to automate the entire process of optimized analogical reasoning with the use of appropriate constraints. But, how ever artistic and skillful our use of analogy may be, the latter should never be confused with the real thing. For the mapping is not the terrain and the analogy is not the reality.

Acknowledgment

The author thanks Dr. Ramon Carbó of the University of Girona, Spain for financial assistance, which made possible the oral delivery in S'Agaró, Spain of an early and much shorter version of the present chapter.

References

1. Rouvray, D. H., *J. Chem. Inf. Comput. Sci.* **32**, 580 (1992).
2. Rouvray, D. H., *J. Chem. Inf. Comput. Sci.* **34**, 446 (1994).
3. Rouvray, D. H., *New Sci.* **138**, 35 (1993).
4. Johnson, M. A. and Maggiora, G. M. (eds), *Concepts and Applications of Molecular Similarity* (Wiley-Interscience, New York, 1990).
5. Tversky, A., *Psychol. Rev.* **84**, 327 (1977).
6. Davies, G. M., Shepherd, J. W., and Ellis, H. D., *Am. J. Psychol.* **92**, 507 (1979).
7. Rosch, E. and Mervis, C. B., *Cognit. Psychol.* **7**, 573 (1975).
8. Oden, G. C. and Lopes, L. L., in *Fuzzy Set and Possibility Theory: Recent Developments*, R. R. Yager (ed.), Pergamon, New York (1982), p. 75.
9. Medin, D. L., *Am. Psychol* **44**, 1469 (1989).
10. Hampton, J. A., *Memory and Cognition* **9**, 149 (1981).
11. Mervis, C. B. and Pani, J. R., *Cognit. Psychol.* **12**, 496 (1980).
12. Dunbar, R., *New Sci.* **121**, 48 (1989).
13. Faith, D. P., *J. Classification* **3**, 257 (1986).
14. Hubert, L. J. and Baker, F. B., *J. Math. Psychol.* **16**, 233 (1977).
15. Omohundro, J., *Memory and Cognition* **9**, 324 (1981).
16. Bailey, D., Thompson, D., and Feinstein, J., *PC AI* **2**, 29 (1988).
17. Dalton, J., *A New System of Chemical Philosophy*, Vol. 1 (Bickerstaff, Manchester, U.K., 1808), Chap. 3.
18. Mendeleev, D. I., *Zhur. Russk. Khim. Obschch.* **1**, 60 (1869).
19. Gerhardt, C. F., *Ann. Chim.* **14**, 117 (1845).
20. Laurent, A., *Méthode de Chimie* (Mallet-Bachelier, Paris, 1854), p. 373; English translation by W. Odling, *Chemical Method* (Harrison, London, 1855), p. 308.
21. de Maupertuis, P. L. M., On Different Laws of Nature (In French), Mém. Acad. Roy. Sci. Paris 564–573 (1744).
22. Le Chatelier, H., *Ann. Mines* **13**, 157 (1888).
23. Caldin, E. F., *The Structure of Chemistry* (Sheed and Ward, London, 1961), p. 17.
24. Feynman, R. P., *The Character of Physical Law*, Penguin Books, London, 1992.

25. Miiniluoto, I., *Philos. Sci.* **45**, 431 (1978).
26. Harnad, S., in *Categorical Perception: The Groundwork of Cognition*, S. Harnad (ed.), Cambridge Univ. Press, Cambridge, U.K. (1987), Chap. 19.
27. Bateson, G., *Mind and Nature: A Necessary Unity* (Dutton, New York, 1980), Chap. 2, 39.
28. Medin, D. L. and Barsalou, L. W., in *Categorical Perception: The Groundwork of Cognition*, S. Harnad (ed.), Cambridge Univ. Press, Cambridge, U.K. (1987), p. 455.
29. Otto, M., *Anal. Chim. Acta* **235**, 169 (1990).
30. Zadeh, L. A., *Inform. and Control* **8**, 338 (1965).
31. Sangalli, A., *New Sci.* **134**, 36 (1992).
32. Cooper, L. (Transl.), *The Rhetoric of Aristotle* (Appleton-Century-Crofts, New York, 1932), Section 1411, 208.
33. Farber, E., *Isis* **41**, 20 (1950).
34. Wold, S. and Sjöjström, M., in *Chemometrics: Theory and Application*, B. R. Kowalski (ed.), ACS Symposium Series No. 52, American Chemical Society, Washington, D.C., (1977), Chap. 12, p. 243.
35. Wipke, W. T. and Hahn, M. A., *Tetrahedron Comput. Methodol.* **1**, 141 (1988).
36. Okada, T. and Kawai, T., *Tetrahedron Comput. Methodol.* **2**, 327 (1989).
37. Okada, T. and Kawai, T., *Tetrahedron Comput. Methodol.* **2**, 337 (1989).
38. Rouvray, D. H., *J. Mol Struct. (THEOCHEM)* **259**, 1 (1992).
39. Fischer, E., *Chem. Ber.* **27**, 2985 (1894).
40. Rebek, J., *Science* **235**, 1478 (1987).
41. Raevskii, O. A., *Teor. Eksp. Khim.* **22**, 431 (1986).
42. Gómez-Jeria J. S., in *Molecules in Physics, Chemistry, and Biology* (Kluwer, Dordrecht, The Netherlands, 1989), Vol. 4, Chap. 10, p. 215.
43. Schneider, H. J., *Angew. Chem. Int. Ed. Engl.* **30**, 1417 (1991).
44. Suckling, C. J., *Sci. Progress Edinburgh* **75**, 323 (1991).
45. Willett, P., *Three-Dimensional Chemical Structure Handling* (Research Studies Press, Taunton, U.K., 1991), Chap. 6, p. 194.
46. Weissbuch, I., Addadi, L., Lahav, M., and Leiserowitz, L., *Science* **253**, 637 (1991).
47. Joanny, J. F., Leibler, L., and Ball, R., *J. Chem. Phys.* **81**, 4640 (1984).
48. Primas H., *Chimia* **36**, 293 (1982).
49. Folse, H. J., *The Philosophy of Niels Bohr: The Framework of Complementarity* (North-Holland, Amsterdam, 1985), Chap. 1, p. 9.
50. Rouvray, D. H., in *Concepts and Applications of Molecular Similarity*, M. A. Johnson and G. M. Maggiora (eds.), Wiley-Interscience, New York (1990), Chap. 2, p. 15.
51. Spinoza, B., *The Ethics (1678)*, Transl. R. H. M. Elwes, Joseph Simon Co., Malibu, California (1981), Part 2, Proposition 13, Lemma 5, p. 77.
52. Tolman, R. C., *Phys. Rev.* **3**, 244 (1914).
53. Claverie, P., in *Symmetries and Properties of Non-Rigid Molecules: A Comprehensive Survey*, J. Maruani and J. Serre (eds.), Elsevier, Amsterdam, (1983), p. 13.
54. Spalding, D. B., The Art of Partial Modeling, *Ninth International Symposium on Combustion (1962)* (Academic Press, New York, 1963), p. 833.
55. Luce, R. D., *Philos. Sci.* **38**, 157 (1971).
56. Buckingham, E., *Phys. Rev.* **4**, 345 (1914).
57. Bruk, Y. and Stasenko, A., *Quantum* **2**, 34 (1992).
58. Gleick, J., *Chaos: Making a New Science*, Viking Press, New York, 1987.
59. Mandelbrot, B. B., *Fractals: Form, Chance and Dimension* (Freeman, San Francisco), 1977.
60. Musès, C., in *The Mathematical Heritage of C. F. Gauss*, G. M. Rassias (ed.), World Scientific Publ. Co., Singapore (1991), p. 526.
61. Wagner, G. C., Colvin, J. T., Allen, J. P., and Stapleton, H. J., *J. Am. Chem. Soc.* **107**, 5589 (1985).
62. Rouvray, D. H. and Pandey, R. B., *J. Chem. Phys.* **85**, 2286 (1986).

63. Avnir, D., Farin, D., and Pfeifer, P., *New J. Chem.* **16**, 439 (1992).
64. Farin, D. and Avnir, D., *J. Phys. Chem.* **91**, 5517 (1987).
65. Kopelman, R., *Science* 241, 1620 (1988).
66. Kaye, B. H., *A Random Walk Through Fractal Dimensions* (VCH Press, Weinheim, Germany, 1989), Chaps. 5, 7, and 9.
67. Peleg, M., *Crit. Rev. Food Sci. Nutrit.* **33**, 149 (1993).
68. Vosniadov, S. and Ortony, A. (eds), *Similarity and Analogical Reasoning* (Cambridge Univ. Press, Cambridge, U.K., 1989).
69. Thagard, P., *Conceptual Revolutions* (Princeton Univ. Press, Princeton, New Jersey, 1992), Chap. 3, p. 34.
70. Priestley, J., *Experiments and Observations Relating to Various Branches of Natural Philosophy* (Johnson, London, 1779), Vol. 1, pp. x–xi.
71. van't Hoff, J. H., *Phil. Mag.* [5] **26**, 81 (1888).
72. Baeyer, A., *Chem. Ber.* **18**, 2269 (1885).
73. Hückel, W., *Fortschr. Chem., Phys. und Phys. Chem.* **19**, 96 (1927).
74. Rosen, J., *Symmetry* **1**, 19 (1990).
75. Sarlemijn, A. and Kroes, P. A., in *Technology and Contemporary Life*, Philosophy and Technology, Vol. 4, P. T. Durbin (ed.), Reidel, Dordrecht, The Netherlands (1988), p. 237.
76. von Neumann, J., *Mathematical Foundations of Quantum Mechanics* (Princeton Univ. Press, New Jersey, 1955), Chap. 5, p. 347.
77. Schrödinger, E., *Proc. Cambridge Phil. Soc.* **31**, 555 (1935).
78. Einstein, A., Podolsky, B., and Rosen, N., *Phys. Rev.* **47**, 777 (1935).
79. Jammer, M., *The Philosophy of Quantum Mechanics*, Wiley, New York, 1974.
80. Youssef, S., *Mod. Phys. Lett.* **A6**, 225 (1991).
81. Lee, P. M., *Bayesian Statistics: An Introduction* (Oxford Univ. Press, New York, 1989).
82. Fisher, R. A., *Statistics Methods and Scientific Inference* (Hafner, New York, 1956).
83. Jaynes, E. T., in *Maximum Entropy and Bayesian Methods*, J. Skilling (ed.), Kluwer, Dordrecht, The Netherlands (1989), p. 1.
84. Bretthorst, G. L., in *Maximum-Entropy and Bayesian Methods in Science and Engineering*, Vol. 1, G. J. Erickson and C. Ray Smith (eds.), Kluwer, Dordrecht, The Netherlands (1988), p. 75.
85. Klimov, G. P. and Marshall, G. R., *Computers Math. Applic.* **26**, 1 (1993).
86. Willett, P., *Similarity and Clustering in Chemical Information Systems* (Research Studies Press, Letchworth, U.K., 1987).
87. Carbó, R., Leyda, L., and Arnau, M., *Int. J. Quant. Chem.* **17**, 1185 (1980).
88. Gower, J. C. and Legendre, P., *J. Classification* **3**, 5 (1986).
89. Johnson, M. A., *J. Math. Chem.* **3**, 117 (1989).
90. Döbereiner, J. W., *Ann. Phys. Chem.* **15**, 301 (1829).
91. Pettenkofer, M., *Bayer. Akad. Wiss. (Munich)* **32**, 261 (1850).
92. Newlands, J. A. R., *Chem. News (London)* **7**, 70 (1863).
93. Karpenko, V., *Ambix* **27**, 77 (1980).
94. Haas, A, *Chem. Zeitung* **106**, 239 (1982).
95. Godovikov, A. A. and Hariya, Y., *J. Fac. Sci., Hokkaido Univ.*, Ser. 4, **22**, 357 (1987).
96. Allen, L. C., *J. Am. Chem. Soc.* **114**, 1510 (1992).
97. Sakka, S., *Kinzoku* **58**, 46 (1988).
98. Suzuki, T., *Kinzoku* **58**, 5 (1988).
99. Spevak, N. D., Spevak, A. N., and Prutskov, D. V., *Ukr. Khim. Zhur. (Russ. Ed.)* **57**, 914 (1991).
100. Dias, J. R., *Acc. Chem. Res.* **18**, 241 (1985).
101. Kong, F. A., *J. Mol. Struct.* **90**, 17 (1982).
102. Hefferlin, R. A., Zhuvikin, G. V., Caviness, K. E., and Duerksen, P. J., *J. Quant. Spectrosc. Radiat. Transfer* **32**, 257 (1984).
103. Dias, J. R., *J. Mol. Struct. (THEOCHEM)* **230**, 155 (1991).

104. Hefferlin, R., *Periodic Systems and their Relation to the Systematic Analysis of Molecular Data* (Edwin Mellen Press, Lewiston, New York, 1989), Chap. 1.
105. Szabó, A., *Anfänge der Griechischen Mathematik*, Oldenburg, Munich, 1969.
106. Tweney, R. D., Doherty, M. E., and Mynatt, C. R. (eds.), *On Scientific Thinking* (Columbia Univ. Press, New York, 1981), p. 285.
107. Warrington, J. (Transl.), *Plato's Timaeus* (Dent, London, 1965), Sect. 31C, p. 22.
108. Gentner, D. and Jeziorski, M., in *Psychology of Science: Contributions to Metascience*, B. Gholson *et al.* (eds.), Cambridge Univ. Press, Cambridge, U.K. (1989), p. 296.
109. Hesse, M. B., *Models and Analogies in Science* (Univ. Notre Dame Press, Notre Dame, Indiana, 1966), p. 101.
110. Shive, J. N. and Weber, R. L., *Similarities in Physics* (Wiley, New York, 1982).
111. Reed, S. K., Ackinclose, C. C., and Voss, A. A., *Memory and Cognit.* **18**, 83 (1990).
112. Bunge, M., *Method, Model and Matter* (Reidel, Dordrecht, The Netherlands, 1973), p. 130.
113. Weitzenfeld, J. S., *Philos. Sci.* **51**, 137 (1984).
114. Clement, J., *Cognit. Sci.* **12**, 563 (1988).
115. Gick, M. L. and Holyoak, K. J., *Cognit. Psychol.* **12**, 306 (1980).
116. Holyoak, K. J., in The *Psychology of Learning and Motivation*, Vol. 19, G. H. Bower (ed.), Academic Press, New York (1985), p. 59.
117. Ortony, A., *Psychol. Rev.* **86**, 161 (1979).
118. Clement, C. A. and Gentner, D., *Cognit. Sci.* **15**, 89 (1991).
119. North, J. D., in *On Scientific Discovery*, M. D. Grmek, R. S. Cohen and G. Cimino (eds.), Reidel, Dordrecht, The Netherlands (1981), p. 115.
120. Falkenhainer, B., Forbus, K. D., and Gentner, D., *Artif. Intell.* **41**, 1 (1989).

Graph Theoretical Approach to Structural Changes in Chemistry

JACQUES-EMILE DUBOIS and GILLES SICOURI
Institut de Topologie et de Dynamique des Systèmes (ITODYS), Université de Paris 7. Denis Diderot, 1 rue Guy de la Broose, 75005 Paris, France

Abstract. A general methodology is proposed to define the various chromatic graphs required for handling structures and their transformation. A transfostructure concept and a chromatic transfograph concept are proposed, and their use is investigated. The transfostructure is considered as a pair of variations on an invariant loose graph (loose structure) of a chemical reaction and its connectability to the two opegraphs linked to the changes.

All of the chromatic graph descriptions are obtained through their systematic generation or step by step rule-based construction. The generation processes described are controlled by priority rules monitoring some elementary information and deal with various types of adjunction or ablation of sites. Ordered generation of the chromatic transfograph, definition of a connection transfograph, identification of structural isomorphism, obtaining bridging transfographs useful for transformation – are all carried out. The generation process for creating the various chromatic graphs implies that they be embedded in general hyperstructures or hypergraphs created in a synchronous process leading simultaneously to graph strucutres and their own hyperstructures (HS). These order-based processes constitute the central feature and controlled power of our graph theory of structural representation and transformation.

This presentation focusses on the algebraic requirements of chromatic families of molecular graphs and their embedding (HS) hyperstructure graphs. In artificial intelligence, these are ordered "spaces of states" composed here of well defined and ordered pathways of theoretical importance. Concentric generation with priority rules of the **DARC** System is used to generate a canonical ordering of a transfograph and its HS. This leads to potential applications mainly concerning organic reactions both for databank management and CAD reactions.

1. Introduction

Several systems have been devised over the last twenty years to express graph transformations for their computer handling. These systems were created in a chemical context (nomenclature, classification, computer-aided design), or in a more mathematical one (graph grammars). In chemistry, names, codes or "parachemical" languages proposed have made it possible to identify the nature of transformations implied in chemical reactions [1, 2, 7]. But they more often use directly an imprecise (and in reality rarely available) mechanistic knowledge, or arbitrary classifications in generic reactions. Moreover theoretical deficiencies of the formalism deprive these descriptions of two fundamental properties: the univocity that stipulates that a descriptor describes only one reaction, and unicity which associates with one reaction a single descriptor. Graph grammars propose generalising usual chain grammars to non linear arrangements of symbols. A rigorous formalism clearly enhances a

R. Carbó (ed.), Molecular Similarity and Reactivity: From Quantum Chemical to Phenomenological Approaches, 203–241.

key problem in a graph transformation definition [3–5]: the relation between the connections of the departed and new elements of the transformation to the invariant subgraph (called "the embedding transformation"). However the formal language derived from such a graph grammar appears heavy and cumbersome when it has to be adapted to complex chromatic graph transformations, and does not take into account equivalence classes of such graphs.

In this article, we propose an original graph theoretical method, and a new structural type for modelling, representing, generating and describing structural changes in chemistry in view of their computer handling. The applications bear mainly on reactions in organic chemistry, and underlay operational procedures regarding storage and interrogation of specific reaction data banks, and automatic generation of synthetic pathways. The starting point of this work are the fundamental concepts regarding generation and description of chemical compounds. But these concepts have been specified and developed to take into account chromatic graph transformations in general, and structural aspects of chemical reactions in particular. The ambition of the method is to respect the dynamic conception of reaction admitted in chemistry, while removing the insufficiencies of previously studied models.

The new structural type of chromatic transfograph [10] – and the transfostructure which is derived from it by isomorphism [11, 12] – is presented after preliminary definitions regarding chemical structure. Its interpretation as the localisation of two variations on an invariant is given. Ordered generation of the chromatic transfograph and its consequences to classes of isomorphism are then presented. Synchronism between generation and description is especially enhanced [13].

2. The transfostructure concept

2.1. Preliminary definitions

What follows are basic definitions (taken mainly from refs. [2, 4]) bearing on a chemical structure and its representation by a chromatic graph.

2.1.1. *Chromatic graph*

With a given node referential **X**, a chromatic graph is defined as an extension of an ordinary graph:

DEFINITION – A *chromatic graph* (undirected and without loops) is a quadruplet $G = (X, U, \chi_N, \chi_L)$ where:

- X is the set of nodes, a subset of referential **X**;
- U is the set of links, a subset of $A(x) = \{\{x, y\} \mid x, y \in X, x \neq y\}$;
- χ_N is the chromatic node function, an application of X in K_N, whereby to every node $x \in X$ is associated a chromaticity $\chi_N(x) \in K_N$;
- K_N is the node chromaticity table,

- χ_L is the chromatic link function, an application of U in K_L whereby to every link $u \in U$ is associated a chromaticity $\chi_L(u) \in K_L$;
- K_L is the link chromaticity table.

A particular choice of chromatic tables defines the graphs representing the structural formulas for chemical compounds, irrespective of charge or stereochemistry.

DEFINITION – A *chemical chromatic* graph is a chromatic graph expressed by $G = (X, U, \chi_N, \chi_L)$ where:

- $\chi_N = \chi_{NA}$ is the atom nature function, an application of X in K_{NA};
- K_{NA} is the atom nature table that contains the list of chemical elements from the Mendeleiev table;
- $\chi_L = \chi_{LI}$ is the bond type function, an application of U in K_{LI};
- $K_L = K_{LI}$ is the bond type table containing the list of chemical bond types; it is limited to *si*ngle, *ta*utomer, *ar*omatic, *do*uble and *tr*iple bonds: K_{LI} = {si, ta, ar, do, tr};
- the nodes of X satisfy the valency constraint (cf. hereafter).

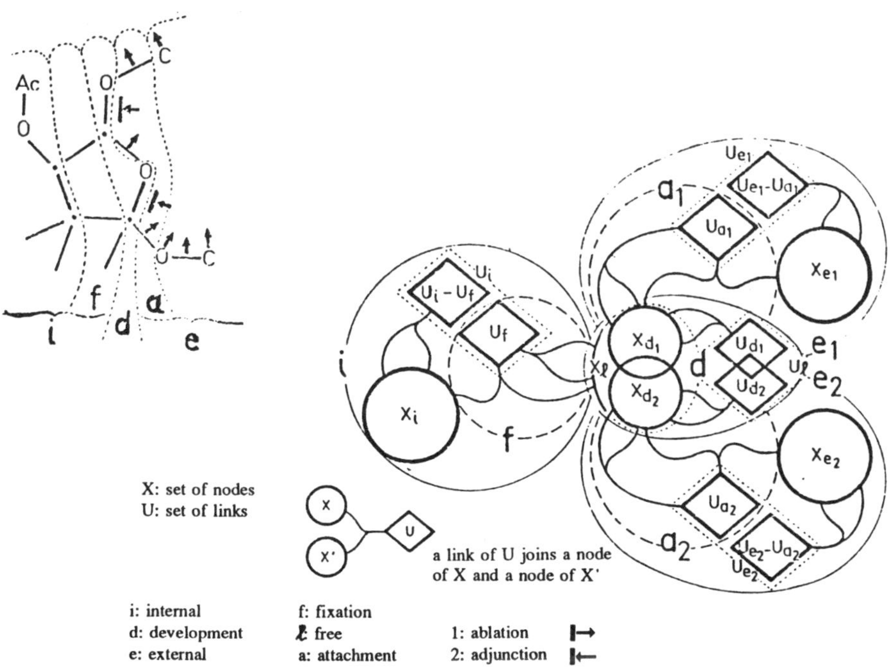

Figure 1. Definition and example of a chromatic transfograph. A chromatic transfograph can be considered as the union of four disjunct modules: free (l), internal (i), ablation (e_1) and adjunction (e_2). This presentation shows the complexity of the topological structure of the chromatic transfograph compared to that of the chromatic graph.

2.1.2. *Valency constraint*

The structural theory of valency bonds imposes a constraint on chromatic graphs that limits the range of topochromatic possibilities. Definitions of usual valency chromaticity, bond multiplicity chromaticity and occupied valency chromaticity serve as an introduction to this valency constraint.

a) Valency and multiplicity functions.

V: $K_{NA} \rightarrow K_V$ is the valency function that, to every chemical element k of K_{NA}, associates the usual valency $V(k)$ of this chemical element.

K_V is the usual valency table for atoms; valencies are whole numbers between 0 and 10.

M: $K_{LI} \rightarrow K_M$ is the formal bond multiplicity function that, to every type of bond k, associates $M(k)$ the multiplicity of this bond.

K_M designates the bond multiplicity table that is limited for four elements: $K_M = \{1/2, 1, 2, 3\}$.

b) Chromatic valency functions, chromatic multiplicity functions and chromatic occupied valency functions.

$\chi_V = V \circ \chi_{NA}$: $X \rightarrow K_M$ associates to every node x representing an atom of nature $\chi_{NA}(x)$, the valency chromaticity of this atom $\chi_V(x) = V(\chi_{NA}(x))$

$\chi_M = M \circ \chi_{LI}$: $U \rightarrow K_M$ associates, to every link u representing a bond of type $\chi_{LI}(u)$, the valency chromaticity of this link $\chi_M(u) = M(\chi_{LI}(u))$.

$\chi_{VO} = X \rightarrow K_V$ associates, to every node x a chromatic designating a valency occupied by a heavy atom link to that node. The sum of multiplicities contributed by bonds connecting heavy atoms to the atom represented by node x is expressed by $\chi_{VO}(x) = \Sigma_{\{x, y\} \in U}\, \chi_M(\{x, y\})$

c) Valency constraints.

On each node x of chromatic graph (X, U, X_{NA}, X_{LI}), the occupied valency chromaticity should be less than or equal to the valency chromaticity:

$$\text{for } x \in X,\ \Delta V(x) = -\chi_V(x) + \chi_{VO}(x) \leq 0 \qquad (1)$$

2.1.3. *Chemical structure*

A chemical structure is derived from a chemical chromatic graph by applying isomorphism relationship f, a binary relationship for the set of all chromatic graphs **G**.

DEFINITION – Two chromatic graphs G and G' are *isomorphic* if:

- there is a bijective mapping f of X onto X',
- the links are preserved by f:
  for $\{x, y\} \in U$, $\{f(x), f(y)\} \in U'$;
- f preserves the node chromaticities.
  for $x \in X$, $\chi'_{NA}(f(x)) = \chi_{NA}(x)$
- f preserves the link chromaticities.
  for $\{x, y\} \in U$, $\chi'_{LI}(\{f(x), f(y)\}) = \chi_{LI}(\{x, y\})$

Since the isomorphism relationship is an equivalence relationship, chromatic graphs can be grouped into classes of equivalence:

DEFINITION – A *chemical structure* is a class of isomorphism for all chromatic graphs respecting the valency constraint Equation (1).

2.2. CHROMATIC TRANSFOGRAPH CONCEPT

A chromatic transfograph is presented as a generalization of a chromatic graph for a given node referential **X**.

Chromatic transfograph F is formally distinguished from chromatic graph G by partitioning sets of nodes and sets of links, and by introducing new bond chromaticities. For simplicity, the elementary concept of a transfograph is given first, and is followed by a presentation of new and usual bond chromaticities and a definition of a chromatic transfograph.

2.2.1. *Transfograph*

The use of two topological elements in a transfograph – X for nodes and Y for links – is the same as in a graph but their structure is more complex in a transfograph than in a graph.

DEFINITION – An undirected *transfograph* is a pair expressed by $F = (X, U)$ where:

- $X = (X_i, X_l, X_e)$ is composed of:
  a) the set of internal nodes X_i, itself a subset of **X**,
  b) the set of loose nodes X_l, itself a subset of **X**,
  c) the pair $X_e = (X_{e1}, X_{e2})$ associating e_{e1} the set of external nodes of ablation with x_{e2} the set of external nodes of adjunction;

  sets X_i, X_l, X_{e1} and X_{e2} are disjoined two by two;
- $U = (U_i, U_e)$ is composed of:
  a) the set of internal links U_i, itself a subset of $A(X_l \cup X_i)$,
  b) the pair $U_e = (U_{e1}, U_{e2})$ associating U_{e1} the set of external links of ablation, subset of $A(X_{d1} \cup X_{e1})$, with U_{e2} the set of external links of adjunction, subset of $A(X_{d2} \cup X_{e2})$.

  Every loose node is an element of an external link.

REMARKS

It should be noted that the disjoint nature of node sets X_i, X_l, X_{e1} and X_{e2} implies the disjoint nature of link sets U_i, U_{e1} and U_{e2}.

It should also be noted that set X_l is the internal equivalent of the pair $X_d = (X_{d1}, X_{d2})$ grouping X_{d1}, the set of ablation nodes of development, with X_{d2}, the set of adjunction nodes of development. Indeed, X_d is obtained from X_l by:

$$X_{dj} = \{x \in X \mid \exists u \in U_{ej}, x \in u\} \text{ for } j \in J.$$

Hereafter, X_l or X_d is used indiscriminately in accordance with its context.

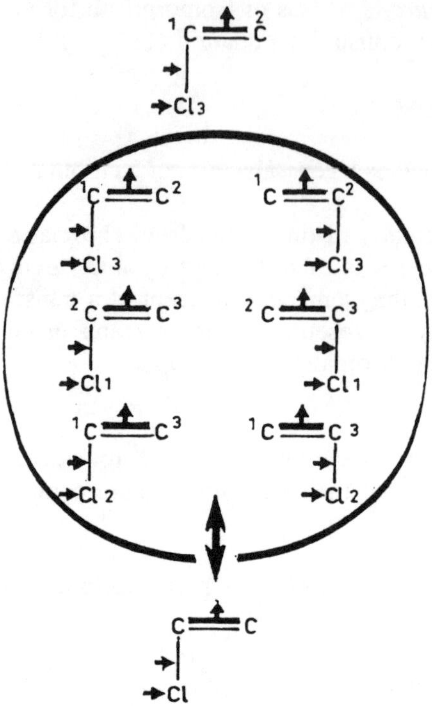

Figure 2. Chromatic transfograph and transfostructure.

2.2.2. *New chromaticity tables*

The expression of chemical bond transformations formalized by loose bonds requires the introduction of new link chromaticity tables: one in which initial and final bond types are gathered, another in which the initial or final type is specified from the generic chromaticity.

a) The generic aspect of loose bonds is expressed by the table K_{LI} indicating the *looseness* for a bond type. Said table is composed of a set of parts, here limited to two elements, from table K_{LI} for the types of bonds:

$$K_{LII} = \{\{k_1, k_2\} \mid k_1, k_2 \in K_{LI}, k_1 \neq k_2\}.$$

b) Passage from the generic to the specific is expressed by the table K_{GL} specifying a bond type. Said table is composed of a set of pairs associating a loose chromaticity with a specific chromaticity:

$$K_{\Gamma LI} = \{\{k_l, k_s\} \mid k_l \in K_{LII}, k_s \in k_l\}.$$

Hereafter, chromaticity $k_{\Gamma LI} = \{\{k_1, k_2\}, k_1\}$ is designated simply by $\{k_1, k_2\}$.

2.2.3. *Chromatic transfograph*

A chromatic transfograph is defined as a transfograph wherein original chromaticities are carried by links called "loose links", and usual chromaticities designating the nature of an atom or the type of bond are carried by all the other sites.

DEFINITION – An undirected *chromatic transfograph* is a quadruplet expressed by $F = (X, U, \chi_N, \chi_L)$ where:

– $X = (X_i, X_l, X_e)$ is composed of:
  a) the set of internal nodes X_i, itself a subset of **X**;
  b) the set of loose nodes X_l, itself a subset of **X**;
  c) the pair $X_e = (X_{e1}, X_{e2})$ associating X_{e1} the set of external nodes of ablation with x_{e2} the set of external nodes of adjunction.

Sets X_i, X_l, X_{e1} and X_{e2} are disjoined two by two;

– $U = (U_i, U_l, U_e)$ is composed of:
  a) the set of internal links U_i, itself a subset of $A(X_l \cup X_i)$;
  b) the set of loose links U, itself a subset of $A(x_l)$;
  c) the pair $U_e = (U_{e1}, U_{e2})$ associating U_{e1} the set of external links of ablation, subset of $A\ (X_{d1} \cup X_{e1})$, with U_{e2} the set of external links of adjunction, subset of $A\ (X_{d2} \cup X_{e2})$.

Every loose node belongs to a loose or external link;

– χ_N is composed of the atom nature function only. Said function is an application of $X_i \cup X_l \cup X_{e1} \cup X_{e2}$ in usual table K_{NA}.

– $\chi_L = (\chi_{LI}, \chi_{LII}, \chi_{\Gamma LI})$ is composed of
  a) χ_{LI} the bond type function; said function is an application of $U_i \cup U_{e1} \cup U_{e2}$ in usual table K_{LI};
  b) χ_{LI} the bond type looseness function; said function is an application of U_l in looseness table K_{LI};
  c) $\chi_{\Gamma LI} = (\chi_{\Gamma LI1}, \chi_{\Gamma LI2})$ the pair of bond type specification functions; said functions are applications of U_l in specification table $K_{\Gamma LI}$. For $j \in J = \{1, 2\}$, if

$$\chi_{\Gamma LIj}(u) = (k_l, k_s), \text{ then } \chi^{l}_{\Gamma Ij}(u) = k_l \text{ and } \chi^{s}_{\Gamma LIj}(u) = k_s.$$

REMARKS

Given the disjoint nature of sets X_i, X_l, X_{e1} and X_{e2}, link sets U_i, U_l, U_{e1} and U_{e2} are disjoint.

The pair $(X_l, (U_i \cup U_l, U_e))$ is a transfograph constructed from the chromatic transfograph (X, U, χ_N, χ_L), this also holds true for the pair $(X_l, (U_i, U_e))$.

Chromaticities on development links present a certain redundancy: the loose chromaticity is repeated in each specification chromaticity of ablation or adjunction, and the same type of bond is mentioned twice in a specification chromaticity. The first type of redundancy is justified by a deliberate will to distinguish invariant elements from variant elements; the second type is related to a desire to generalize the specification of a generic chromaticity to the case where it has more than two specific chromaticities. By convention, redundancy is curtailed when expressed in writing or graphically.

2.2.4. *Modules of a chromatic transfograph*

To clarify the design of the chromatic transfograph and facilitate the presentation of its properties, topological and chromatic elements corresponding to the same category (loose, internal, or external) are grouped into modules.

- The loose module is composed of loose nodes (X_l), their chromaticities (χ_{NA}/X_l), loose links (U_l) and their loose chromaticities (χ_{LIl}).
- The internal module is composed of internal nodes (X_i), their chromaticities (χ_{NA}/X_i), internal links (U_i) and their chromaticities (χ_{LI}/U_i).
- The external module with subindex j is composed of external nodes (X_{ej}), their chromaticities (χ_{NA}/X_{ej}), external links (U_{ej}), their chromaticities (χ_{LI}/U_{ej}), and finally, specification chromaticities on development links with subindex j ($\chi_{\Gamma LIj}/U_l$). A chromatic transfograph has two external modules: an ablation module corresponding to index $j = 1$, and an adjunction module corresponding to index $j = 2$.

Thus, a chromatic transfograph can be viewed as an "assemblage" of four disjoint modules: loose, internal, ablation and adjunction. This facilitates its interpretation as a transformation of chromatic graphs, considered as an ablation-adjunction sequence localized on an invariant.

- The invariant corresponds to the loose and internal modules,
- the ablation variation corresponds to the external ablation module and the adjunction variation corresponds to the external adjunction module.

Within internal and external modules, links involving at least one loose node are characterized in a more specific manner, thereby making it possible to highlight their original role in each category.

- The set of fixation links $U_f = U_i \cap (X_l \times (X_l \cup X_i))$ groups together internal links between a loose node and an internal node, or between two loose nodes.
- The pair $U_a = (U_{a1}, U_{a2})$ associates $U_{a1} = U_{e1} \cap (X_{d1} \times X_{e1} \cup X_{d1}))$ the set of ablation attachment links with $U_{a2} = U_{e2} \cap (X_{d2} \times X_{e2} \cup X_{d2}))$ the set of adjunction attachment links.

Fixation links (U_f), as well as their chromaticities indicating the type of bond (χ_{LI}/U_f), belong to the internal module and constitute the fixation sub-module. Likewise, attachment links U_{aj}, their chromaticities indicating the type of bond (χ_{LI}/U_{aj}), and the specification chromaticities ($\chi_{\Gamma LIj}$) on the development links of U_l, belong to the external module with subindex j, and constitute the attachment sub-module with subindex j.

2.3. Representative graph concept

To establish the tie with the chromatic graph theory, an attempt has been made to associate, to every chromatic transfograph F, a unique chromatic graph said to be a representative graph of F and designated by $G_r(F)$ or, simply, G_r. The complexity of the topological structure of the chromatic transfograph is thus transferred to its chromatic structure.

The representative graph of chromatic transfograph F is chromatic graph G_r associated to F by one-to-one pairing. The existence and unicity of G_r is established hereafter by constructing the elements of G_r from the elements of F. Inversely, reconstructing F from G_r demonstrates the unambiguity of the representative graph.

2.3.1. *Existence and unicity*

To each and every chromatic transfograph $F = (X, U, \chi_N, \chi_L)$ is associated the chromatic graph $G_r = (X_r, U_r, \chi_{Nr}, \chi_{Lr})$ where:

- $X_r = X_i \cup X_l \cup X_{e1} \cup X_{e2}$ groups together the following types of nodes: internal, loose, external ablation and external adjunction;
- $U_r = U_i \cup U_l \cup U_{e1} \cup U_{e2}$ groups together the following types of links: internal, loose, external ablation and external adjunction;
- χ_{Nr} is an application of X_r in K_{NA}, defined simply by

$$\chi_{Nr}(x) = \chi_{NA}(x) \text{ for } x \in X_r;$$

- χ_{Lr} is an application of U_r in:

$$K_{Lr} = \{\{k_i, k_l, k_1, k_2\} \mid k_i \in K_{LI} \cup \{b\}, k_l \in K_{LII} \cup \{b\},$$
$$k_1 \text{ et } k_2 \in K_{LI} \cup K_{\Gamma LI} \cup \{b\}\},$$

(empty chromaticity b designates an absence of chromaticity). It is defined by:

$$\begin{aligned}\chi_{Le}(u) &= (\chi_{LI}(u), b, b, b) \text{ if } u \in U_i \\ &= (b, \chi_{LII}(u), \chi_{\Gamma LI1}(u), \chi_{\Gamma LI2}(u)) \text{ if } u \in U_l \\ &= (b, b, \chi_{LI}(u), b) \text{ if } u \in U_{e1} \\ &= (b, b, b, \chi_{LI}(u)) \text{ if } u \in U_{e2}\end{aligned}$$

If $\chi_{Lr}(u) = (k_i, k_l, k_1, k_2)$ then $\chi^j_{Lr} = k_j$ for $j \in \{i, l, 1, 2\}$.

2.3.2. *Unambiguity*

At this point, no attempt is made to seek the conditions that any chromatic graph whatever must meet to qualify as the representative graph of a chromatic transfograph. Only the unique reconstruction of $F = (X, U, \chi_N, \chi_L)$ from $G_r = (X_r, U_r, \chi_{Nr}, \chi_{Lr})$ is established, because this is enough to demonstrate the unambiguity of representing F by G_r.

- Constructing sets of links.

The link character is deduced directly from link chromaticities.

$$U_i = \{u \in U_r \mid \chi^i_{Lr}(u) \neq b\};$$

$$U_l = \{u \in U_r \mid \chi^l_{Lr}(u) \text{ and } \chi^j_{Lr}(u) \neq b\} \text{ for } j \in J;$$

$$U_{ej} = \{u \in U_r \mid \chi^l_{Lr}(u) = b \text{ and } \chi^j_{Lr}(u) \neq b\} \text{ for } j \in J.$$

- Constructing sets of nodes.
  The node category is deduced from the link character.

$$X_{dj} = \{x \in X_r \mid \exists u \in U_f \cup U_l \cup U_{aj}, x \in u\} \text{ for } j \in J;$$

$$X_{ej} = \{x \in X_r \mid \exists u \in U_{ej}, x \in u\} - X_{dj} \text{ for } j \in J;$$

$$X_i = X_r - X_l - X_{e1} - X_{e2}.$$

- Constructing node chromaticities.

$$\chi_{NA}(x) = \chi_{Nr}(x) \text{ for } x \in X_r.$$

- Constructing link chromaticities.

$$\chi_{LI}(u) = \chi^i_{Lr}(u) \text{ for } u \in U_i;$$

$$\chi_{LI}(u) = \chi^j_{Lr}(u) \text{ for } u \in U_{ej} \text{ and } j \in J;$$

$$\chi_{LI}(u) = \chi^l_{Lr}(u) \text{ for } u \in U;$$

$$\chi_{\Gamma LIj}(u) = \chi^j_{Lr}(u) \text{ for } u \in U \text{ and for } j \in J.$$

2.4. Valency constraints in chemistry

The non-independence between node chromaticity and link chromaticity in chemistry defines valency constraints that can be separated into two categories: valency constraints applying to internal or external nodes (involving usual chemical bonds) and valency constraints applying to loose nodes (involving original loose bonds).

2.4.1. *Ordinary constraint*

The constraint on internal and external nodes is the usual constraint applying to chromatic graphs representing chemical compounds. It stipulates that the multiplicity brought by bonds connected to a given atom should not be greater than the usual valency of this atom.

For $c = i, e1, e2$ and for $x \in X_c$,

$$\Delta V_c(x) = -\chi_V(x) + \sum_{\{x, y\} \in U_c} \chi_M(\{x, y\}) \leq 0.$$

2.4.2. *Loose node valency constraint*

The valency constraint for loose nodes in a chromatic transfograph differs from the usual valency constraint in that each type of bond category (internal, loose, external) in the vicinity of the loose nodes is brought into play in a different manner. This constraint stipulates that within the immediate vicinity of a given loose atom, the multiplicity contributed by the internal bonds added to the multiplicity contributed by the loose bonds and the external bonds of each type (ablation, adjunction), should not be greater than the usual valency of this atom.

For $x \in X_l$,

$$\Delta V_l(x) = -\underset{+\max}{\chi V(x)} + \sum_{j \in J} {}_{\{x, y\} \in U_i} \chi_M(\{x, y\})$$
$$(\textstyle\sum_{\{x, y\} \in U_l} \chi_{Mj}(\{x, y\}) + (\textstyle\sum_{\{x, y\} \in U_{ej}} \chi_M(\{x, y\})) \leq 0,$$

(where $\chi_{Mj}(\{x, y\}) = M(\chi^s_{\Gamma LIj}(\{x, y\})$ for $j \in J$ and $\{x, y\} \in U_l$).

2.5. Transfostructure concept

The transfostructure concept is derived from the chromatic transfograph concept by applying an isomorphism relationship to a set of chromatic transfographs.

2.5.1. *Isomorphism relationship*

The isomorphism relationship is defined on the basis of bijection between the node sets of each transfograph.

DEFINITION – Two chromatic transfographs $F = (X, U, \chi_N, \chi_L)$ and $F' = (X', U', \chi'_N, \chi'_L)$ are *isomorphic* if there is a bijection f of $X_r = X_i \cup X_l \cup X_{e1} \cup X_{e2}$ in $X'_r = X'_i \cup X'_l \cup X'_{e1} \cup X'_{e2}$ so that:

a) the node character is preserved
   - for $x \in X_i$, $f(x) \in X'_i$;
   - for $x \in X_l$, $f(x) \in X'_l$;
   - for $j \in J$ and for $x \in X_{ej}$, $f(x) \in X'_{ej}$;

b) node chromaticities are preserved
   - for $x \in X_r$, $\chi'_{NA}(f(x)) = f(\chi_{NA}(x))$;

c) links are preserved
   - for $\{x, y\} \in U_i$, $\{f(x), f(y)\} \in U'_i$;
   - for $\{x, y\} \in U_l$, $\{f(x), f(y)\} \in U'_l$;
   - for $j \in J$ and for $\{x, y\} \in U_{ej}$, $\{f(x), f(y)\} \in U'_{ej}$;

d) link chromaticities are preserved
   - for $\{x, y\} \in U_i \cup U_{e1} \cup e_{e2}$,

   $$\chi'_{LI}(\{f(x), f(y)\}) = \chi_{LI}(\{x, y\});$$

   - for $j \in J$ and for $\{x, y\} \in U_l$,

   $$\chi'_{\Gamma LIj}(\{f(x), f(y)\}) = \chi_{\Gamma Ij}(\{x, y\});$$

   - for $j \in J$ and for $\{x, y\} \in U_l$,

   $$\chi'_{LIl}(\{f(x), f(y)\}) = \chi_{LIl}(\{x, y\}).$$

This binary relationship for the entire set **F** of chromatic transfographs is an equivalence relationship: the reflexivity, symmetry and transitivity properties are easily deduced from the usual set properties. Said relationship can be used to group chromatic transfographs into classes of equivalence.

REMARKS

Other types of more or less refined equivalence relationships will be envisioned for the constitution of classes of transfographs better suited to other uses. They will make it possible to group

- transfographs leading from the same initial graph to the same final graph,
- transfographs leading from a graph belonging to the same initial structure to a graph belonging to the same final structure.

2.5.2. *Transfostructure*

The restriction of classes of isomorphism to chromatic transfographs respecting the valency constraint yields a transfostructure.

DEFINITION – A *transfostructure* T is a class of isomorphism containing chromatic transfographs respecting the valency constraints.

Three particular cases must be pointed out.

a) A transfostructure with chromatic transfographs bearing an empty set of loose nodes is isomorphic to a triplet of three chemical structures, an invariant structure, an ablation structure and an adjunction structure.
b) A transfostructure with chromatic transfographs bearing an empty set of loose nodes and an empty set of external nodes is isomorphic to a pair of chemical structures.
c) A transfostructure with chromatic transfographs bearing an empty set of loose nodes and empty sets of external nodes is isomorphic to a single chemical structure, an invariant structure.

2.6. INVARIANT AND VARIATIONS

A transfostructure is initially interpreted as the localization of a pair of variations on the invariant of a chemical reaction. A study on the derivation of an invariant loose structure and of a pair of operating structures from a transfostructure is followed by a discussion on the establishment of the conditions under which an invariant loose structure and a pair of operating variation structures are connected in order to build a set of transfostructures.

2.6.1. *Loose graph and opegraphs derived from a transfograph*

When attempting to extract a loose graph G_0 and two opegraphs Ω_1 and Ω_2 from transfograph F, no condition need be imposed on transfograph F to authorize these derivations.

RECONSTRUCTING G_0 FROM F

- Sets of nodes and links gather the internal and loose nodes and links of the transfograph:

$$X^0 = X_i \cup X_l;\ U^0 = U_i \cup U_l.$$

- atom natures are obtained directly:

$$\chi^0_{NA}(x) = \chi_{NA}(x) \text{ for } x \in X_i \cup X_l.$$

- bond types are obtained directly from internal and external bonds and require calculating a "minimal bond type" when they originate from a loose bond:

$$\begin{aligned}\chi^0_{LI}(u) &= \chi_{LI}(u) \text{ for } u \in U_i \cup U_{e1} \cup U_{e2};\\ &= \chi^m_{LI}(u) \text{ for } U \in U.\end{aligned}$$

If $\chi_{LIl}(u) = \{k_1, k_2\}$, then the minimal bond type is expressed by: $\chi^m_{LI}(u) \leq k_1, k_2$, (where < designates the adjacent succession relationship in K_{LIl}).

- The lower bounds of loose chromaticity are nil for nodes originating from internal nodes; these bounds are strictly negative for nodes originating from development nodes. Calculation of these bounds depends on the multiplicity contributed by the attachment bonds and the loose bonds of the transfograph.

$$\begin{aligned}\inf \chi^0_{VL}(x) &= 0 \text{ for } x \in X_i;\\ &= \max_{j \in J} \textstyle\sum_{\{x,\,y\} \in U} \chi_M(\{x, y\}) + \sum_{\{x,\,y\} \in U} \chi^m_M(\{x, y\})\\ &\quad \text{for } x \in X_{d1} \cup X_{d2}\\ &= \textstyle\sum_{\{x,\,y\} \in U} \chi_M(\{x, y\}) + \sum_{\{x,\,y\} \in U} \chi^m_M(\{x, y\})\\ &\quad \text{for } x \in X_{d1} - X_{d2};\\ &= \textstyle\sum_{\{x,\,y\} \in U} \chi_M(\{x, y\}) + \sum_{\{x,\,y\} \in U} \chi^m_M(\{x, y\})\\ &\quad \text{for } x \in X_{d2} - X_{d1}.\end{aligned}$$

$$\begin{aligned}\inf \chi^0_{ML}(u) &= 0 \text{ pour } u \in U_i;\\ &= \max_{j \in J} \textstyle\sum_{u \in U} \chi^s_{Mj}(u) + \sum_{u \in U} \chi^m_M(u)\\ &\quad \text{for } u \in U.\end{aligned}$$

(with $\chi^m_M(u) = M(\chi^m_{LI}(u))$)

RECONSTRUCTING Ω_1 AND Ω_2 FROM F

The elements of each opegraph $\Omega_j (j \in J)$ are built from the development elements and the external elements of the transfograph.

- external elements are deduced directly from the external elements of the transfograph.

$$X^j_e = X_{ej} \text{ and } \chi^j_{NA}(x) = \chi_{NA}(x) \text{ for } x \in X_{ej} \text{ and } j \in J;$$

$$U^0_e = U_{ej} \text{ for } u \in U_{ej} \text{ and } j \in J;$$

- Development elements belong to the loose elements of the transfograph. Should the minimal bond type of a loose bond find itself in the specification, the opegraph makes no mention of this bond.

For $u \in U_{dj}$, if $\chi^{m}_{LIl}(u) \neq \chi^{j}_{\Gamma LI}(u)$, then

$U^{j}_{d} = \{u \in U_{dj} \mid \chi_{LIl}(u) \neq \chi^{s}_{\Gamma LIj}(u)\}$ and

$\chi^{j}_{\Delta LI}(u) = \{\chi^{m}_{LIl}(u), \chi^{m}_{\Delta LIj}(u)\}$.

$X^{j}_{d} = \{x \mid \exists u \in U^{j}_{a}, x \in u\} \cup \{x \mid \exists u \in U^{j}_{d}, x \in u\}$

2.6.2. *Connectability of a loose graph to two opegraphs*
The conditions for the connection of two opegraphs to a loose graph simultaneously bring into play relationships tying each opegraph to the loose graph and relationships between each of the two opegraphs.

DEFINITION – Two opegraphs Ω_1 and Ω_2 *are connectable* to a loose chromatic graph G_0, if they meet the following three sets of conditions:

a) development element conditions stipulate that nodes and links with loose non-empty chromaticities are the development nodes and development links of the opegraphs; said conditions are expressed by

$$X^{1}_{d} \cup X^{2}_{d} = \{x \in X^{0} \mid \inf \chi^{0}_{VL}(u) \neq 0\};$$

$$U^{1}_{d} \cup U^{2}_{d} = \{u \in U^{0} \mid \inf \chi^{0}_{ML}(u) \neq 0\}.$$

b) external element conditions stipulate that the external nodes and external links of each opegraph are disjoint from each other and from the nodes of the loose graph; said conditions are expressed by

$$X^{1}_{e} \cap X^{2}_{e} = \varnothing \text{ and } U^{1}_{e} \cap U^{2}_{e} = \varnothing;$$

$$X^{j}_{e} \cap X^{0} = \varnothing \text{ and } U^{j}_{e} \cap U^{0} = \varnothing \text{ for } j \in J.$$

c) link chromaticity conditions stipulate that the attachment bonds and attachment chromaticities of at least one of the opegraphs completely fill in the loose valencies and loose multiplicities of the loose graph; said chromaticities are expressed by

$$\chi^{0}_{LI}(u) = \chi^{j}_{\Delta LI}(u) \text{ for } u \in u_l \text{ and } j \in J;$$

$$\begin{aligned}|\inf \chi^{0}_{ML}(u)| &= \max(\Delta M(\chi^{1}_{\Delta LI}(\{x, y\})), \Delta M(\chi^{2}_{LI}(\{x, y\}))) \\ &\quad \text{for } u \in U^{1}_{d} \cap U^{2}_{d}; \\ &= \Delta M(\chi^{1}_{\Delta LI}(\{x, y\})) \text{ for } u \in U^{j}_{d} \text{ and } j \in J.\end{aligned}$$

$$\begin{aligned}\inf \chi^{0}_{VL}(x) &= \max\nolimits_{j \in J} \\ &\quad (\textstyle\sum_{\{x, y\} \in U^{j}_{e}} M(\chi_{LI}(\{x, y\})) + \\ &\qquad \textstyle\sum_{\{x, y\} \in U^{j}_{d}} \Delta M(\chi^{j}_{\Delta LI}(\{x, y\})) \\ &\quad \text{for } x \in D_{d1} \cap X_{d2}; \\ &= (\textstyle\sum_{\{x, y\} \in U^{j}_{e}} \chi_{M}(\{x, y\}) + \textstyle\sum_{\{x, y\} \in U^{j}_{d}} \chi^{j}_{m}(\{x, y\})) \\ &\quad \text{for } x \in X_{dj} \text{ and } j \in J.\end{aligned}$$

2.6.3. *Connection transfograph*

The dual role of the connection transfograph is to bring together, with no loss of information, a loose graph and two opegraphs that complete it, and to establish their connection on their common elements of development.

If the triplet $(G_0, \Omega_1, \Omega_2)$ is composed of loose graph G_0 and two opegraphs Ω_1 and Ω_2, then Ω_1, Ω_2 completes G_0. The connection transfograph $F = Cn\ (G_0, \Omega_1, \Omega_2)$ is a unique transfograph built from this triplet in such a way that G_0, Ω_1 and Ω_2 can be reconstructed from F with no ambiguity. The equations for constructing the elements of F from the elements of G_0, Ω_1, and Ω_2 given hereafter demonstrate the existence and unicity of the connection transfograph.

CONSTRUCTING F FROM $(G_0, \Omega_1, \Omega_2)$

– The internal elements are the atoms and bonds of G_0 with loose chromaticities equal to zero.

$$x_i = \{x \in X^0, \inf \chi^0_{VL}(x) = 0\};$$

$$\chi_{NA}(x) = \chi^0_{NA}(x) \text{ for } x \in X_i;$$

$$U_i = \{u \in U^0, \inf \chi^0_{ML}(u) = 0\};$$

$$\chi_{LI}(u) = \chi^0_{LI}(u) \text{ for } u \in U_i.$$

– The development nodes and links originate in Ω_1 and Ω_2:

$$X_d = (X^1_d, X^2_d);$$

$$U_d = (U^1_d; U^2_d).$$

– loose chromaticities come from G_0:

$$\chi_{MA}(x) = \chi^0_{NA}(x) \text{ for } x \in X_l;$$

$$\chi_{LIl}(u) = \{\chi^1_{LI}(u), \chi^2_{LI}(u)\} \text{ for } u \in U_l.$$

– The external elements are the external elements of Ω_1 and Ω_2:

$$\chi^s_{\Gamma LIj}(u) = \chi^i_{\Delta LI^+} \text{ for } u \in U^j_d \text{ and for } j \in J;$$

$$X_e = (X^1_e, X^2_e);$$

$$\chi_{NA}(x) = \chi^j_{NA}(x) \text{ for } x \in X^j_e \text{ and for } j \in J;$$

$$U_e = (U^1_e, U^2_e);$$

$$\chi_{LI}(u) = \chi^j_{LI}(u) \text{ for } u \in U^j_e \text{ and for } j \in J.$$

2.6.4. *Extension of isomorphism classes*

The use of a transfostructure for the connection between a loose structure and two operating structures derives from the connection between two representative graphoids of each of these structures.

DEFINITION – A transfostructure $T = Cn\ (L_0, W_1, W_2)$ is the *connection structure* of a loose structure L_0 and of two operating structures W_1 and W_2, provided that G_0 is a loose chromatic graph of L_0 and that Ω_1 is an opegraph of W_1 and Ω_2 is an opegraph of W_2, so that the transfograph $F = Cn(G_0, \Omega_1, \Omega_2)$ exists and belongs to W.

It should be noted that several transfostructures $T = Cn(L_0, W_1, W_2)$ can be the connection structures of the same triplet. There can be several connection transfographs constituted from elements G_0, Ω_1 and Ω_2. The associated transfostructures of these connection transfographs are not necessarily isomorphic.

DEFINITION – The population $P_{Cn}(L_0, W_1, W_2) = \{T \mid T = Cn(L_0, W_1, W_2)\}$ is the population of connection structures between loose structure L_0 and two operating structures W_1 and W_2.

2.7. Initial and final structures

A second interpretation of the transfostructure is presented independently of the first interpretation, although related to it. Said interpretation involves establishing the tie between the transformation structure and the initial and final structures of a chemical reaction. Two chemical structures are first derived from a given transfostructure before proceeding with the reciprocal determination of the transformation structures constructed from two given chemical structures.

2.7.1. *Initial and final graphs derived from a transfograph*

The way to construct a pair of graphs (G_1, G_2) from any F transfograph whatever is given hereafter.

CONSTRUCTING G_1 AND G_2 FROM F

a) nodes regroup internal nodes, development nodes and external nodes.

$$X^j = X_i \cup X_{dj} \cup X_{ej} \text{ for } j \in J.$$

b) links regroup internal links, development links and external links.

$$U^j = U_i \cup U_l \cup U_{ej} \text{ for } j \in J.$$

c) atom natures are deduced directly.

$$\chi^i_{NA}(x) = \chi_{NA}(x) \text{ for } x \in X_i \cup X_{d1} \cup X_{e1} \text{ and for } j \in J;$$

d) Bond types are easily deduced.

$$\chi^j_{LI}(u) = \chi_{LI}(x) \text{ for } u \in U_i \cup X_{d1} \cup U_{e1} \text{ and } j \in J;$$

$$\chi^j_{LI}(u) = \chi^s_{\Gamma LIj}(u) \text{ for } u \in U \text{ and } j \in J.$$

2.7.2. *Conditions for bridging from one graph to another*
The conditions on graphs G_1 and G_2, for passing from one graph to the other by means of a transfograph concern node chromaticities. Said conditions stipulate that if a node of G_1 and a node of G_2 are identical, they necessarily have the same chromaticity.

$$\text{For } x \in X^1 \cap X^2, \chi^1_{NA}(x) = \chi^2_{NA}(x).$$

This limitation stems from the very definition of a chromatic transfograph, which disregards transformations in atom nature.

2.7.3. *Bridging transfograph*
A traversal transfograph intended to guarantee the passage from one graph to another is supposed to gather two chemical graphs with no loss of information, and to highlight those elements common to both graphs as well as the distinct elements that are specific to each of these two graphs.

CONSTRUCTING F FROM G_1 AND G_2

If G_1 (called the initial graph) and G_2 (called the final graph) meet the above-stated conditions for passing from one graph to another, then a bridging transfograph

$$F = Pa(G_1, G_2)$$

can be determined from these two graphs by:
– constructing sets of links

$$U_i \subset \{\{x, y\} \mid \{x, y\} \in U^1 \cap U^2, \chi^1_{LI}(\{x, y\}) = \chi^2_{LI}(\{x, y\}), \chi^1_{NA}(x) = \chi^2_{NA}(x), \chi^1_{NA}(y) = \chi^2_{NA}(y)\};$$

$$U_l \subset \{\{x, y\} \mid \{x, y\} \in U^1 \cap U^2, \chi^1_{LI}(\{x, y\}) \neq \chi^2_{LI}(\{x, y\}), \chi^1_{NA}(x) = \chi^2_{NA}(x), \chi^1_{NA}(y) = \chi^2_{NA}(y)\};$$

$$U_{ej} = U^j - U_i - U_l \text{ for } j \in J.$$

– constructing sets of nodes

$$X_{dj} = \{x \mid \exists u \in U_{ej}, x \in X^1 \cap X^2, x \in u\} \cup \{x \mid \exists u \in U_l, x \in u\};$$

$$X_i = X_1 \cap X_2 - X_{d1} \cup X_{d2};$$

$$X_{ej} = X_j - X_{dj} - X_i \text{ pour } j \in J.$$

– constructing chromaticities

$$\chi_{NA}(x) = \chi^j_{NA}(x) \text{ for } x \in X^j \text{ and } j \in J;$$

$$\chi_{LI}(u) = \chi^j_{LI}(u) \text{ for } U \in U_j - U_l \text{ and for } j \in J;$$

$$\chi_{LIl}(u) = \{\chi^1_{LI}(u), \chi^2_{LI}(u)\} \text{ for } u \in U;$$

$$\chi^s_{\Gamma LIj}(u) = \chi^j_{LI}(u) \text{ for } u \in U \text{ and } j \in J.$$

Each choice of U_i and of U_l determines a single traversal transfograph. The case where both inclusions are equalities corresponds to the maximal transfograph regarding anteriority between transfographs; redundancy between the two graphs is thus minimized by maximalizing internal and loose modules.

2.7.4. *Extension to classes of isomorphism*
The relationship for bridging these two structures is the relationship for the passage between two graphs belonging to each of these two structures.

DEFINITION – A transfostructure T is the bridging structure from an initial structure S_1 to a final structure S_2 ($T = Pa(S_1, S_2)$), if G_1 is a graph of S_1 and G_2 is a graph of S_2 so that the existence of transfograph $F = Pa(G_1, G_2)$ and its membership in T are ensured.

DEFINITION – $P_{Pa}(S_1, S_2) = \{T \mid T = Pa(S_1, S_2)\}$ is the population of bridging structure from an initial structure S_1 to a final structure S_2. Said definition applies to
- two graphs connected by transfographs, whether isomorphic or not,
- two isomorphic transfographs connected by distinct pairs of graphs.

3. Generation-description of a transfostructure

3.1. Generation of a (chromatic) transfograph

The generation of a transfograph F involves six stages: an adjacent anteriority relationship a is introduced throughout an entire set of transfographs **F**, thereby inducing a partial order on this set (3.1.1); a is defined by means of a collection of elementary adjunction operators grouped into ten categories (3.1.2) according to their conditions of applicability (3.1.3) and their activation procedure (3.1.4); a restriction is then applied to this relationship to define the generation network or hyperstructure $HS(F)$ of a particular transfograph F (3.1.5); $HS(F)$ then gives out onto all the generation pathways of all the transfographs in **F** (3.1.6).

3.1.1. *Adjacent anteriority relationship*
The adjacent anteriority relationship a sought for the entire set of transfographs **F** can be viewed in two different ways:
- by $a = \{(F_1, F_2) \mid F_1\ a\ F_2\}$ the set composed of all pairs of transfographs in **F** joined together by adjacent anteriority;
- by $+a$ the multivocal many-to-one operator of elementary topochromatic adjunctions which, when applied to an initial transfograph F_1, produces a population $F_2{+}a$ of final transfographs F_2 linked to F_1 by anteriority.

Moreover, the breakdown of set a into a partition $\hat{a}$ of subsets of a can result in the breakdown of operator $+a$ into a set a of univocal operators of elementary topochromatic adjunctions; when applied to an initial F_1 transfograph, each univocal operator $+ \alpha$ produces a single final transfograph $F_2 = F_1 + \alpha$ (deterministic operator as defined for example in [2] or [4].

By using the latter process to define *a*, each elementary topochromatic adjunction operator + of set *a* (simply called an "elementary adjunction") is characterized by a topochromatic element. (It should be recalled that a topochromatic element involves no more than a single node and its adjacent links.)

The type of topochromatic elements which can be adjoined is, of course, constrained by the internal structure of operand transfograph F_1; however, many possibilities corresponding to a more or less sizeable chromatic or topological density can be envisaged.

Conformity to structure generation principles initially applied to structures of chromatic graphs representing chemical entities and a wish to facilitate subsequent chemical and computational applications served as guidelines for our choice of topochromatics elements. From a chemical standpoint, the aim is to prepare an interpretation of anteriority links as significant neighboring relations between chemical transformations intended to guide the establishment of transformation-behavior relationships. From a computational standpoint, the aim is to clarify, simplify and parametrize the establishment of algorithms used in building a transfograph and in selecting a single generation pathway.

The formal approach proposed hereafter meets these aims

- by highlighting maximal topochromatic elements (i.e. nodes and links that always have chromaticity, . . .);
- by avoiding the generation of non-transfograph intermediate entities, and by curtailing the number of intermediate entities that cannot be interpreted as transformations;
- by authorizing a concentric anteriority order around the "natural" focus, composed of all development nodes and links, that privilege attachment to these nodes.

What qualifies as a topochromatic element?

- A node assigned a chromaticity corresponding to the nature of an atom;
- A link assigned a corresponding chromaticity e.g. a bond;
- A node and one of its adjacent links with their assigned chromaticities e.g. an atom and one of its adjacent bonds;
- a chromaticity specifying the type of bond.

3.1.2. *Elementary adjunctions*

To define adjacent anteriority relationship *a* with accuracy, associated multivocal operator + *a* is broken down into a set of elementary adjunction univocal operators **a*:

$$F_1 \; a \; F_2 \Leftrightarrow \exists \alpha \in \hat{a}, \; F_2 = F_1 + \alpha.$$

Each elementary adjunction $+ \alpha$ should specify the nature of the adjoined topochromatic element as well as its character and its subsequent index; $+ \alpha$

should also indicate the topochromatic elements needed to carry out the adjunction. A redundant information item identifying the type of adjoined topochromatic element (node, link, chromaticity, . . .) is also needed to express ordering criteria as a lexicographical classification order for elementary adjunctions.

It is suggested that each elementary adjunction be designated by a finite set of attributes wherein each attribute is taken from distinctly separate finite sets called reference sets defined hereafter.

REFERENCE SETS

Three types of reference sets can be distinguished: sets of non-topochromatic references, topological references, and chromatic tables.

1. Sets of non-topochromatic references are:
   - set $\{+\}$ reduced to + to symbolize adjunction (at this stage, there are no plans to use the minus sign – to designated ablations);
   - the set of characters $CR = \{l, i, e\}$: l for loose, i for internal, e for external;
   - the set of topochromatic types $CA = \{K, L, N', N\}$: in which K designates bond chromaticity, L a link with its chromaticity, N a node with its chromaticity, N' a node and its adjacent link of attachment with their chromaticities;
   - the set of indices $J = \{1, 2\}$: in which 1 designates ablation and 2 designates adjunction.
2. Topological references are:
   - node referential $\mathbf{X}$;
   - the reference for unoriented links built from $\mathbf{X}$:

   $$\mathbf{U} = \{\{x, y\} \mid x, y \in \mathbf{X}, x \neq y\}$$

   - the reference for oriented links built from $\mathbf{X}$:

   $$\mathbf{U}' = \{(x, y) \mid x, y \in \mathbf{X}, x \neq y\}$$

3. chromatic tables are all tables involved in the transfograph definition, i.e. tables giving the atom natures K_{NA}, the bond types K_{LI}, bond type looseness K_{LIl} and bond type specifications $K_{\Gamma LI}$

An adjunction is expressed by a given combination of elementary attributes derived from these reference sets:

RULE – Each and every adjunction of a is designated by a plus sign +; a character from set CA; an index from set J if the character is external; a topological element from $\mathbf{X}$, $\mathbf{U}$ or $\mathbf{U}'$; two chromaticities if the topochromatic type is N' (node and link), a single chromaticity in every other case.

TYPES OF ELEMENTARY ADJUNCTIONS

Elementary adjunctions in the collection of items constituting a are separated into ten categories so as to group together adjunctions with a shared meaning and thereby simplify the definition of their conditions of applicability and of their application procedure.

In these ten categories (numbered from 01 to 010) listed below, specific elements (elements that can take on a certain given value only) are printed in boldface to distinguish them from generic elements (each member element that can take on any value whatsoever from a set of possible values)

01. +, *l*, *L*, *x*, *y*, K_{LII}: designates the input of a new loose link $\{x, y\}$ (between two existing nodes *x* and *y*) with an assigned looseness chromaticity indicating the bond type K_{LII}.

02. +, *l*, *N′*, (*x*, *y*), k_{LII}, k_{NA}: designates the input of a new loose link $\{x, y\}$ (from existing node *x*) with an assigned looseness chromaticity indicating the bond type k_{LI}, and of a new loose node *y* with an assigned chromaticity indicating the atom nature k_{NA}.

03. +, *l*, *N*, *y*, k_{NA}: designates the input of a new free node *y* lone with an assigned chromaticity to indicate the nature of the atom k_{NA}.

04. +, *i*, *L*, $\{x, y\}$, k_{LI}: designates the input of a new internal link (between two existing nodes *x* and *y*) with an assigned chromaticity indicating the bond types k_{LI}.

05. +, *i*, *N′*, (*x*, *y*), k_{LI}, k_{NA}: designates the input of a new internal link (*x*, *y*) (starting from an existing node *x*) with a chromaticity indicating the bond type k_{LI}, and of a new internal node *y* with a chromaticity indicating the atom nature k_{NA}.

06. +, *i*, *N*, *y*, k_{NA}: designates the input of a new internal node *y* with a chromaticity indicating the atom nature k_{NA}.

07. +, *e*, *j*, *K*, $\{x, y\}$, $k_{\Gamma LI}$: designates the input of a chromaticity specifying the bond type $k_{\Gamma LI}$ with index *j* on an existing loose link $\{x, y\}$.

08. +, *e*, *j*, *L*, $\{x, y\}$, k_{LI}: designates the input of a new external link $\{x, y\}$ with an index *j* (between two existing nodes *x* and *y*) and with a chromaticity indicating the bond type k_{LI}.

09. +, *e*, *j*, *N′*, (*x*, *y*), k_{LI}, k_{NA}: designates the input of a new external link *x*, *y* bearing an index *j* (between two existing nodes *x* and *y*) with an assigned chromaticity indicating the bond type k_{LI}, and of a new external node *y* with an assigned chromaticity indicating the atom nature k_{NA}.

10. +, *e*, *j*, *N*, *y*, k_{NA}: designates the input of a new external node *y* bearing index *j* and with an assigned chromaticity indicating the atom nature k_{NA}.

So that the use of these ten types of elementary adjunctions can be accurately characterized, the way in which the operations for the transformation of transfograph F_1 into transfograph F_2 are actually conducted is dealt with hereafter via a definition of the conditions of applicability as well as of the application procedure for each of the ten types.

3.1.3. *Applicability conditions*

The conditions of applicability associated with each type of elementary adjunction are those conditions under which an initial transfograph F_1 (the operand) can be transformed into another transfograph F_2 (the result of the operation) by using a given type of adjunction $+\alpha$; these conditions mark off the field of applicability of adjunction $+\alpha$, i.e. the set of transfographs to which adjunction $+\alpha$ can be applied.

The applicability conditions (described with set theory symbols) corresponding to each type of operator are given in Table I. In deciding whether a given specific elementary adjunction applies to a transfograph, the type of adjunction must be determined and the applicability conditions corresponding to the type of adjunction must be tested. These conditions are classified into four groups arranged in descending order of importance. The meaning of each group of conditions is briefly explained hereafter.

C1. The conditions in this first group are the most fundamental: they are directly related to the disjunction conditions resulting from the very *definition* of a *chromatic transfograph* and account for the fact that an elementary adjunction can occur only by respecting the disjoint nature of the sets of internal, loose and external nodes, and by using authorized relationship connecting nodes from a given character.

For example, in the case of a 01 type of adjunction, the conditions stipulate that a loose link can be added only if this link is not already in the transfograph as an internal or external link. This stems from the disjoint nature of the sets of internal, loose and external links.

C2. The conditions in this second group are essential because they reflect the adopted *system of adjunctions* and the nature of adjoined topochromatic elements. These conditions stipulate that: to add a link between two nodes, each of the two nodes must already exist in operand transfograph F; to adjoin a node and an adjacent link starting from another node, this other node must already lie in F_1; to adjoin a loose chromaticity specification, the loose chromaticity must already be given in F_1.

C3. The conditions in the third group stem from the valency constraints that need to be checked for each node in a chromatic transfograph if it is to receive a chemical interpretation. The addition of an isolated node is not subject to this type of condition; however, a bond adjunction can take place only if its addition does not result in an overextension of the maximum authorized valency.

Together, conditions C1, C2 and C3 guarantee the internal nature of the adjunctions: the result of carrying out an adjunction on transfograph F_1 (a chromatic transfograph obeying the valency constraint) will also be a transfograph.

C4. The conditions in the fourth group are in a category by themselves because they are optimization constraints that are less indispensable than conditions C1–C3: they serve mainly to curtail the triggering of temporary operations that will be repeated or replaced. As the sequence of adjunctions is intended to build a target transfograph in the most efficient possible way, repeated adjunction of the same topochromatic element and replacement of one chromaticity by another chromaticity are

TABLE I
Applicability conditions associated with the ten types of elementary adjunctions

Adjunction \ Conditions		C1	C2	C3	C4
O_1	$+, l, L, \{x, y\}, k_{LIl}$	$\{x, y\} \notin U_i \cup (\bigcup_j U_{ej})$	$x, y \in X_l$	$M^+(k_{LIl}) \leq -\chi_{\Delta V}(x), -\chi_{\Delta V}(y)$	$\{x, y\} \notin U_l$
O_2	$+, l, N', \{x, y\}, k_{LIl}, k_{NA}$	$y \notin X_i \cup (\bigcup_j X_{ej})$	$x \in X_l$	$M^+(k_{LIl}) \leq -\chi_{\Delta V}(x), V(k_{NA})$	$y \notin X_l$
O_3	$+, l, L, y, k_{NA}$	$y \notin X_i \cup (\bigcup_j X_{ej})$			$y \notin X_l$
O_4	$+, i, L, \{x, y\}, k_{LI}$	$\{x, y\} \notin U_l \cup (\bigcup_j U_{ej})$	$x, y \in X_i \cup X_l$	$M(k_{LI}) \leq -\chi_{\Delta V}(x), -\chi_{\Delta V}(y)$	$\{x, y\} \notin U_i$
O_5	$+, i, N', \{x, y\}, k_{LI}, k_{NA}$	$y \notin X_l \cup (\bigcup_j X_{ej})$		$M(k_{LI}) \leq -\chi_{\Delta V}(x), V(k_{NA})$	$y \notin X_i$
O_6	$+, i, N, y, k_{NA}$	$y \notin X_l \cup (\bigcup_j X_{ej})$			$y \notin X_i$
O_7	$+, e, j, K, \{x, y\}, k_{\Gamma LI}$		$\{x, y\} \in U_l;\ k^l_{\Gamma LI} = \chi_{LIl}(\{x, y\})$		$\chi_{LIl}(\{x, y\})$ non defini
O_8	$+, e, j, L, \{x, y\}, k_{LI}$	$\{x, y\} \notin U_i \cup U_l$	$x, y \in X_l \cup X_{ej}$	$M(k_{LI}) \leq -\chi_{\Delta V}(x), -\chi_{\Delta V}(y)$	$\{x, y\} \notin U_{ej}$
O_9	$+, e, j, N', \{x, y\}, k_{LI}, k_{NA}$	$y \notin X_i \cup X_l \cup X_{ej'}\ (j' \neq j)$	$x \in X_l \cup X_{ej}$	$M(k_{LI}) \leq -\chi_{\Delta V}(x), V(k_{NA})$	$y \notin X_{ej}$
O_{10}	$+, e, j, N, y, k_{NA}$	$y \notin X_i \cup X_l \cup X_{ej'}\ (j' \neq j)$			$y \notin e_j$

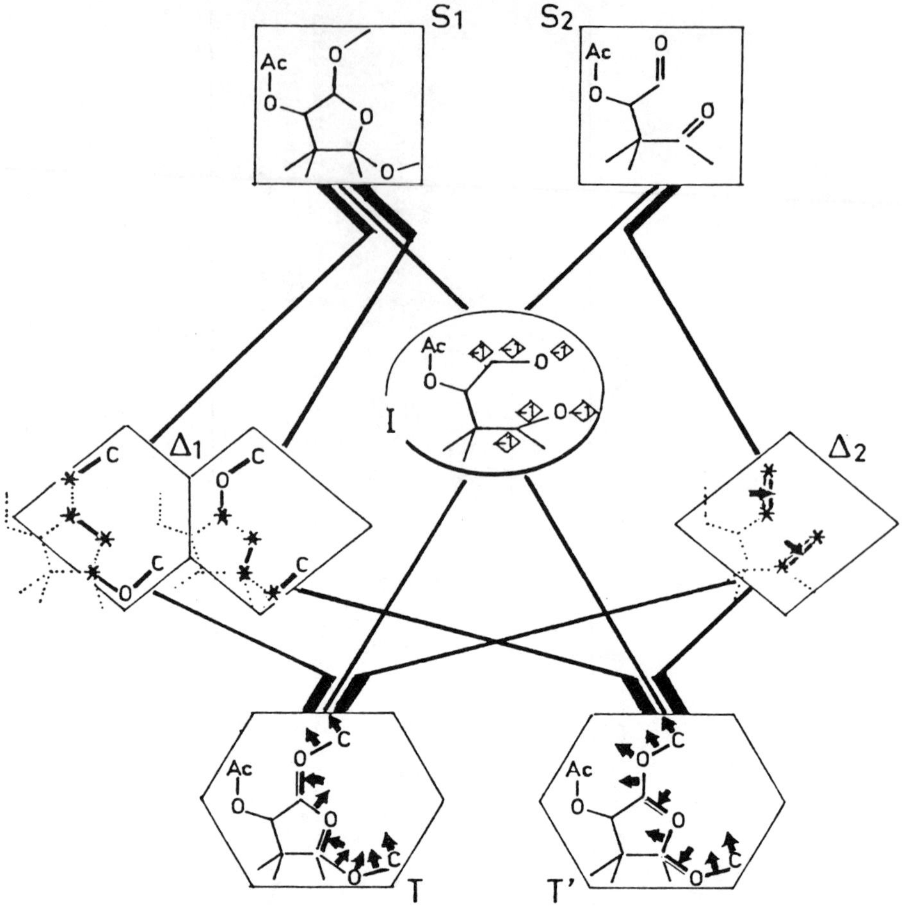

Figure 3. Transfostructures as chemical reactions: The localization problem.

avoided. By using conditions C4 it is possible to remain in the strict realm of pure adjunction operations: no topochromatic ablation (not even implicit ablation), no substitution of chromaticity, no transition to another chromaticity.

3.1.4. *Application procedure*

The application procedure associated with each type of elementary adjunction describes the way to obtaining final transfograph F_2 from an initial transfograph F_1 by enumerating the topochromatic changes (adjunctions) to be made in F_1 to build F_2.

Table II shows how to actually build F_2 from F_1 by using a given type of elementary adjunction $+\alpha$. The usual set theory symbols for assignment and "union" have been used to describe the construction of the elements of F_1 from the elements in F_2. For simplicity, only the way to obtain those elements of

TABLE II
Applicability procedure associated with the ten types of adjunctions

	Adjunction type	Aplication procedure
O_1	$+, l, L, \{x, y\}, k_{LIl}$	$U_l \leftarrow U_l \cup \{\{x, y\}\}; \chi_{LIl}(\{x, y\}) \leftarrow k_{LIl}$
O_2	$+, l, N', (x, y), k_{LI}, k_{NA}$	$U_l \leftarrow U_l \cup \{\{x, y\}\}; \chi_{LIl}(\{x, y\}) \leftarrow k_{LIl}$ $X_l \leftarrow X_l \cup \{y\}; \chi_{NA}(y) \leftarrow k_{NA}$
O_3	$+, l, N, y, k_{NA}$	$X_l \leftarrow X_l \cup \{y\}; \chi_{NA}(y) \leftarrow k_{NA}$
O_4	$+, i, L, \{x, y\}, k_{LI}$	$U_i \leftarrow U_i \cup \{\{x, y\}\}; \chi_{LI}(\{x, y\}) \leftarrow k_{LI}$
O_5	$+, i, N', (x, y), k_{LI}, k_{NA}$	$U_i \leftarrow U_i \cup \{\{x, y\}\}; \chi_{LI}(\{x, y\}) \leftarrow k_{LI}$ $X_i \leftarrow X_i \cup \{y\}; \chi_{NA}(y) \leftarrow k_{NA}$
O_6	$+, i, N, y, k_{NA}$	$X_i \leftarrow X_i \cup \{y\}; \chi_{NA}(y) \leftarrow k_{NA}$
O_7	$+, e, j, L, \{x, y\}, k_{\Gamma LI}$	$\chi_{\Gamma LIj}(\{x, y\}) \leftarrow k_{\Gamma LI}$
O_8	$+, e, j, L, \{x, y\}, k_{LI}$	$U_{ej} \leftarrow U_{ej} \cup \{\{x, y\}\}; \chi_{LIj}(\{x, y\}) \leftarrow k_{LI}$
O_9	$+, e, j, N', (x, y), k_{LI}, k_{NA}$	$U_{ej} \leftarrow U_{ej} \cup \{\{x, y\}\}; \chi_{LIj}(\{x, y\}) \leftarrow k_{LIl}$ $X_{ej} \leftarrow X_{ej} \cup \{y\}; \chi_{NA}(y) \leftarrow k_{NA}$
O_{10}	$+, e, j, N, k_{NA}$	$X_{ej} \leftarrow X_{ej} \cup \{y\}; \chi_{NA}(y) \leftarrow k_{NA}$

F_1 unlike those in F_2 is dealt with; the other unspecified elements of F_1 are equal to those elements of F_2 which remain unchanged by adjunction.

A hierarchical classification by descending order of importance of construction operations follows.

M1. Major topological or chromatic modifications corresponding to the addition of a node, a link or a chromaticity from a specified character;

M2. Secondary chromatic modifications, so-called because borne by the adjoined topological element;

M3. Consequences on the assignment of loose nodes or loose links to specific development sets.

DISJOINT NATURE OF ADJUNCTIONS

Another fundamental property of elementary adjunctions, aside from univocity (which means that a single entity results from an adjunction) and the internal nature of adjunctions (the resulting entity is a transfograph), is proposed here. They are all disjoint

$$F_2 = F_1 + \alpha \text{ and } F_2 = F_1 + \alpha' \Rightarrow \alpha = \alpha'.$$

This property is easily demonstrated by the fact that:

a) two adjunctions of $*a$ differ either by the adjoined topochromatic element, or by the set (loose, internal, or external with index J specified) to which it is adjoined;

b) an adjunction of $*a$ actually modifies a F' transfograph to which a can apply, i.e. this adjunction of a produces a final transfograph F that is unlike transfograph F' and that has more topochromatic elements than transfograph F' (thanks to applicability conditions C4).

So, two different adjunctions $+\alpha$ and $+\alpha'$ applied to the same initial transfograph F' will necessarily yield two different transfographs.

3.1.5. *Generation hyperstructure*

Adjacent anteriority relationship a defined via a, the collection of elementary adjunction operators, puts into set **F** all chromatic transfographs abiding by the valency constraint of a partially ordered structure. An overall anteriority relationship designated by A, the transitive closure of a, is used hereafter to define this order. It should be remembered that A is recursively defined by transitivity:

$$F_1\ a\ F_2 \Rightarrow F_2\ A\ F_1;\ (F_1\ A\ F_2 \text{ and } F_2\ A\ F_3) \Rightarrow F_1\ A\ F_3$$

in which F_1, F_2 and F_3 designate the transfographs in **F**.

By isolating $P_A(F_2) = \{F_1 \in \mathbf{F} \mid F_1\ A\ F_2\}$ the population of all the transfographs linked to F_2 by A, from those populations marked off by anteriority relationship A, we obtain the population of anteriors of F. The ordering relationship designated by $<$ is defined by the inclusion of populations of "anteriors":

$$F_1 < F_2 \Leftrightarrow P_A(F_1) \subset P_A(F_2)$$

So the hyperstructure (**F**, a) can be represented by an oriented chromatic graph that is connected and circuitless in which: each node represents a transfograph of **F**. The chromaticity of the chromatic graph designates the structure of this graph; an arc joins two nodes when they represent two transfographs related by anteriority a; arc chromaticity specifies the adjunction (or the type of adjunction) actually used in building each arc.

Actually, hyperstructure (**F**, a) regroups all generation networks $HS(F)$ (not necessarily disjoint) of all F transfographs in **F**.

The restriction of anteriority relationship a to $P_A(F)$, simply designated by a, determines the generation hyperstructure $HS(F)$ of transfograph F.

DEFINITION – Hyperstructure $HS(F) = (P_A(F), a)$, associating anterior transfograph population $P_A(F)$ of F and the restriction of anteriority relationship a to $P_A(F)$, is the *generation hyperstructure* of transfograph F.

Generation hyperstructure $HS(F)$ can be represented by a connected and circuitless chromatic graph that has two specific nodes: the source node representing empty transfograph F_0 and the target node representing F, the transfograph to be generated. In the $HS(F)$ network, F_0 is the only minimal element of $P_A(F)$ ordered according to A and F is the only maximal element: the graph of the hyperstructure is therefore a lattice.

3.1.6. *Generation pathways*

In the lattice representing generation hyperstructure $HS(F)$ of transfograph F, every pathway from F_0 to F is a generation pathway.

DEFINITION – A sequence $C(F) = (F_0, F_1, \ldots, F_n = F)$ in which two successive transfographs are related by adjacent anteriority a is a transfograph F *generation pathway* from empty transfograph F_0.

The result of this definition and of the partial ordering structure for an anteriority graph is that a generation pathway constitutes a strictly ascending monotonic sequence of transfographs. Moreover it should be noted that the generation pathways of the same F transfograph share the source transfograph F_0 and the target transfograph F; the pathways differ by the number and nature of the intermediate transfographs $F_1, \ldots, F_{n-1}$.

Another way of characterizing a generation pathway $C(F)$ is to enumerate, in order to use, the elementary adjunctions of a used in building each of the arcs connecting the successive intermediate transfographs of the pathway. Indeed, every arc $(F_j, F_{j'})$ of the pathway is such that F_j is anterior to $F_{j'}$; therefore, there is one and only one adjunction $+\alpha_{j'}$ and only one which fits $F_{j'} = F_j + \alpha_{j'}$ (because the adjunctions of $*a$ are disjoint) . . .

PROPERTY – To every generation pathway $C(F)$ is associated the sequence $(+\alpha_1, \ldots, +\alpha_n)$ which fits $F_j = F_{j-1} + \alpha_j$ when $j \in \{1, \ldots, n\}$.

Conversely, one must envision the conditions under which a generation pathway is determined by a finite series of elementary adjunctions, i.e. the conditions under which the finite series of elementary adjunctions is "generative".

DEFINITION – A *generative series* is defined recursively as:

a) series $(+\alpha_1)$ is generative if $+\alpha_1$ applies to F_0; if so, then $F((+\alpha_1)) = F_1 = F_0 + \alpha_1$;

b) series $(+\alpha_1, \ldots, +\alpha_n)$ wherein $n > 1$ is a generative series if $(+\alpha_1, \ldots, +\alpha_{n-1})$ is a generative series and if $+\alpha_n$ applies to $F_{n-1} = F((+\alpha_1, \ldots, +\alpha_{n-1}))$;

if so, then $F((+\alpha_1, \ldots, +\alpha_n)) = F_n = F_{n-1} + \alpha_n$.

CHARACTERISTIC PROPERTY: A generative sequence $(+\alpha_1, \ldots, +\alpha)$ characterizes a generation pathway $(F_0, F_1, \ldots, F_n)$ of F in which $F_j = F((+\alpha_1, \ldots, +\alpha_j))$ for $j \in \{1, \ldots, n\}$.

It should be noted that two generation pathways from the same final transfograph F can differ in the number, nature and order of elementary adjunctions characterizing them.

There are several possibilities:

- two pathways characterized by a different number of adjunctions;
- two pathways characterized by the same number of adjunctions, but of a different type, for each pathway;
- two pathways characterized by the same number of adjunctions of the same type, with the only difference lying in the order of adjunctions for each pathway.

In other, simpler, systems of adjunction, generation pathways differ neither in the number nor in the nature of adjunctions, but only in the order of adjunctions.

It should also be noted that, in a generative series, elementary adjunctions are not only different, they are also independent. In the same generative series,

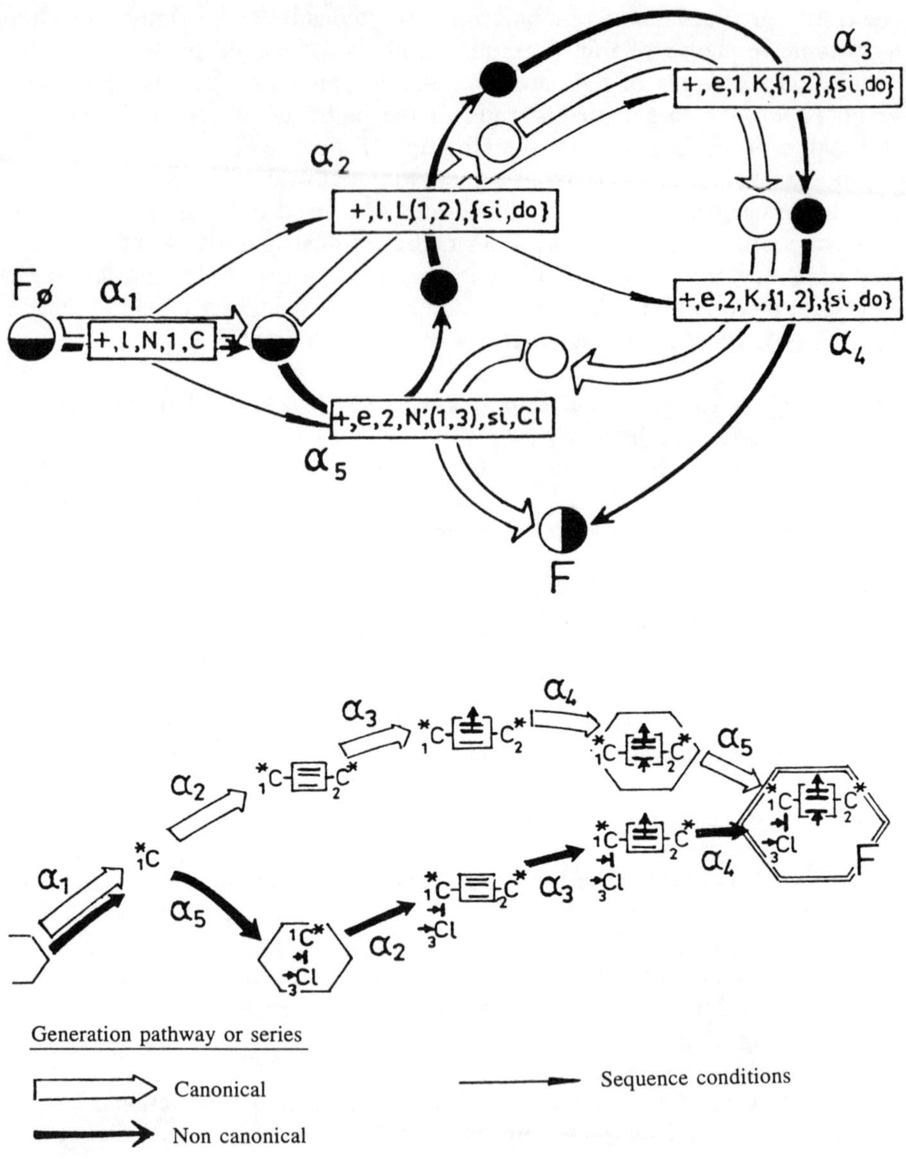

Figure 4. Elementary adjunctions and generation of a chromatic transfograph *F*.

one either uses a straightforward type N' adjunction to build a given node x and a given attachment link $\{x, y\}$, or one uses a type N adjunction to build x and a type L adjunction to build $\{x, y\}$. Use of type N', N and L adjunctions for the same (x, y) pair would constitute a dependency relationship between elementary adjunctions.

3.2. Ordered generation of a chromatic transfograph

The ordered generation of a F transfograph should result in the establishment of the canonical pathway of F. The presentation of this ordered generation in four parts – ordering criteria for elementary adjunctions (3.2.2), ordered set of elementary adjunctions (3.2.3), canonical pathway (3.2.4) and synchronism condition (3.2.5) – is preceded by a preliminary consideration (3.2.1) introducing the methodology adopted in approaching the problem of ordered generation.

3.2.1. *The ordered generation in its context*

To solve the problem of choosing a canonical generation pathway, it is necessary to establish a general systematic rule that makes it possible to:

- reach every transfograph F of $\mathbf{F}$ via a single generation pathway, canonical generation pathway $C^o(F)$;
- reach every intermediate transfograph F_j on the canonical pathway $C^o(F)$ of F via a sub-pathway of $C^o(F)$ that is the canonical generation pathway $C^o(F_j)$ of intermediate transfograph F_j.

More specifically, an attempt is being made to establish the adjacent anteriority arboresence restriction on the adjacent anteriority hyperstructure network ($\mathbf{F}$, a) bearing:

- all the generation pathways of each of the transfographs in $\mathbf{F}$
- the canonical generation pathway of all the transfographs in $\mathbf{F}$.

The proposed solution to this problem involves four stages:

1. Seeking a total order for the entire set a of elementary adjunctions;
2. Using the lexicographical order deduced from the order for $*a$ to classify all generation pathways of $\mathbf{F}$;
3. Restricting this order to classify all generation pathways of the same F transfograph;
4. Selecting the first pathway as the canonical generation pathway $C^o(F)$.

Once this procedure is set into motion, the only authorized latitude in choice is in stage 1 (determining the order for the adjunctions in a); the other three stages are purely deductive, although determinant.

In establishing an order for elementary adjunctions, an attempt has been made to abide by the general principles of focalization and concentric generation around a focus, while taking into account the specificity of a chromatic transfograph as the mode of expression for a chemical transformation. A particular effort has been made to give priority to the loose elements of development constituting the dynamic center of transformation (reaction sites of a chemical reaction) and to the external elements of attachment characterizing the active dynamics of the transformation (valency bonds formed, broken or altered by a chemical reaction).

First, a loose module of the transfograph is built; then the three other modules are built around it. Hence, the loose module acts as a generation focus; given its particularity of bearing origins of development only, it is called a transfograph pivot focus (FOP). The other three modules constitute three environments around the pivot focus: internal environment E_i, ablation environment E_{e1}, adjunction environment E_{e2}. Global environment $E = (E_i, E_{e1}, E_{e2})$ encompasses these three environments. They have been generated independently and successively for simplicity and to isolate the invariant and the ablation and adjunction variations.

The principle of generation via the propagation of an Environment that is Limited, Concentric and Ordered (ELCO) has been obeyed for each environment, E_i, E_{e1} and E_{e2}; however, no use was made of the law for generating by propagation of an environment limited to two layers (e_B) governed by priorities P_0, P_A and P_B that defines the major order of the DARC system. A two layer environment is better suited to graphs of chemical compounds motivated by considerations of a chemical nature (several properties of mono- and bifunctional compounds vary little beyond the second layer), and by a desire to optimize the description procedure (decision-making speed for classifying positions in the same environment, ease in passing to a descriptor of existence). Instead, for the sake of simplicity and to highlight the role of attachment and fixation bonds, the chosen generation procedure is via propagation of an environment E_a limited to a single concentric LAYER.

A strict description of ordering criteria and their impact on the choice of canonical pathway is preceeded by an informal summary of the way to generate transfograph F.

1. modules are generated in the following order: loose, internal, ablation and, finally adjunction E_{e2}.
2. Once the referential node is entirely ordered, the loose module is generated concentrically around the first loose node and each of the three remaining modules is generated successively and concentrically around the loose module while respecting the order of the nodes (classification according to the order of the node predecessors, then according to their node numbers.
3. Within each module, adjunction priority is given as soon as possible to lone chromaticities, to atoms connected by a single bond to existing atoms, to bonds between two existing atoms, and, lastly, to isolated atoms. Consequently, it should be noted that within a module a connected component is generated in its entirety before the next subsequent connected component is generated.

3.2.2. *Ordering criteria*

A total order for $*a$ the entire set of elementary adjunctions is defined by two levels of ordering criteria.

It should be recalled that a set of attributes, each of which is taken from a distinct finite reference set, has been associated to every adjunction in $*a$ (cf. 3.1.2). An adjunction in a is designated by:

– a plus sign +,

- a character of CA,
- an index of J if the character is external,
- a topological element of $\mathbf{X}$, $\mathbf{U}$, $\mathbf{U}'$,
- two chromaticities if the topochromatic type is N' (node and link),
- a single chromaticity in every other case.

Two types of ordering criteria are discussed: inter-reference ordering criteria for the classification of all reference sets and intra-reference ordering criteria for the classification of elements within each reference set. Intra-reference orders are inter-dependent. Orders in $\mathbf{U}$ and $\mathbf{U}'$ are deduced from the order in $\mathbf{X}$; orders in K_{LII} and $K_{\Gamma LI}$ are deduced from the order in K_{LI}.

E^o designates reference set E with its intra-reference total order; j_E designates the jth element in E^o.

1. INTER-REFERENCE CRITERIA – Reference sets involved in elementary adjunctions are classified in ascending order:

$$(CR, J, CA, \mathbf{U}, \mathbf{U}', \mathbf{X}, K_{\Gamma LI}, K_{LII}, K_{LI}, K_{NA}).$$

It should be noted that the non-topochromatic references (CR, J and CA) are at the head of these reference sets, and that they are followed by the topological references ($\mathbf{U}$, $\mathbf{U}'$ and $\mathbf{X}$), before ending with the chromatic tables.

2. INTRA-REFERENCE CRITERIA – A distinction is made between non-topochromatic sets, topological references and chromatic tables.

a) Non-topochromatic reference sets are ordered totally as follow:

$$CR^o = (l, i, e);\ J^o = (1_J, 2_J);\ CA^o = (K, L, N', N).$$

b) topological references are ordered totally on the basis of an arbitrary total order used for node reference $\mathbf{X}$:
   - the order in $\mathbf{X}$ defines $\mathbf{X}^o = (1_\mathbf{X}, 2_\mathbf{X}, \ldots)$;
   - the order in $\mathbf{U}$ follows therefrom: since $x1 < y1$ and $x2 < y2$, then $\{x1, y1\} < \{x2, y2\} \Leftrightarrow x1 < x2$ or ($x1 = x2$ et $y1 < y2$);
   - the order in $\mathbf{U}'$ also follows therefrom:

$$(x1, y1) < (x2, y2) \Leftrightarrow x1 < x2 \text{ or } (x1 = x2 \text{ et } y1 < y2).$$

c) Chromatic tables are ordered totally:
   - chromaticities of K_{NA} are classified in the usual order of their ascending atomic numbers in a Mendeleiev table:

$$K^o_{NA} = (LI, B, Be, \ldots);$$

   - chromaticities of K_{LI} are first classified by ascending multiplicity, then, when multiplicity is the same, by the arbitrary order $ta < ar$:

$$K^o_{LI} = (si, ta, ar, do, tr);$$

   - chromaticities of K_{LI} are first classified by using an ascending multiplicity variation, then by using the lexical order derived from the order in K_{LI}: since $K_1 < k'_1$ and since $K_2 < K'_2$,

$$k_1, k'_1 < k_2, k'_2 \Leftrightarrow \Delta M(\{k_1, k'_1\}) < M(\{k_2, k'_2\}) \text{ or}$$
$$(\Delta M(\{k_1, k'_1\}) = M(\{k_2, k'_2\}) \text{ and } (k_1 < k_2 \text{ or } (k_1 = k_2 \text{ and } k'_1 < k'_2)))$$

– chromaticities of K_{LI} are first classified by ascending loose chromaticity, then by ascending specified chromaticity:

$$k_l, k_s < k'_l, k'_s \Leftrightarrow k_l < k'_l \text{ or } (k_l = k'_l \text{ and } k_s < k'_s).$$

3.2.3. *Ordered set of adjunctions*

These ordering criteria make it possible to define the total order for the set of elementary adjunctions a as a lexicographical order obtained from an order for all the reference sets together taken as one.

The total order for all the reference sets taken as one is obtained by successively applying inter- and intra-reference criteria.

DEFINITION – An element e' from a reference set E' is smaller than an element e from another reference set $E(e' < e)$ if, and only if, one of the two following propositions is true:

a) $E' < E$ in abidance with inter-reference criterion 1;

b) $E' = E$ and $e' < e$ in abidance with intra-reference criterion 2a, 2b or 2c.

Each adjunction is designated by a strictly ascending series: $+\alpha = (e_1, \ldots, e_j, \ldots, e_n)$. The definition of lexicographical order for a is simplified by the fact that the start of each series determines the length of the series.

DEFINITION – Adjunction $+\alpha'$ is smaller than adjunction $+\alpha$ $(+\alpha' < +\alpha)$ if one of the two following propositions is true:

a) the first element in $+\alpha'$ is smaller than the first element in $+\alpha$: $e'_1 < e_1$;

b) the first j elements in both series are equal and the $(j + 1)$th element of $+\alpha'$ is smaller than the $(j + 1)$th element of $+\alpha$:

$$j \in [1, \min(n, m)],$$
$$(ee'_1, \ldots, e'_j) = (e_1, \ldots, e_j) \text{ et } e'_{j+1} < e_{j+1}.$$

DEFINITION – Set $*a$ associated with a total order relationship is the ordered set $*a^o$ of elementary adjunctions.

3.2.4. *Canonical generation pathway*

The lexicographical order for the set of adjunction series deduced from the order for $*a^o$ is a total order. The restriction of this order to hyperstructure $HS(F)$ regrouping all the generation pathways of an F transfograph makes it possible to classify these pathways and to choose the canonical generation pathway on the list.

DEFINITION – A series of adjunctions $s' = (+\alpha'_1, \ldots, +\alpha'_n)$ is smaller than a series of adjunctions $s = (+\alpha_1, \ldots, +\alpha_m)$ if, and only if, one of the following three propositions is true:
a) $+\alpha'_1 < +\alpha_1$;
b) $j \in [1, \min(n, m)]$, $(+\alpha'_1, \ldots, +\alpha'_j) = (+\alpha_1, \ldots, +\alpha_j)$ and $+\alpha'_{j+1} < +\alpha_{j+1}$;
c) $n < m$ and $(+\alpha'_1, \ldots, +\alpha'_n) = (+\alpha_1, \ldots, +\alpha_n)$.

Given the canonical association of a generative series to every generation pathway, the classification of the generative pathways automatically follows from the classification of the series.

DEFINITION – Generation pathway C' is smaller than generation pathway $C(C' < C)$ if, and only if, generative series s and s' characterizing these two pathways are such that $s' < s$.

The canonical generation pathway is a sub-product of the total order for all the generation pathways.

DEFINITION – The *canonical generation pathway* $C^o(F)$ of an F transfograph is the first in the list when classifying the generation pathways of F in ascending order.

3.2.5. *Synchronism condition*

The synchronism condition, one of the fundamental constraints on ordered generation, is given and verified hereafter.

PROPERTY – Every intermediate transfograph F_j on the canonical generation pathway $C^o(F)$ of transfograph F is reached via a sub-pathway of $C^o(F)$. $C^o(F_j)$ is the canonical generation pathway of transfograph F_j.

DEMONSTRATION

a) Let $F = F_n$ be a transfograph of $\mathbf{F}$, $C^o_n = (F_0, F_1, \ldots, F_j, \ldots, F_n)$ its canonical generation pathway, and $s^o_n = (+\alpha_1, \ldots, +\alpha_j, \ldots, +\alpha_n)$ the generative series (called "canonical series") characterizing this pathway.
b) Let F_j be the jth transfograph on the canonical pathway of F_n. By definition of the generative series, $F_j = F((+\alpha_1, \ldots, +\alpha_j)$, i.e. F_j is determined by the generative series $s_j = (+\alpha_1, \ldots, +\alpha_j)$.
c) To show that s_j is the canonical series s^o_j of F_j, let it be assumed that there is another series $s'_j = (+\alpha'_1, \ldots, +\alpha'_j)$ that is the canonical series of F_j. Then $S'_n = (+\alpha'_1, \ldots, +\alpha_{j+1}, \ldots, +\alpha_n)$ is a generation series of F_n. Since $s^o_n < S'_n$, and, hence, $(+\alpha_1, \ldots, +\alpha_j) < (+\alpha'_1, \ldots, +\alpha'_j)$, in contradiction with the initial hypothesis that s'_j is the canonical series of F_j, it follows that s_j is the canonical series, thereby demonstrating the validity of the property.

The restriction of anteriority hyperstructure $(\mathbf{F}, a)$ to the canonical generation pathways of all the transfographs in $\mathbf{F}$ is the ordered anteriority hyperstructure $(\mathbf{F}, a^o)$. It follows from the synchronism property that this restriction is represented by a rooted tree F_0: every F transfograph of **F** has only a single direct predecessor on its canonical generation pathway derived from F_0. This rooted arborescence has a more stronger ordering structure than an ordinary tree: the order for adjunctions attached to each arc makes it possible to classify all siblings from the same father.

Whereas anteriority hyperstructure (**F**, a) guarantees the exhaustivity of the generation pathways thanks to the adjunction links ordered anteriority hyperstructure (**F**, a^o), guarantees non-redundancy in the generation of the transfographs of **F**: only a single copy of a given F transfograph is generated. Hence, this property is used in choosing a transfograph from among transfographs belonging to the same class of isomorphism or to different classes.

3.3. Extension to classes of isomorphism

Hereafter described generation and description laws for the transformation structure – a class of isomorphism for transfographs (chromatic transfographs obeying the valency constraint) – are derived from above-established laws (3.1 and 3.2) concerning the transfograph. As for the transfograph, two stages are involved.

1) The generation pathways of transformation structure T are easily determined by the generation pathways of all F transfographs in T.
2) The canonical pathway for the ordered generation $C^o(T)$ of transformation structure T is chosen from among the canonical generation pathways for the transfographs in T; this choice is based on the same ordering criteria as those used in selecting a single canonical pathway from among the generation pathways of the same transfograph. This pathway determines a target transfograph called the canonical transfograph $F^o(T)$.

Key definitions bearing on these two stages are given hereafter. Algorithmic rules for directly determining a canonical transfograph in line with rules for determining the ordered canonical chromatic graph of a chemical structure were given previously by J.-E. Dubois in 1976 [2].

3.3.1. *Generation of a transformation structure*

The generation of the class of isomorphism defining a transformation structure is proposed as a simple extension of the generation of a transfograph (3.2); so, by simple analogy, recourse is made to the series of definitions that led to the establishment of the generation pathways of a transfograph (cf. essentially 3.1.5 and 3.1.6).

As for transfograph F, the generation of transformation structure T is defined with the help of a partial order relationship also referred to as an "adjacent anteriority" relationship and designated by a. This relationship can be used to organize the entire set of transformation structures **T** into an hyperstructure network (**T**, a) grouping all the generation pathways of all transformation structures T in **T**.

Adjacent anteriority between two transformation structures means adjacent anteriority between transfographs (cf. 3.2.1) belonging to each of the two transformation structures.

DEFINITION – Transformation structure T_1 is adjacent anterior to transformation structure T_2 (T_1 a T_2) if there is a F_1 transfograph in T_1 and a F_2 transfograph in T_2 so that F_1 is anterior to F_2:

$$T_1 \; a \; T_2 \Leftrightarrow \exists F_1 \in T_1, \exists F_2 \in T_2, F_1 \; a \; F_2.$$

Overall adjacent anteriority relationship A, defined as the transitive closure of a, makes it possible to regroup all anterior structures into a single transformation structure.

DEFINITION – The population of anterior transformation structure of transformation structure T is the set of transformation structures related to T by anteriority: $P_A(T) = \{T' \in \mathbf{T} \mid T' \; A \; T\}$.

The restriction of adjacent anteriority relationship a to this population of anterior transformation structures constitutes a characteristic hyperstructure of T.

DEFINITION – Hyperstructure $HS(T) = (P_A(T), a)$ associating anterior transformation structure population $P_A(T)$ to T, together with the restriction of adjacent anteriority relationship a to $P_A(T)$, constitutes the generation hyperstructure of transformation structure T.

As for the generation hyperstructure $HS(F)$ of a particular F transfograph, $HS(T)$ has a partially ordered structure with unique minimum and maximum elements and can therefore be represented by a lattice: T_0 is the minimum element and T the maximum element.

The exhaustive traversal between T_0 and T in this lattice establishes all the generation pathways of T.

DEFINITION – A series $C(T) = (T_0, T_1, \ldots, T_0 = T)$ in which two successive transformation structures are related by adjacent anteriority relationship a is a *generation pathway* of transformation structure T starting from empty transformation structure T_0.

So, to build a generation pathway of T, it suffices to build a generation pathway $C(F)$ of an F transfograph in T. Since each pathway $C(F)$ is characterized by a unique set of adjunctions $s(F)$ (cf. 3.1.6), it is also possible to determine a pathway $C(T)$ from such a series.

Since, in practice, the network of all the generation pathways of all the transfographs of T is never handled in its entirety, no detailed study of its properties is undertaken here, despite any theoretical interest it might hold.

3.3.2. *Ordered generation of a transformation structure*
The canonical generation pathway of transformation structure T is deduced by counting, then classifying the canonical generation pathways of all the transfographs constituting T. For simplicity, and in the absence of other constraints, the same classification order used in selecting the canonical pathway $C^o(F)$ of a F transfograph from among all the generation pathways in T (3.3) has also been used in determining the canonical generation pathway.

Determining the canonical generation pathway relies on the lexicographical order for adjunction series (3.3.3 and 3.3.4) derived from the total order for the set $*a$ of elementary adjunctions (3.3.2) for building transfographs.

DEFINITION – The *canonical generation pathway* $C^o(T)$ of transformation structure T is the first pathway on the list when classifying the canonical generation pathways of all transfographs in ascending order (cf. 3.3.4).

It should be noted that the same result is obtained if all the generation pathways (and no longer the canonical pathway only) of all the transfographs in T are classified in ascending order.

The canonical transfograph of T is simply the impact of the canonical generation pathway on the class of isomorphism constituting T.

DEFINITION – *Canonical transfograph* $F^o(T)$ of transformation structure T is the target transfograph of canonical generation pathway $C^o(T)$.

Canonical transfograph $F^o(T)$ is representative of class T: all the transfographs of T can be obtained from the canonical transfograph. Indeed, since a node order imposed on each of the loose (X), internal (X_i) and external (X_{e1} and x_{e2}) sets of nodes determines a single F transfograph of class T, it suffices to carry out all the permutations on the nodes of $F^o(T)$ to obtain all the other transfographs.

It should be underscored that the canonical transfograph itself can be generated along various pathways: if no rules for choosing a canonical pathway are adjoined to the canonical transfograph, the transfograph structure will have several descriptions.

3.3.3. *Determining a canonical transfograph*
For the sake of comparison, the same methodology as used for the DARC major order has been used for the canonical order of a transfograph: definition of a size of limited environment, establishment of an order for this environment and propagation of this order to induce a linear order throughout the environment. The result of this process is the canonical generation pathway, as well as the canonical transfograph $F^o(T)$.

No formal development of the establishment of intermediate transfographs of the pathway, nor of the establishment of elementary adjunctions of the generative associated series is given here. By analogy with the DARC major order, the nature and mode of internal ordering of each limited environment are mentioned, and the techniques for numbering all the nodes of transformation structure T in order to select the canonical transfograph representing T is given.

The three main characteristics involved in building a canonical transfograph are given hereafter. They are followed by the priority rule p_0 (which applies to development origins only, and which should be compared to p_0, p_A, and p_B applying to development origins, to positions A and to positions B) that can be used within a given module.

1. The generation of canonical transfograph F^o is considered as the successive ordered generation of loose, internal, ablation and adjunction modules; so nodes are numbered in ascending order (loose, internal, ablation, adjunction) according to the nature of the module to which they belong.
2. The generation of each module is conducted by propagation of an environment that is concentric, ordered and limited to a single layer (ELCOa). An ELCOa is made up of node rank Aj originating from a single development node and from the links connecting these nodes to already generated existing nodes. The origin, or focus, of the generation numbered first is a node from the loose module, whereas internal, ablation and adjunction modules are generated from all loose nodes.
3. A single priority rule p_0 expressed by four sets of criteria simultaneously determines the internal generation of each ELCOa (directly), the generation sequence for ELCOa's (by propagating the order) and the loose node that is the generation focus.

PRIORITY p_0

(In each case, the existence of a topological or chromatic element takes priority over its absence and a lower node number takes priority over a higher node number; cf. 3.3.2 for the classification of the types of node and link chromaticities).

Nodes Aj originating from a single development node D are classified according to:

a) the existence of a link between D and Aj, and thereafter by the nature of the chromaticity of the link (D, Aj)
b) the nature of the atom chromaticity on node Aj;
c) the existence of a specification chromatism between Aj and another existing node C, the order number of the other end C, then the nature of the chromaticity on the link (Aj, C).

Recursive application of p_0 to all the nodes of the same module of transformation structure T thereby results in the classification of these nodes according to a total order. A particular transfograph of transformation structure T, i.e. its canonical transfograph $F^o(T)$ is obtained by adding the ordering criteria between the modules (loose, internal, ablation, adjunction).

From a practical standpoint, it should be mentioned that the reduced size of the chosen elementary module of the environment ELCOa means that in a great many cases, the classification decision cannot be taken at the level of a single ELCOa and that it requires using the propagation criteria.

Conclusion

The new structural type of "transfostructure" proposed for modeling structural transformations, and its ordered generation by elementary adjunction operators provide fundamental graph theoretical tools for efficient computer applications. They lead to structural abstract data types and to manipulation algorithms, directly applicable for data base management and for definition and traversal of adapted search spaces.

The transfostructure is related to schematic structures used in several fields of chemical transformation studies: transition states, arrowed graphs representing rearrangement mechanisms, fragmentation graphs in mass spectrometry, generic reaction diagrams with more or less explicit indication of broken and formed bonds. All these transformation diagrams are directly representable with the transfostructure. Moreover using a single structure to define with this level of accuracy what takes part in a transformation has never been proposed to generally model every chemical transformation given by its initial and final structures.

The ordered generation process has several kinds of applications. The most straightforward is a linear unambigous and unique descriptor which makes explicit the synchronism between canonical description and ordered generation. This descriptor of a structural transformation consists of totally ordered concatenation of elementary adjunction descriptors as in the DARC code. Some other applications are: the selection of a unique transfostructure for modelling a chemical transformation, the determination of steochiometric coefficients of a chemical reaction, the structural adaptation of chemical compound structures to reaction representation, the appropriate automatic reaction editing [14]. Paramaterizing generation permits building reaction spaces where vicinity relations reflect transformational criteria. It thus allows for an intelligent search of significant analogies and differences between reactions, and to the establishment of generic reactions with strictly controlled procedures.

This presentation of a global Transformation Graph Theory or Model corresponds to the need felt during the years of implementing the **DARC** System, both for compounds and their transformation. At first, static chromatic graphs were used to describe structural changes, but gradually chromatic transfographs seemed better adapted to describing the transformation operation itself. In the **DARC** System (Description, Acquisition, Retrieval, CA Design), these two different reaction management languages are called CO and RO (Compound Oriented and Reaction Oriented) languages or codes. The CO language, based on the ordered generated graph descriptors, handles "pairs of initial and final structures" of a reaction that is dealt with as a black box.

The RO transfograph provides the means for more in-depth investigation of the complexity or structural dynamics of a reaction and for clarifying reaction similarities. To date, most reaction files are handled with CO concepts, but for CA elucidation and synthesis, as well as for patent search, more complex graph descriptions must be developed [15].

Both theoretical and similarity approaches still require more tools to define and code transformation stages. At the confluence of structural and mechanistic chemistry, compound transformation is at the heart of chemistry. Computer treatment of graphs, transfographs and T-hyperspaces should help to improve our knowledge of reactions. Transformation studies still present many, very exciting challenges.

References

1. Dubois, J.-E., Laurent, D., and Viellard, H., *C.R. Acad. Sci. Paris, Série A* **270**, 228 (1970).
2. Dubois, J.-E., Bonnet, J.-C., Laurent, D., and Viellard, H., *Bull. IRIA (Institut de Recherches Informatique Appliquées)*, suppl. IRIA N°6 (1971), pp. 1–16.
3. Dubois, J.-E., *Isr. J. Chem.* **14**, 17 (1975).
4. Dubois, J.-E., in *The Chemical Applications of Graph Theory*, A. T. Balaban (ed.), Academic Press, London (1976), pp. 333–370.
5. Dubois, J.-E., Picchiottino, R., Sicouri, G., and Sobel, Y., *C. R. Acad. Sci. Paris*, Série I, **294**, 251 (1982).
6. Dubois, J.-E., Panaye, A., Picchiottino, R., and Sicouri, G., *C. R. Acad. Sci. Paris*, Série II, **295**, 1081 (1982).
7. Attias, R., *J. Chem. Inf. Comput. Sci.*, **23**, 102 (1983).
8. Picchittino, R., Sicouri, G., and Dubois, J.-E., in *Data for Science and Technology*, P. S. Glaeser (ed.), North Holland (1983), pp. 229–334.
9. Picchiottino, R., Georgoulis, G., Sicouri, G., Panaye, A., and Dubois, J.-E., *J. Chem. Inf. Comput. Sci.*, **24**, 241 (1984).
10. Dubois, J.-E., Sicouri, G., Sobel, Y., and Picchiottino, R., *C. R. Acad. Sci. Paris*, Série II, **298**, 525 (1984).
11. Picchiottino, R., Sicouri, G., and Dubois, J.-E., in *Computer Science and Data Banks*, Z. Hippe (ed.), Polish Academy of Sciences (1984), pp. 183–193.
12. Sicouri, G., Sobel, Y., Picchiottino, R., and Dubois, J.-E., *C. R. Acad. Sci. Paris*, Série (1984), pp. 523–528.
13. Sicouri, G., Sobel, Y., Picchiottino, R., and Dubois, J.-E., in *The Role of Data in Scientific Progress*, P. S. Glaeser (ed.), North-Holland (1985), pp. 573–580.
14. Dubois, J.-E. and Messabih, B., *Communication and Information for Progress in Science and Industry – Codata International Conference* (1984) (to be published – look 1995).
15. Dubois, J.-E., Sicouri, G., and Picchiottino, R., in *Modern Approaches to Chemical Reaction Searching*, P. Willet (ed.), Gower Publishing Co. Ltd. (1986), pp. 240–256.

Detection of the Largest Patterns Common to a Set of Molecules, Using 2D Graphs and 3D Skeletons

DENIS M. BAYADA and A. PETER JOHNSON
School of Chemistry, University of Leeds, Leeds LS2 9JT, U.K.

Abstract. We present an algorithm, the multiple algorithm for comparing a set of chemical structures in two and three dimensions. Based on pairwise comparisons of structures, it finds all the largest substructures common to a set of chemical structures. For 2D molecular graphs, a backtracking algorithm for pairwise comparisons of graphs is presented. This algorithm together with the multiple algorithm is refined to find different types of solutions, connected, disconnected and augmented. For 3D structures, a clique detection algorithm is used, within the multiple algorithm instead of a backtrack. Some results for 2D graphs as well as for 3D structures are presented.

1. Introduction

The similarity of a set of chemical compounds can be assessed by finding the largest pattern common to the set. This pattern can be a subgraph if the set is composed of molecular graphs or a three-dimensional substructure if 3D skeletons of molecules are considered. The size of this pattern relative to the size of the molecules of the set has been used as a similarity value. The Tanimoto index, for instance, is such a similarity index for sets of two molecules [1]. However, one is sometimes more interested in the pattern itself than in the similarity measure it has generated. In QSAR studies the pattern common to a set of molecules can indicate a reason for a similar behaviour [2, 3]. When clustering a large set of compounds, there is no unique set of clusters. Knowing the pattern common to a set of molecules, different clusters can be organised on a hierarchical basis, where a cluster C1 is higher in the hierarchy than a cluster C2 if the pattern associated with C2 is included in the pattern of C1 [4].

Several algorithms have been designed to solve the problem of finding a common pattern to a set of molecules. For 2D molecular graphs, Varkony and co-workers [5] as well as Takahashi and co-workers [6] have developed such an algorithm. In 3D space, Crandell and Smith [7] and also Brint and Willett [8] have published algorithms able to find a 3D substructure common to a set of molecular skeletons. This last algorithm has been successfully used to map groups of pharmacophores [2].

In this article, after defining some terms, a general algorithm for matching a set of objects and some heuristics to enhance the performance are presented. We will then focus on applying this algorithm to 2D molecular graphs and show that it is flexible enough to find different types of solutions. Finally

R. Carbó (ed.), Molecular Similarity and Reactivity: From Quantum Chemical to Phenomenological Approaches, 243–265.

the multiple algorithm is applied to 3D molecular skeletons and some results are presented.

DEFINITIONS

A *labelled graph* $G(V, E, Lv, Le)$: V is the set of vertices corresponding to the atoms and E is the set of edges corresponding to the bonds. Lv is a function $Lv : V \rightarrow \mathbb{N}^+$ and Le is a function $Le : \mathrm{E} \rightarrow \mathbb{N}^+$. Lv (respectively Le) denotes the positive integer labels assigned to the vertices (respectively edges) (Figure 1). In the following paragraphs the terms graph and structure are interchangeable, though structure is often used to mean three-dimensional molecular skeleton.

A *tree* is a graph with no rings. This means that between any two nodes a and b there exists only one path $a_0a_1a_2 \ldots a_p$, where $a = a_0$, $b = a_p$ and there is an edge between a_i and a_{i+1}, $\forall i, 0 \leq i < p$.

The *degree* of a node of a graph is the number of edges that touch this node.

A *structure* is a set of points in 3D space representing the atoms of a molecule (Figure 1).

A *subgraph* $g(v, e)$ of a graph $G(V, E)$ is a graph where: $v \subseteq V$ and $e \subseteq E$. This is sometimes called a partial subgraph of a graph.

The *2D isomorphism* problem can be defined as follow:

Instance: Two labelled graphs $G1(V, E, Lv, Le)$ and $G2(V', E', Lv', Le')$, with $|V| = |V'|$ and $|E| = |E'|$.

Question: Are $G1$ and $G2$ isomorphic, i.e. does there exist a one-to-one onto function f: $V \rightarrow V'$ / $\{x, y\} \in E$ iff $\{f(x), f(y)\} \in E'$, $Lv(x) = Lv'(f(x))$, $Lv(y) = Lv'(f(y))$ and $Le(\{x, y\}) = Le'(\{f(x), f(y)\})$?

The *3D isomorphism* problem is defined as follow:

Instance: Two structures $G1(V)$ and $G2(V')$ with $|V| = |V'|$ and a threshold k.

Question: Are $G1$ and $G2$ isomorphic, i.e. is it possible to superimpose

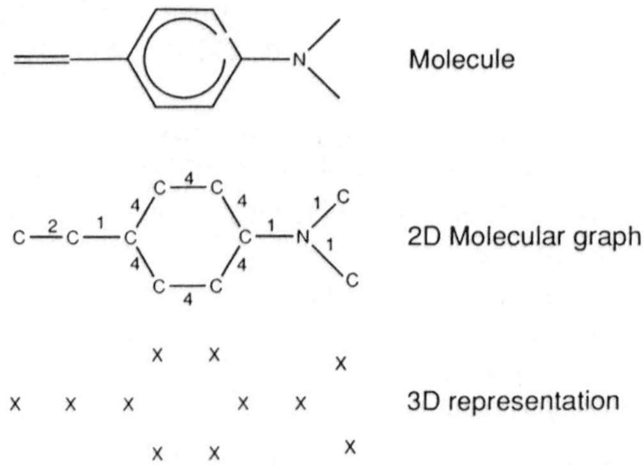

Figure 1. Representation of a molecule, its 2D graph and 3D structure.

$G1$ and $G2$, such that there exists a one-to-one onto function f: $V \rightarrow V'$ and the Euclidian distance between any two nodes, x from $G1$ and $f(x)$ from $G2$, is smaller or equal to k?

2. The multiple algorithm

Introduction

Let us first define the problem for 2D graphs (the definition is very similar for 3D structures):

The Multiple Largest Common Subgraphs Problem:

Instance: A set of n labelled graphs ($G1$, $G2$, . . . , Gn) and a positive integer k where $k \leq$ minimum number of nodes of ($G1$, $G2$, . . . , Gn).

Question: Find all the largest graphs H isomorphic to a subgraph g_i of each Gi, $\forall i$ ($1 \leq i \leq n$), such that the number of nodes of $H \geq k$ and all graphs H have the same number of nodes. All graphs H are maximal with respect to graph isomorphism.

The multiple algorithm is based on the repeated use of two algorithms that compare a pair of structures. The first algorithm, called *larger*, finds all the substructures common to the two input graphs, of size larger or equal to a given limit. These substructures are maximal, i.e. any two substructures are different and no substructure is the subgraph of another one. The definition of this problem, below, is given for 2D graphs:

The Maximal Common Subgraphs Problem:

Instance: Two labelled graphs $G1$ and $G2$ and a positive integer k where $k \leq$ minimum number of nodes of $G1$ and $G2$.

Question: Find all the graphs H isomorphic to a subgraph of $G1$ and a subgraph of $G2$, such that the number of nodes of $H \geq k$. All graphs H are maximal with respect to subgraph and graph isomorphism.

The second one, called *largest*, finds only the largest substructures common to the two input graphs, these substructures being of size larger or equal to a given limit. The largest substructures are all the same size and are all different. Here is the definition for 2D graphs:

The Largest Common Subgraphs Problem:

Instance: Two labelled graphs $G1$ and $G2$ and a positive integer k where $k \leq$ minimum number of nodes of $G1$ and $G2$.

Question: Find all the largest graphs H isomorphic to a subgraph of $G1$ and a subgraph of $G2$, such that the number of nodes of $H \geq k$ and all graphs H have the same number of nodes. All graphs H are maximal with respect to graph isomorphism.

The algorithm *Larger* is called a maximal common subgraphs (or substructures) algorithm and the algorithm *Largest* is called a largest common subgraphs (or substructures) algorithm [9]. These two pairwise algorithms are very similar and it is often easy to modify one of them into the other one. The algorithms *larger* and *largest* in 2D and 3D will be presented in paragraphs 2D matching and 3D matching respectively.

THE ALGORITHM

The organisation of the multiple method is based on a tree where at each node, *larger* or *largest* is used. It is explored in depth-first search and its depth is $n - 1$ for n graphs. Figure 2 shows an example of an execution of the multiple algorithm on five graphs. At the root of the tree, level 1, *larger* is run on the two first input graphs (step 1 in Figure 2). It returns a list (A1, A2, . . . , Am) of m solutions, that is sorted, the largest solution first. The root will therefore have m children although some of them will be pruned away later. Down one level in the tree, to level 2, at the first child of the root, *larger* is run on A1, largest solution of the list of level 1, and graph 3 (step 2 in Figure 2). Solution A1 is then deleted from the list. Down one level again, the best solution found at level 2 is matched with graph 4 using

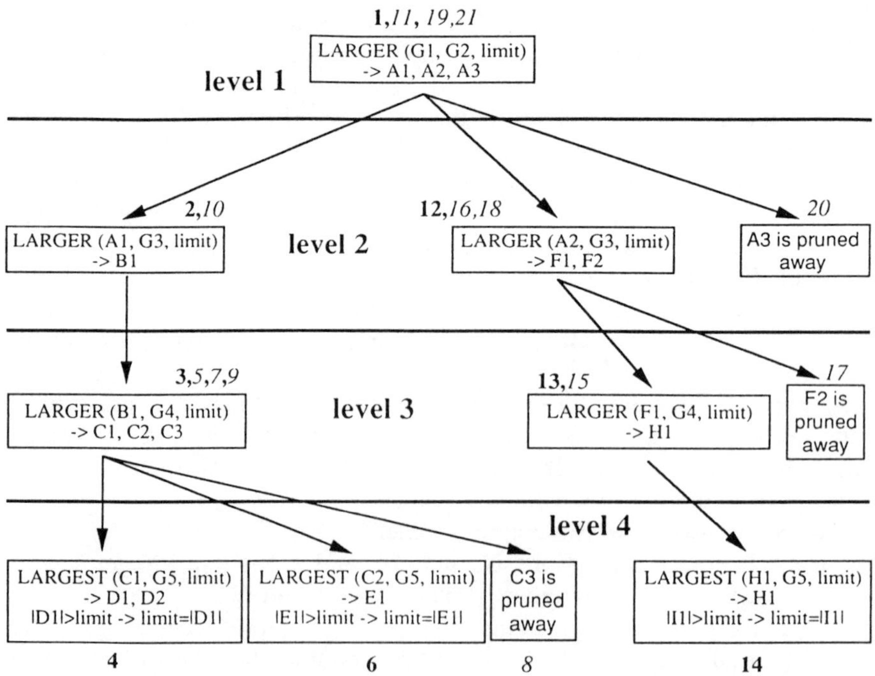

Figure 2. Tree of an execution of the multiple algorithm on the five input graphs G_1, G_2, G_3, G_5 and G_4. The numbers in bold and italic indicate the order in which the nodes of the tree are explored. The numbers in italic indicate backtracking steps or elimination of intermediate solutions and the numbers in bold indicate executions of the pairwise algorithms. The final solution is I_1.

larger (step 3 in Figure 2), etc. . . . At level $n - 1$ (level 4 in Figure 2), *largest* is used to match the largest solution found at level $n - 2$ with graph n (step 4 in Figure 2). Solutions, say of size *nb*, are found that are substructures common to all the *n* graphs, though they are not necessarily the largest solutions to the problem. At this point, it must be noticed that the largest solutions to the multiple problem must be at least of size *nb*. The limit is set to *nb* and all the intermediate solutions at levels 1 to $n - 2$ that are smaller than *nb* will be discarded when they are reached, i.e. the branches that they have created will be pruned (steps 8, 17 and 20 in Figure 2). Then, the other solutions found at level $n - 2$ and graph number n are matched using *largest* (step 6 in Figure 2), setting the limit higher again if solutions larger than the limit are found. The next step consists in backtracking to level $n - 2$ (step 9 in Figure 2) and matching the first solution of the list at level $n - 3$, if any exists, with graph $n - 1$ using *larger*. If no solution exists at this level, a backtracking step to level $n - 3$ is done (step 10 in Figure 2), etc. . . . The search is stopped when level 2 is reached and there are no more solutions unexplored at level 1 (step 21 in Figure 2). This multiple algorithm works well if and only if a limited number of solutions are found at each branch. Thus it is important to produce no more than the necessary intermediate solutions.

Finding a proper limit

As *larger* and *largest* find only solutions that are larger than a given limit, the two pairwise algorithms can be forced to find fewer solutions the larger the limit is, in other words the limit should be set as close as possible to the size of the final solutions. Figure 3 shows that calls to *larger* with increasing limits reduce the cpu time and the number of solutions. We have developed a quick way of finding a reasonably good limit, still smaller or equal to the size of the final solutions but quite close to it. This is achieved by applying *largest* on graphs 1 and 2 and recording only one of the best solutions, say S1. Then *largest* is applied on S1 and graph 3 and solution S2 is recorded, etc. . . . At the end, *largest* is applied on solution $Sn - 2$ and graph *n* and the size of the best solutions it has produced is recorded. They are solutions to the multiple problem, though not necessarily the best. It is probable that the size of these solutions is close to the size of the optimum solutions to the problem and it can be used as a starting limit. If the smallest input graph is composed of less than 8 nodes, the limit is set to 2 and this heuristic is not executed.

Sorting the input graphs

Another way of increasing the performance of the algorithm is to order the graphs at the beginning. We know that when the multiple algorithm goes down the tree, the intermediate solutions can only decrease in size. Thus, it is better to find right from the start small solutions and to do this the multiple algo-

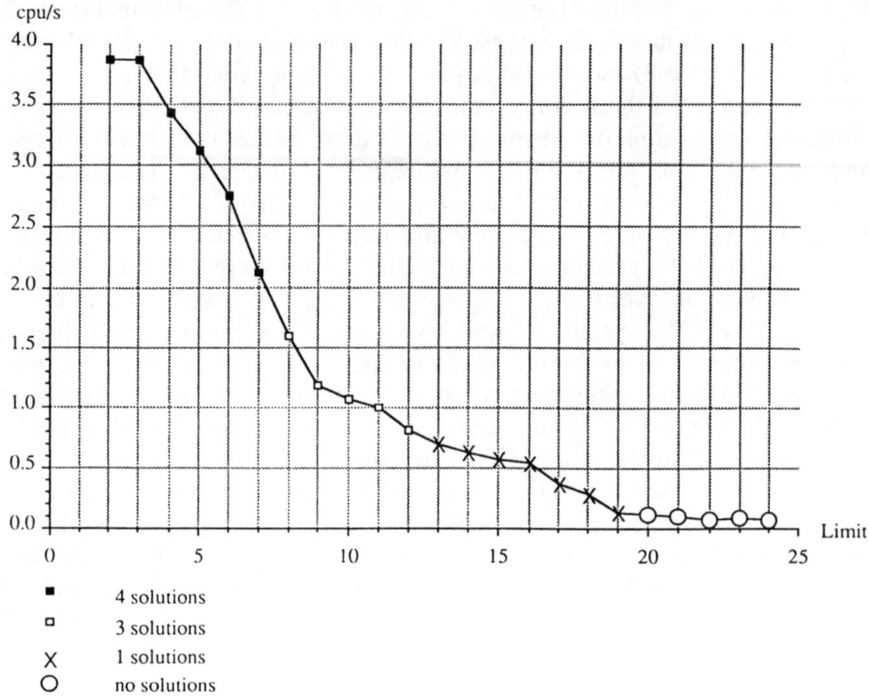

Figure 3. Two graphs of 22 and 29 nodes and different limits are used as parameters to the algorithm *larger*. It is clear that the larger the limit, the faster the execution and the fewer the solutions.

rithm has to start from the smallest graphs. This can also be applied to the complexity of the graphs. Especially in two dimensions, the graphs having several rings are more difficult to match than simple trees because they produce more solutions. Thus it is better to start from the simplest graphs.

The output

There are two different types of output from the multiple algorithm. The first type of output is a list of two-dimensional graphs or three-dimensional structures that are solutions to the problem. The other type of output consists of lists of nodes from the input graphs for each graph solution. For each graph or structure S that is a solution to the multiple algorithm and for each input graph G_i, all the lists of the nodes of G_i that match S are given in the output. There can be more than one list, because S can match different parts of G_i (see Figure 4).

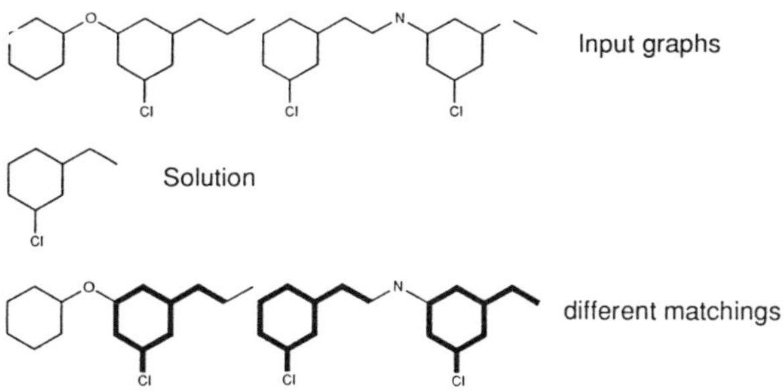

Figure 4. There is only one largest common subgraph to the two input graphs. The solution matches the first input graph once (bold bonds) and the second input graph twice (bold bonds).

3. 2D matching

Introduction

Several algorithms exist to find the maximal subgraphs common to two graphs, as well as their largest common subgraphs. Some algorithms are strongly based on chemical knowledge like the fragment approach [10], others on clique detection algorithms [11, 12] or backtracking methods like McGregor's algorithm [13, 14]. Few algorithms [5, 6] have been published for the detection of a subgraph common to a set of molecules. The method of Varkony and co-workers [5] finds all the common subgraphs of size k and tries to add a node to the subgraphs to create all the subgraphs of size $k + 1, \ldots$ The worst cases for this algorithm occur when all the molecules are very similar. Due to the nature of our problem, we choose to adopt the backtracking approach. It has the advantage over the others of being fast when the two graphs to compare are similar, but also to work in any circumstance [10]. This is an important advantage as this algorithm is often used on similar molecules and especially when measuring the validity of a cluster of molecules. A cluster has been created by putting together some molecules that share some properties or fragments. Thus it is probable that these molecules have similar molecular graphs. Moreover, at level i of the multiple algorithm, an intermediate solution S is a structure common to graphs 1 to $i - 2$. Thus S is likely to be also similar to graph $i - 1$ with which it is going to be matched.

The backtrack algorithm

The main difference between our backtracking method and McGregor's is that it explores the graphs in breadth-first search, whereas McGregor's algorithm explores the graphs without considering the connections between the nodes. The two graphs, $G1$ and $G2$ are explored in parallel, following their spanning trees. First, a node from $G1$, called *beginning node*, together with the nodes from *G*2 that it can match (they have the same atom type for instance) are stored in a list of pairs, at level 1. A pair of nodes (*a*, *b*) means that node *a* from *G*1 can match node *b* from *G*2. The algorithm takes the first member of the list, say (*a*, *b*) and considers all the children of these two nodes. It creates and stores some combinations of children of *a* and their matching children of *b*, that is all the possible pairs of nodes, one a child of *a*, and the other a child of *b*, that are compatible are explored and combined in different combinations of pairs at level 2. We make a distinction between pairs that have one of their members attached by a ring bond to its parent and pairs whose members are attached to their respective parent by chain bonds. For instance, for the pairs of nodes (*aa*, *bb*) and (*cc*, *dd*), whose parents are *a* in *G*1 and *b* in *G*2, if the bond between *a* and *aa* or the bond between *b* and *bb* is in a ring, then the pair (*aa*, *bb*) is considered as a ring pair, and if the bond between *a* and *cc* and the bond between *b* and *dd* are not in a ring, (*cc*, *dd*) is a chain pair. This distinction is necessary because, it is possible to attain *aa* from *a* by at least two ways, whereas *cc* and *dd* can be reached from *a* and *b* respectively by a unique path. Thus all possible combinations of ring pairs must be created to which some combinations of chain pairs are added, knowing that the final lists of pairs must be maximal with respect to the chain pairs. To illustrate this, let us assume that we have two ring pairs: (*aa*, *bb*) and (*cc*, *dd*) and three chain pairs: (*ee*, *ff*), (*gg*, *ff*) and (*gg*, *hh*). For instance, all the nodes from *G*1, namely *aa*, *cc*, *ee*, *gg*, have the same parent, *a* and the nodes from *G*2, *bb*, *dd*, *ff*, *hh*, have *b* for parent. All the combinations of ring pairs are:

(*aa*, *bb*)
(*cc*, *dd*)
(*aa*, *bb*) (*cc*, *dd*)

To which chain pair combinations are added:

1. (*aa*, *bb*) + (*ee*, *ff*) (*gg*, *hh*)
2. (*aa*, *bb*) + (*gg*, *ff*)
3. (*cc*, *dd*) + (*ee*, *ff*) (*gg*, *hh*)
4. (*cc*, *dd*) + (*gg*,*ff*)
5. (*aa*, *bb*) (*cc*, *dd*) + (*ee*, *ff*) (*gg*, *hh*)
6. (*aa*, *bb*) (*cc*, *dd*) + (*gg*, *ff*)

Figure 5 shows why it is necessary to create all the ring pairs combinations. After having created the two first levels, the largest combination from level 2, combination 5 in the example above, is chosen and the children of the nodes involved in this combination, are combined etc. . . . When the children of the nodes of a combination cannot be combined into a new combination, a new solution has been found. In our example the solution is S = (*a*, *b*) (*aa*, *bb*) (*cc*, *dd*) (*ee*, *ff*) (*gg*, *hh*) . . . The solutions are stored in *sol* that is a list of solutions. If the length of S is larger than the limit, it is compared with solutions in *sol* and eliminated if it is included in a solution of *sol*. At the same time, if a solution S′ of *sol* is included into S, S′ is eliminated. A

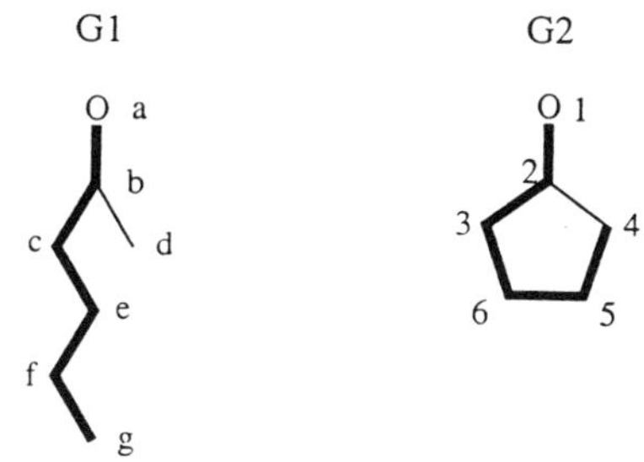

Level 1	(a 1)		
Level 2	(b 2)		
Level 3	(c 3) (d 4) *	(c 3) *	(d 3) *
Level 4	(e 6)	(e 6)	End
Level 5	(f 5)	(f 5)	
Level 6	End	(g 4)	
		End	

* 3 and 4 = equivalent nodes

Figure 5. This example shows the necessity of combining differently ring pairs and chain pairs. Graphs *G*1 and *G*2 have two largest solutions. The solution shown with the bold bonds can only be found if the non maximal combination (c 3) at level 3 is produced. Some equivalent nodes, 3 and 4, reduce the number of combinations at level 3.

solution S′ is included into a solution S if each pair (a, b) of S′ is also a pair of S and S′ is smaller than S. If S has not yet been eliminated, it is stored in *sol*. Eventually, the algorithm backtracks to the previous level and tries to grow the new combination, etc. . . . When there is no more combination available, it means that all possible solutions larger than the limit involving the *beginning node* have been explored. Thus this node can be eliminated and the search starts again from a new *beginning node*. If by eliminating the *beginning node* from *G*1, then *G*1 has been disconnected into several components, the components smaller than the limit are eliminated. The graph *G*1 is now smaller than at the beginning and if it is smaller than limit there cannot be any new solution larger than the limit and the search is concluded All the lists of pairs in *sol* are converted into graphs that are tested for subgraph and graph isomorphism as only maximal graphs are kept. To sum up the process of elimination of solutions, first non maximal matchings are eliminated (when S′ is included into S) and at the end non maximal graphs are eliminated.

Difference between larger and largest

The differences between *larger* and *largest*, for this algorithm, are that the value of limit is unchanged in *larger* whereas it is always set to the size of the best solution found so far in *largest* and the list *sol* is updated to contain only solutions of size equal to limit.

EQUIVALENT NODES

In Figure 6, at level 3, we had only three combinations, (c3)(d4), (c3) and (d3), and not the six possible combinations that are the three above plus the following three: (c4)(d3), (c4) and (d4). This is because in graph $G2$, nodes 3 and 4 are equivalent and it is then useless to record more than three combinations at level 3. We use an algorithm based on the Morgan algorithm [15] for graph isomorphism to find equivalent nodes [9]. At the end of the algorithm, nodes with identical marks are equivalent to each other. We are interested only in equivalent nodes with a common parent. However this algorithm does not always work but is still used because it is fast and examples where it does not work are relatively rare. A correct and relatively fast algorithm for finding equivalent nodes has been published by Rucker and Rucker [16].

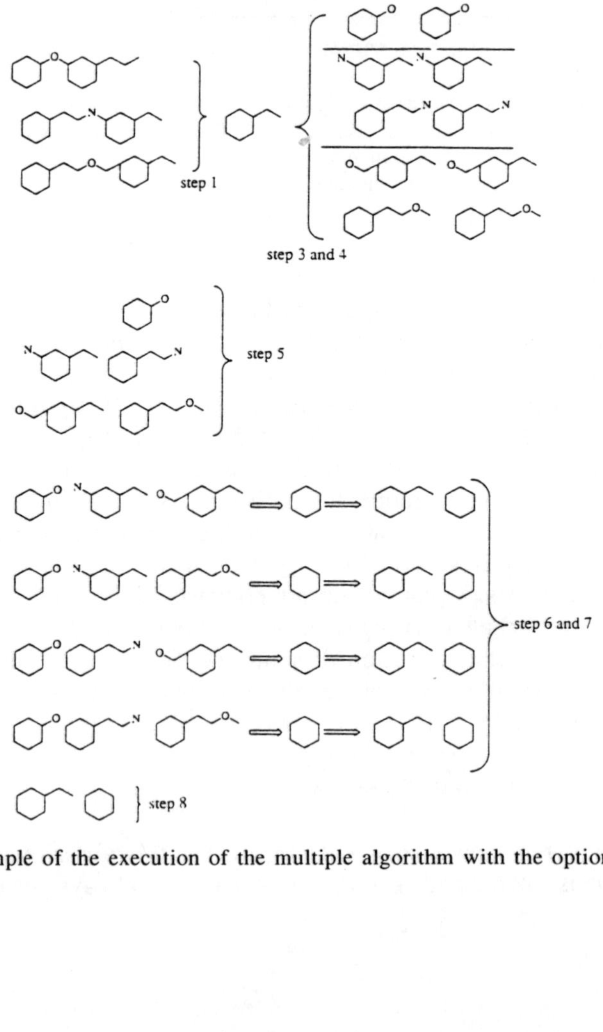

Figure 6. Example of the execution of the multiple algorithm with the option disconnected set on.

Renumbering the nodes of the graphs

To improve the performance of the algorithm *largest*, *beginning nodes* must be chosen so as to lead quickly to the best solutions, thus the nodes that are likely to be in the best solutions have to be guessed. Moreover, for both *larger* and *largest* we want the algorithm to stop as soon as possible. *Beginning nodes* that have a high degree must be chosen first because once they have been eliminated from $G1$, the graph will be simpler and it might also be disconnected into components smaller than the limit. We have developed a system of marking the nodes in that respect [9]. Our hypothesis is that if the label of a node or the bond type of a bond touching this node appears in the two graphs a small number of times, usually hetero-atoms and double and triple bonds, then this node is likely to be in the best solutions. Thus each node, X, is given a mark, $M(X)$, according to the number of times its atom type and the bond type of the bonds it is connected with appear in the graphs, and its degree:

$$M(X) = 1000 \times AT(X) + 10 \times BT(X) + 5\text{-deg}(X). \tag{1}$$

$AT(X)$ is the frequency of the atom type of X in the graph containing X. $BT(X)$ is the frequency of the least frequent bond type of all bonds touching X. $\deg(X)$ is the degree of X. The lower the mark, the more likely the node is to be in the solutions. The rarity of an atom type is considered more important than the rarity of a bond type that is itself considered more important than the degree of the node. This is a heuristic and there is no guarantee that it will work each time.

Subgraph check

Let us now suppose that we are at level i of the multiple algorithm. It is about to run *larger* between intermediate solution S and graph $i - 1$. We know that S is a subgraph of graphs 1 to $i - 2$. We also know that the graphs have been ordered by increasing size and we can assume that they are all quite similar. This means that it is probable that S is a subgraph of graph $i - 1$. Thus the first step of *larger* and *largest* is to check if graph $G1$ is a subgraph of graph $G2$. The Ullmann algorithm [17] is used for this check. If the answer is yes, *larger* and *largest* return the list of matching nodes returned by the Ullmann algorithm as well as $G1$. As this subgraph algorithm is much faster than *larger* and *largest*, there is no significant increase in cpu time when $G1$ is not a subgraph of $G2$. The same Ullmann algorithm is used for two other purposes. It is used to check that the solutions are maximal, at the end of the algorithm. It is also used to create all the possible matchings of a subgraph to a graph. This is done for instance at the end of the program, when the matchings of the solutions to the input graphs are required (see above, paragraph The output).

OPTIONS

1. *Disconnected solutions*

As the solutions are found following a spanning tree of the graphs, the graphs of the solutions are connected. It would be interesting to find solutions whose graphs are disconnected. Such solutions are sometimes more meaningful in chemistry than connected solutions. One problem however is that looking for disconnected solutions can lead to an explosion in the number of solutions explored and stored, thus leading to a substantial increase in cpu time and space. Another problem is that some disconnected solutions can consist of a large number of small components with no importance. To avoid all this, the algorithm only tries, at the end, to add new components to the best solutions it has found so far. Figure 6 shows how this is done. The following steps are executed:

1. Find the largest common subgraphs, $S_1 \ldots S_p$, to the input graphs using the multiple algorithm.
2. For each solution S do steps 3 to 7.
3. Create all the possible matchings of S to the input graphs, $G1 \ldots Gn$, using the Ullmann algorithm.
4. Subtract these matchings from the input graphs. For each Gi, there is a list of graphs Gij that correspond to Gi–S.
5. For each Gi, eliminate Gik if Gik is isomorphic to G_{ij}.
6. Create the lists of all combinations of Gij, $0 \leq i \leq n$. Such a list has n members, one for each input graph.
7. For each list, L, run recursively the multiple algorithm on L and add the new solutions to S.
8. Eliminate the non maximal solutions.

Let us assume that for each graph Gi, two graphs, $Gi1$ and $Gi2$ are created during the subtraction of S from Gi. Then 2^n combinations will be produced at step 6. This exponential explosion cannot be avoided and if too many combinations are produced, the program either quits and outputs the solutions found so far or after reaching a given time limit, stops and returns the solutions it has stored.

2. *Matching options*

There are other options to use for 2D matching. We have implemented three different node-options. A node from $G1$ matches a node from $G2$ when they have the same label, i.e. same atom type (option 1, standard compatibility), but it is also possible to consider that nodes having a hetero-atom label are compatible with any other hetero-atom labelled node (option 2) or that any node is compatible to any other node (option 3). When two nodes are matched together in a combination, the edge between the parent of each node and the node is checked. Both edges must be compatible according to one of three bond-options. The first one is the standard compatibility, accepting the matching of two edges if they are of the same bond type. The second option accepts to match two edges in a ring whatever their bond types are. This option is useful when two aromatic rings are being matched. These rings are sometimes modelled as alternative single and double bonds and it can be a serious problem when matching for instance Single-Double-Single-Double-Single-Double with Double-Single-Double-Single-Double-Single. The third option disregards bond types and thus allows any two edges to match. We can also use a ring-option that concerns rings and chains. This option is used in combination with a bond option. It allows the matching of two compatible bonds only if they are both in a chain or both in a ring. When we are looking for a suitable limit to start with and *largest* is used several times, the options are set to ensure the fastest 2D matching. These options are standard node and bond compatibility, i.e. two nodes (resp. bonds) of different types do not match, and ring bonds are only allowed to match ring bonds and chain bonds are only allowed to match chain bonds.

3. *Starting from a known partial subgraph*
It is sometimes the case that a partial subgraph common to the set of input graphs is known, or that we want to constrain the search to include a particular subgraph. For instance, a cluster of molecules is known to have two aromatic rings in common and we want these two rings to be subgraphs of the best solution common to the set of molecular graphs, although a better solution might exist without these two rings. To achieve this, we apply a similar strategy to the one used for finding disconnected solutions. The input is composed of the set of input graphs plus one graph that is a subgraph of all the input graphs. This last graph is called a starting graph. Using the Ullmann algorithm, all the possible matchings of the starting graph with each of the input graphs are created. At this stage, for each input graph *Gi* there is a list L_i of possible matchings of the starting graph and *Gi*. At the first step of the multiple algorithm *larger* of *G*1 and *G*2 is run with *L*1 and *L*2 passed as parameters. At level 1 of *larger*, instead of matching the *beginning node* of *G*1 with all compatible nodes of *G*2, all the possible combinations of the members of list *L*1 with all the members of list *L*2 are created. It then proceeds normally, until there is no more combinations at level 1. Then instead of eliminating the *beginning node* of *G*1 and starting from a new *beginning node*, the pairwise algorithm is stopped. Each time a solution is found within *larger*/*largest*, a trace is kept of the nodes of the solution that correspond to the starting graph, i.e. the solution is a list of pairs of nodes and the *s* first pairs of the list are the nodes that correspond to the starting graph, where *s* is the number of nodes in the starting graph. The second step of the multiple algorithm involves the matching of one solution S found at the previous step with graph *G*3. The same as between *G*1 and *G*2 is done, except that *Ls* has only one member. If S is a subgraph of *G*3 at level 2 of the multiple algorithm, this level is skipped and *S*(*Ls*) with *G*4(*L*4) are matched at level 3. However, we face a similar drawback as with the disconnected solutions algorithm. At level 1 of *larger*/*largest*, between *G*1 and *G*2, all the possible

combinations of $L1$ and $L2$ have been created. This can lead to a large number of possibilities to explore, with no improvement in speed relatively to having no starting graph. For instance, if the starting graph is a simple six-member ring, there is 12 different ways of matching it on another identical ring. Thus it leads to a list $L1$ with $L2$ matchings and if the same is done with $G2$, there is 12×12 combinations at level 1.

4. *Augmenting a solution*
Figure 7 shows another type of solution that it is possible to look for. By not considering the atom types when the nodes are matched the 2D algorithm can create chemically meaningless solutions, thus something different is needed. We want meaningful solutions to which some "fuzzy pieces" are added. To do this, firstly the algorithm is run with a standard type of node and bond matching. Let us assume that a solution S has been found. S is used as a starting graph and the multiple algorithm is run again but this time the matching of any two nodes and any two bonds is allowed. However, a slight modification to the algorithm is necessary. We want the starting graph and the input graphs to be compared strictly (using standard matchings) and the nodes and bonds in the 2D algorithms to match in a less strict manner (every two nodes and every two bonds are compatible). Moreover, the solutions must be eliminated according to the standard matching. Thus there are two types of matchings: the matching options for the starting graph and the solutions and the matching options for the matching of nodes and bonds in *larger* and *largest*.

Figure 7. The two first graphs produce a solution that is used as a starting graph in the second execution of the algorithm. The final augmented solution has a "fuzzy" atom type and a "fuzzy" bond.

RESULTS

14 Macrolides (Figure 8)
This example is taken from [5]. In previous articles, the largest common connected subgraph to the 14 graphs was found in three hours [5] and 430 seconds [6] respectively. The multiple algorithm needs 4 seconds for the connected solution and 4.8 seconds for the disconnected one. The multiple program has been written in C and runs on a MicroVAX 3100/90. It is difficult to compare different programs written in different languages and running on different machines, but the multiple algorithm seems to be significantly quicker. Figure 9 shows how the multiple algorithm has worked for the 14 macrolides. It can be noticed that due to the important similarity

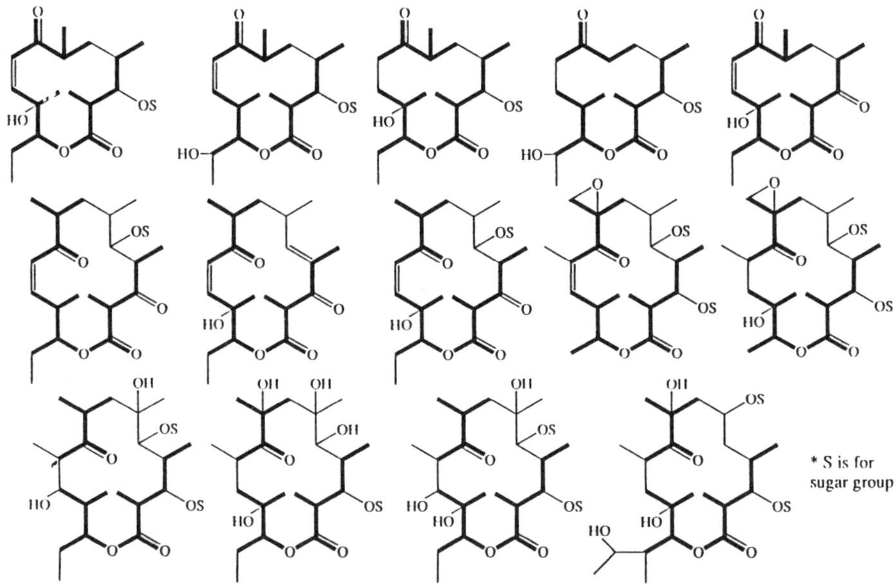

Figure 8. 14 macrolides and their largest common disconnected subgraph shown with the bold bonds. The molecules are shown in the order in which the multiple algorithm has used them, the smallest first (left to right and top to bottom).

between these molecular graphs, the Ullmann algorithm for subgraph isomorphism has been used often (Figure 9).

20 Polycyclic Structures (Figure 10)

These structures have been used as a test case for the multiple algorithm in 2D and 3D. It must be noticed that all the ring bonds are labelled as aromatic bonds. Figure 11 shows the solutions that have been found by the multiple algorithm in two dimensions with different options. In one case, a cpu time of 191.4 seconds is used, because 2048 combinations of graphs have been generated to find the disconnected parts (parts 3–6 of the disconnected algorithm).

A Cluster of 34 Structures (Figure 12)

A cluster of 34 molecules was processed by the multiple algorithm. All the molecules of the set are very similar and their largest common subgraph constitutes a large proportion of each molecular graph. The multiple algorithm needed 2.3 seconds to find the largest subgraph common to the 34 structures, with standard atom and bond matching and the disconnected option off.

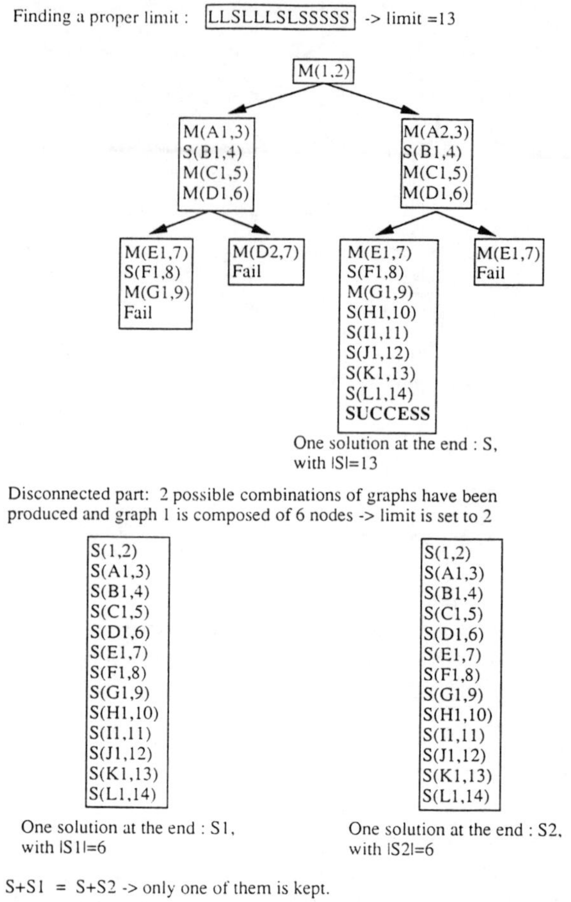

Figure 9. This figure shows how the solution to the 14 macrolides of Figure 8 was found. L is for *largest*, M is for *larger* and S is for subgraph isomorphism algorithm (Ullmann's algorithm).

4. 3D matching

INTRODUCTION

All the techniques for comparing 3D molecular skeletons do not use the bonds between atoms. They compare the positions of the atoms in space. The output is a set of tuples of atoms that can be superimposed and the rotations and the translation that has to be applied to one or several skeletons to superimpose them. The hydrogen atoms are discarded and often the atom types of the atoms are not considered. Kuhl and co-workers [19] have described a method to superimpose two sets of points in 3D space of

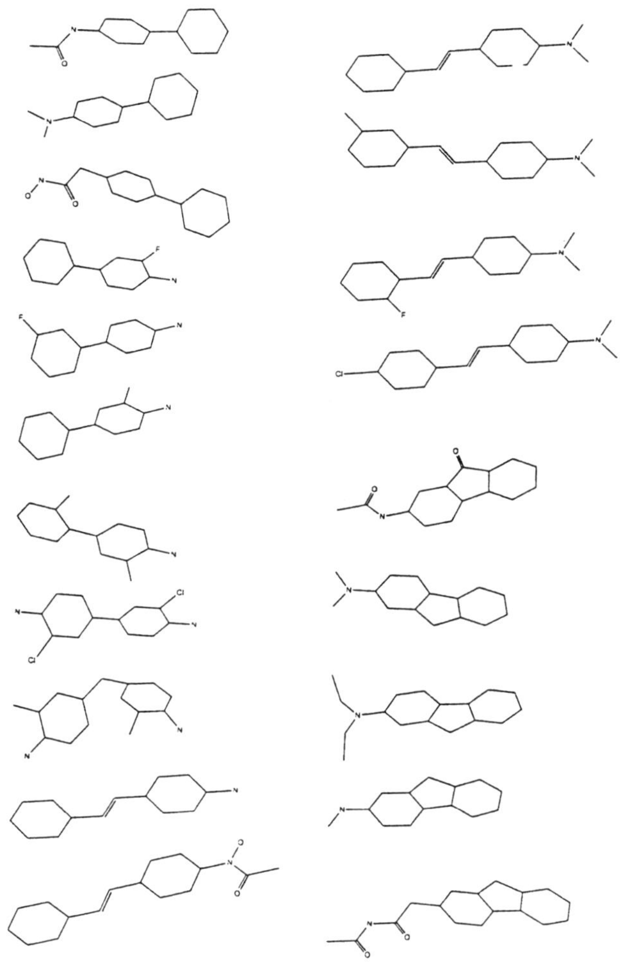

Figure 10. 20 structures for which all rings are aromatic. They are used for 2D and 3D matching. Their conformation has been calculated by the program Concord [18].

polynomial complexity: $O(n^7)$ for two structures of n atoms. This brute force method is inefficient but demonstrates that the problem of finding the best superimposition of two sets of points in space is not an NP-complete problem. Crandell and Smith [7] were the first to present an algorithm to find the maximal common substructures of a set of molecular skeletons. They use a similar technique as Varkony and co-workers [5] and Takahashi and co-workers [6] have used in 2D. It is based on growing common substructures until they cease to be common to the whole set of input structures. Two substructures are considered to be identical if their canonical names derived from their distance matrices are the same. The clique detection method is another technique [19], well explored by Brint and Willett [8]. Brint and Willett have

Result	Bond	Atom	Ring bonds / chain bonds	Discon	cpu/s
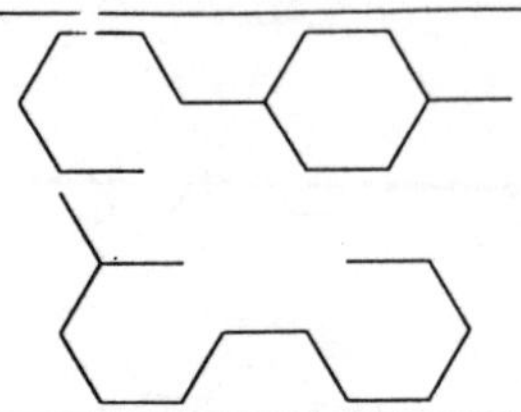	No bond type	No atom type	No distinction between rings and chains	no	26.9
				yes	27.4
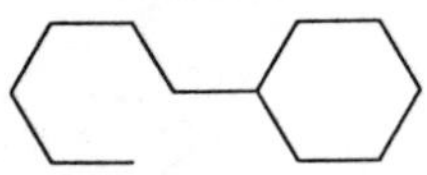	No bond type	Normal atom types	No distinction between rings and chains	no	14.4
				yes	14.9
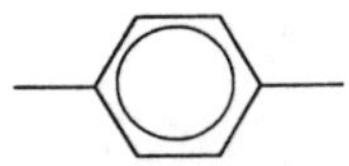	Normal bond types	No atom type	No distinction between rings and chains	no	1.6
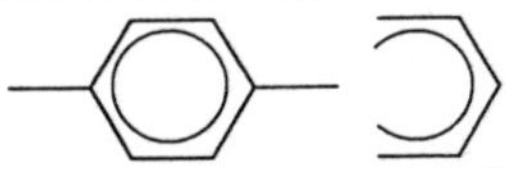	Normal bond types	No atom type	No distinction between rings and chains	yes	2.1
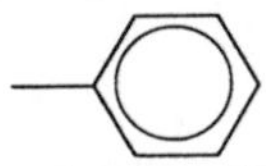	Normal bond types	Normal atom types	No distinction between rings and chains	no	1.8
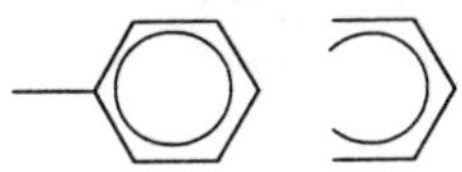	Normal bond types	Normal atom types	No distinction between rings and chains	yes	191.4 (2048 *)
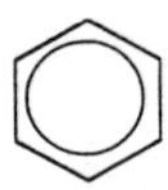	Normal bond types	Normal atom types	Ring bonds with ring bonds and chain bonds with chain bonds	no	2.1
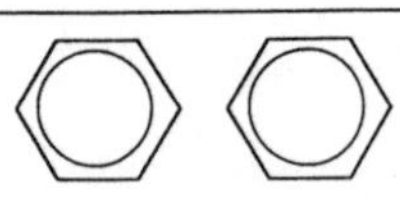	Normal bond types	Normal atom types	Ring bonds with ring bonds and chain bonds with chain bonds	yes	7.3

Figure 11. Solutions found for the structures of Figure 10, with different options used. *2048 combinations of graphs have been found as a result of the subtraction of the solution to the input graphs to find a disconnected part to add to the solution. This explains the important cpu time.

extended this method to matching more than two three-dimensional structures [8]. For N structures, the clique detection method is used to match one chosen structure against each of the $N - 1$ remaining structures in turn. Each execution produced a list of solutions concerning the chosen structure and one of the $N - 1$ other input structures, thus there are $N - 1$ such lists. A solution is itself a list, whose members are nodes from the chosen structure. A growing based algorithm, similar to the Crandell and Smith algorithm, is used to explore these lists, but here a grown structure is accepted if it is included in one solution

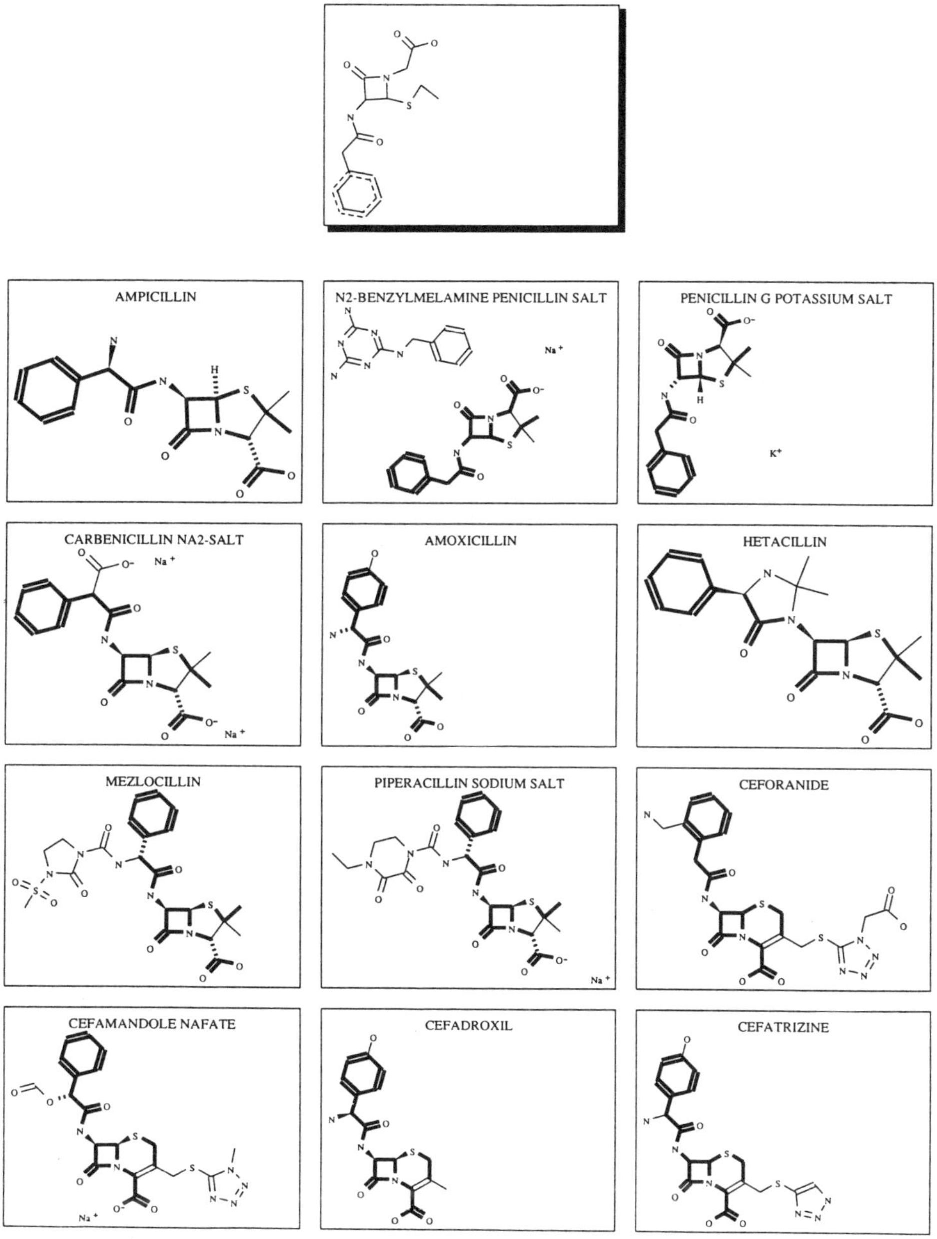

Figure 12. At the top of the figure, the largest common subgraph is shown. It is common to a set of 34 molecular graphs, 12 of which are displayed in the figure.

of each of the $N - 1$ lists. We choose to use the clique detection approach, but only to simulate *larger* and *largest*, i.e. for a pair of structures.

THE CLIQUE DETECTION METHOD

This technique involves creating the product graph of the two input graphs and to detect in this product graph the cliques, that are the maximal common substructures of the two input graphs.

Creation of the product graph

The product graph $GP(N_{GP}, E_{GP})$ of two graphs $G1(N_1)$ and $G2(N_2)$ where N_{GP}, N_1, N_2 are sets of nodes and E_{GP} is a set of edges is:

$N_{GP} = N_1 \times N_2$. The nodes in GP are all pairs (n_1, n_2) of nodes, $n_1 \in G1$ and $n_2 \in G2$.

$E_{GP} \subseteq (N_1 \times N_2) \times (N_1 \times N_2)$.

$(a, b) \in E_{GP} \Leftrightarrow$ a $= (n_1, n_2)$, $b = (n_1', n_2')$, where $n_1, n_1' \in N_1$ and $n_2, n_2' \in N_2$, n_1 is compatible with n_2, n_1' is compatible with n_2' and distance(n_1, n_1') = distance$(n_2, n_2') \pm \varepsilon$.

We consider that any atom n_1 is compatible with any other atom n_2, but it is possible to constrain the graph by making some atom types incompatible to others [3]. We also consider ε as a constant, but it is possible to make clusters of distances between points and to consider two different distances to be equivalent if they are in the same cluster [7]. Finally, the complexity of the creation of this product graph for two skeletons of n and m points, is clearly dependant on the number of edges, and thus $O(n^2m^2)$.

Clique detection

Kuhl and co-workers [19] have shown that the cliques in the product graph are the maximal common substructures to the two input graphs. A clique C in a graph G is a maximal complete subgraph in G, i.e. a subgraph where all nodes are connected to each other and there is no node from G other than those of C that can be added to C to produce a larger complete subgraph. As the nodes of the product graph are pairs of nodes from the two input graphs, a clique in the product graph is a list of pairs of nodes from the two input graphs that can be superimposed within a threshold ε. Several algorithms exist for solving the clique detection problem. This problem involves detecting all the cliques of a given graph. Brint and Willett [8] have reviewed some of them and have shown that the Bron and Kerbosch algorithm [20] is the fastest in most cases. This algorithm keeps a set S of current nodes that are a complete subgraph, a set *Not* of nodes that have been eliminated from S and a set *Candidates* of nodes that are connected to all the nodes of S. S, *Not* and *Candidates* have no node in common. After having moved a node from *Candidates* to S, and having updated the *newNot* and the *newCandidates* sets, if *newCandidates* is empty then S is a clique, otherwise it runs recursively on S, *newCandidates* and *newNot*. The choice of the node of *Candidates* to add to S will determine the order in which the cliques are found and in which the bad candidates are eliminated. After the recursive call, the node from *Candidates* that was put in the set S, is added to *Not*. As we need to find solutions that are larger than a given limit, we modify slightly the Bron and Kerbosch algorithm in two places:

1. When a solution is found, it is kept only if it is larger or equal to the limit. Moreover for *largest*, the value of the limit is changed if a solution larger than the limit is found.
2. It runs recursively on the new sets, only if $|S| + |newCandidates| \geq$ limit.

Some solutions must be eliminated. For instance, let us assume that two solutions $S1$ and $S2$ are $S1 = (1, a)(2, b)(3, c)(4, d)$ and $S2 = (1, b)(2, d)(3,e)$ where 1, 2, 3, 4 are nodes of the first input graph $G1$ and a, b, c, d, e of the second graph $G2$. If only the nodes from $G1$ are considered: $S2_{(G1)} \subset S1_{(G1)}$. This means that in 3D space, $S2$ can be superimposed to a part of $S1$ and as it is smaller than $S1$, $S2$ is not maximal. So $S2$ can be eliminated. However, the Bron and Kerbosch algorithm has not been designed to find the largest cliques only. Other algorithms such as [21] that find the largest cliques only are apparently faster than the Bron and Kerbosch algorithm to simulate *largest*.

Complexity of some algorithms

The clique detection problem is NP-complete for general graphs [20, 22], though the Bron and Kerbosch algorithm is fast in most real cases. The speed of the algorithm depends obviously on the number of edges of the product graph, which itself depends on the threshold ε. The larger ε is, the more edges the product graph will have.

The worst case for the Crandell and Smith algorithm as well as for Brint and Willett's extension of the clique detection method occurs when all the n input structures are the same and have a high degree of symmetry. In this case, if a structure can be mapped onto itself in K different ways, K^n solutions will be found.

The speed of our multiple algorithm for 3D structures depends on the number of cliques found by the clique detection and the number of maximal solutions stored at the end of *larger*. If all the structures are the same and have a high degree of symmetry, a lot of cliques will be found, but only one maximal solution will be kept. Thus it is quite a good case for our algorithm. If the input structures are very dissimilar, the clique detection will be very fast and find a limited number of small cliques. Thus this is also a good case for the multiple algorithm. It would seem that the worst case occurs with a set of structures that are "quite" symmetrical and "quite" similar. In this case a lot of large cliques might be found and it is not possible to eliminate them when they are stored as solutions.

RESULTS

The 20 structures of Figure 10 have been matched together in 3D using the multiple algorithm. Figure 13 shows the different solutions found with different distance thresholds and without considering any atom type. The solutions are shown for the sixth graph, the smallest of the 20 structures, of Figure 10 (top to bottom, left to right).

Solutions	Threshold/Å	cpu/s
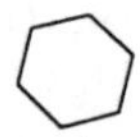	0.05	29.4
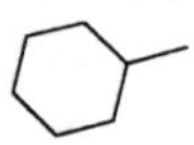	0.1	27.4
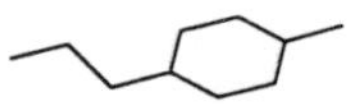	0.5	52.5
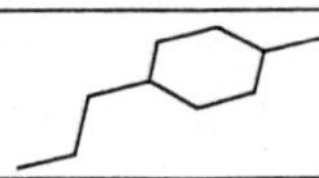		
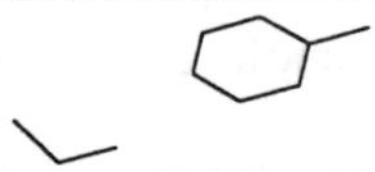	0.8	67.5
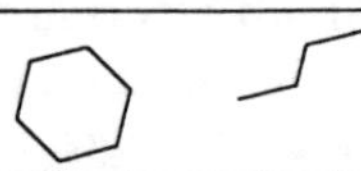		

Figure 13. 3D solutions found for the 20 structures of Figure 10, using different distance threshold. No atom type has been taken into account.

5. Conclusion

We have presented an algorithm that can find the largest substructures common to a set of compounds. The multiple algorithm has been used for 2D molecular graphs, using a backtracking method to compare pairs of graphs and using the clique detection method for pairs of 3D structures. Several options can be set to create different types of solutions. In 2D, disconnected solutions can be found and it is also possible to start from a known substructure common to the input graphs. Augmented solutions are also available, where "fuzzy" atoms and "fuzzy" bonds are added to a common subgraph. In 3D, it is possible to find several different solutions when a distance threshold is modified.

It is possible to extend the scope of the multiple algorithm by applying it to other chemical or non-chemical problems. As long as it is possible to extract a sub-object from an object and to find the largest sub-object common to two objects, it is possible to find the largest sub-objects common to a set of objects using the multiple algorithm. In a future paper, we intend to demonstrate this on molecular surfaces, by finding the largest regions of surfaces common to a set of molecular surfaces.

References

1. Willett, P., *Similarity and clustering in chemical information systems*, (Research Studies Press: Letchworth, Herts., U.K., 1987), p. 54.
2. Richards, W. G. and Hodgkins, E. E., *Chem. Britain* **11**, 1141 (1988).
3. Martin, Y. C., Bures, M. G., Danaher, E. A., DeLazzer, J., Lico, I., and Pavlik, P. A., *J. Comput. Aided Mol. Design* **7**, 83 (1993).
4. Johnson, A. P., Marshall, C., and Cho, A., unpublished report.
5. Varkony, T., Shiloach, Y., and Smith H., *J. Chem. Inf. Comput. Sci.* **3**, 104 (1978).
6. Takahashi, Y., Satoh, Y., and Sasaki, S., *Anal. Sci.* **3**, 23 (1987).
7. Crandell, C. W. and Smith, D. H., *J. Chem. Inf. Comput. Sci.* **23**, 186 (1983).
8. Brint, A. T. and Willett, P., *J. Chem. Inf. Comput. Sci.* **27**, 152 (1987).
9. Bayada, D. M., Simpson, R. W., Johnson, A. P., and Laurenço, C., *J. Chem. Inf. Comput. Sci.* **32**, 680 (1992).
10. Johnson, A. P. and Marshall, C., *J. Chem. Inf. Comput. Sci.* **32**, 426 (1992).
11. Levi, G., *Calcolo*, **9**, 341 (1972).
12. Kvaniscka, V. and Pospichal, J., *Reports Mol. Theo.* **1**, 99 (1990).
13. McGregor, J. J., *Software-Practice Experience* **32**, 23 (1982).
14. McGregor, J. J. and Willett, P., *J. Chem. Inf. Comput. Sci.* **21**, 137 (1981).
15. Morgan, H. L., *J. Chem. Doc.* **5**, 107 (1965).
16. Rucker, G. and Rucker, C., *J. Chem. Inf. Comput. Sci.* **30**, 187 (1990).
17. Ullmann, J. R., *J. ACM.* **23**, 31 (1976).
18. Rusinko, A., Skell, J., Balducci, J. M., McGarity, C. M., and Pearlman, R. S., *CONCORD: a program for the rapid generation of high quality approximate 3-dimensional molecular structures* (University of Texas, Austin and Tripos Associates: StLouis, Missouri, 1988).
19. Kuhl, F. S., Crippen, G. M., and Friesen, D. K., *J. Comput. Chem.* **5**, 24 (1984).
20. Bron, C. and Kerbosch, J., *Commun. ACM.* **16**, 575 (1973).
21. Balas, E. and Yu, C. S., *SIAM J. Comput.* **15**, 1054 (1986).
22. Garey, M. R. and Johnson, D. S., *Computers and Intractability: A Guide to the Theory of NP-Completeness* (Freeman & Co.: New York, 1979), p. 53.

A Proposal Toward the Identification of Substructure Electronic Similarity

BARBARA LEONI and GUIDO SELLO
Universita Degli Studi Di Milano, Dipartimento Di Chimica Organica Ed Industriale, Via Venezian 21, Milano, Italy

Abstract. A new measure of substructural similarity is introduced; it is based on our method for the calculation of molecular electronic energy. The differences obtained comparing complete molecules with the corresponding structure where each atom has been separated can be used as numeric or graphic similarity measures. The application of the methodology to simple functional groups, heteroaromatic compounds, mono and disubstituted benzene derivatives, and nitrogenous bases of nucleic acids, is presented and commented.

1. Introduction

Theoretical models have a constant and fruitful use in many different applications even if their importance is not always explicitly recognised. One of the most important of the possible application fields is given by the possibility to obtain a real representation of a bad defined or vague concept or idea. This aspect of the modelling activity has taken a crucial role mainly since the mathematical procedures of description have increased in all scientific fields and, as for us concerned, in organic chemistry.

One of the most useful such concepts used in organic chemistry is the one concerning the definition of similarity [1]. The scope of studies dealing with similarity is extremely heterogeneous and ranges from the characterisation of similarity in molecular shape [2] to the application of similarity measures in computer aided synthesis [3]. The attempt to relate the similarity in molecular properties to the similarity in the electron structure of corresponding molecules is another application field [4]. (A typical example is the similarity measure of electron density).

The definition of a model for the quantification of any kind of similarity can be shown by two different viewpoints: 1) the choice of a system to measure the similarity with the corresponding definition of its variational capacities; 2) the choice of a similarity index which, being a well defined quantity, allows for a more efficient comparison even if static. Similarity measures can be easily transformed into similarity indexes, therefore the definition of model systems to measure similarity is the determining point [5].

During our work mainly devoted to the modelling of reactivity in organic chemistry [6] and to the planning of organic synthesis [7], we run into the problem of the identification of functional groups present in a molecule. The traditional methodologies, which are based on the search for specific inter

R. Carbó (ed.), Molecular Similarity and Reactivity: From Quantum Chemical to Phenomenological Approaches, 267–289.

atomic relations compared with pre-existing tables, seemed too restrictive and in any case not in line with our modelling approach, which would describe molecules and their behaviour in general terms based only on the intimate properties of atoms and on the interactions between atoms. As a consequence we decided to use a new model able to identify the functional groups of a molecule using the calculated properties of its atoms [8]. The characteristic satisfying this purpose was the perturbation caused by each atom on the molecular electronic energy. The model and the corresponding procedure furnished the searched result and were able to reproduce the classic functional groups and to suggest the presence of complex interactions between atoms and atomic groups not easily describable in standard terms. A further interesting result of this study was the ascertainment of the transferability of the values calculated for an atomic group from one to another molecule; i.e., for example, an isolated ketonic group gives an equal or similar value of the perturbation energy in any molecule, whereas different values are calculated only if its electronic surrounding is really changed (e.g. in presence of a conjugated double bond). It is therefore clearly confirmed the validity, in our model, of the principle of summability of atomic or atomic group quantities [9].

2. Electronic similarity

Excited by the obtained results and by the ideas suggested by the literature in the field, we come to consider the possibility of using the electronic energy differences (EED) for finding a similarity measure. In order to be better understood in the following discussion we need to briefly describe the calculation methodology used in our general model for reactivity [10]. It gives, among other things, the possibility to calculate the molecular electronic energy summing the atomic electronic energies; these last are calculated using the following equation:

$$E_i = k_3 \times (A + B + C) - k_2 \times N_3$$
$$k_3 = -k_1/(Z^{0*\prime} \times R^0_{\text{cov}})$$

where k_1 and k_2 are constants depending on the atom type; $Z^{0*\prime}$ is the effective nuclear charge of the isolated atom for a complete electronic shielding; R^0_{cov} is the atomic covalent radius of the isolated atom. The sum is on all the atoms.

$$A = (N^2 + aN - 2NN_1 - 2bNN_2 + N_1^2 + 2bN_1N_2 - \text{a}N_1 + b^2N_2^2 - abN_2) \times N_3$$
$$\text{B} = 0.5 \times (-2aN + 2aN_1 + 2abN_2 - a^2) \times N_3^2$$
$$\text{C} = (a^{2/3}) \times N_3^3$$

where a, b, c, are Slater's coefficients; N is the atomic number; N_i are the shell occupation numbers.

The variation of the electronic occupation of the atomic shells results in a variation in the energy value and as a consequence in a variation of the molecular electronic energy. A different energy is thus obtained for each molecule, for each perturbation of the molecule, for each transformation of the molecule.

Nevertheless the direct comparison of the molecular electronic energies doesn't give information on molecular similarity; in fact if we take as an example two molecules of linear alkanes (C4 $\leftrightarrow$ C6) their energy constantly increase but can happen that it is casually coincident with the energy of a molecule even much different. On the contrary a much more powerful source of information is furnished by the analysis of the variations of the electronic energy, mainly if it is compared on many points. It is the use of these variations that allowed us to introduce the already described new definition of functional groups. The immediate next step is the application of the just introduced concept to the analysis of the similarity between substructures, similarity which we called "electronic similarity" in order to emphasise the impossibility of its direct application to different properties of each class of molecules (biological activity, reactivity, etc.), but its limited application to the comparison of molecules in terms of their reaction to the perturbation induced on their electronic structure.

An interesting and rich of consequences aspect of our approach is represented by the mechanism of the used calculation. This is based on a very simple idea: if we think that the electronic energy of a molecule is *strictly* in relation with the energy of its atoms as result from their reciprocal interactions, we can affirm that the presence/absence of an atom in a specific position determines an energetic variation which can be thought as the quantification of the difference between that molecule and a hypothetical molecule where the atom is isolated (annihilation principle). Therefore if we consider two identical molecules and we work on atoms in complete correspondence the energy variation will be identical; on the contrary, if we consider two different molecules and we work on similar atoms or if we consider two identical molecules and we work on different atoms the calculated energy differences will be different and they will represent a measure of the "importance" of the atom in the molecule.

The final effect of the calculation mechanism can be viewed as a kind of "wave" of energy variations moving on the molecule touching each atom. The order of move becomes therefore important, giving different "waves" for different routes.

Thus we have available two similarity measures: one direct and straightforward which comparing the energy variations corresponding to specific substructures evaluates the degree of their similarity; the other, strongly depending on the analysis conditions, which gives a measure of the similarity between two substructures comparing their energy waves; this similarity, using a visual approach, can be quantitative and/or qualitative and even difficult to be described by numbers [2].

In this work we will analyse both the possibilities of the model. We will try to identify the substructure electronic similarity both from the viewpoint of numbers and from that of graph behaviour. We think that the amount of information present in this kind of similarity is very high and that the present work only partly describes it.

On the contrary, we will not suggest any kind of similarity index, not just because it is impossible to obtain an index from the measuring system, but because, at present, it is not one objective of ours, being more interested in checking the power of the model.

3. Results and discussion

In the following we will examine some examples of the application of the similarity measuring system just introduced. In order to meet at the same time the requirements of clarity and of utility we decided to present four groups of different examples and to discuss for each of them one or more aspects which could be of interest. The groups are: 1) a set of very simple molecules containing functional groups; 2) two groups of heterocycles with 5 or 6 member cycles at different levels of heteroatom presence; 3) a set of mono and disubstituted benzene derivatives; 4) some bases, both natural and modified, of nucleic acids.

3.1. Set 1. Functional groups

The molecules used in this group are reported in Tables I and I bis together with the calculated values of electronic energy; it is clear that there are

TABLE I
Electronic energy differences of simple organic compounds[a]

Compound	Atom 1	Atom 2	Atom 3	Atom 4	Total
C—C—C	0	0	0	–	0
C=C—C	1.772	1.767	0.303	–	3.842
C#C—C	4.570	4.407	0.879	–	9.883
C=C=C—C	0.721	3.418	0.697	0.299	5.135
C—C—C$^+$	0	0	0.563	–	0.563
C=C—C$^+$	1.418	1.203	0.773	–	3.394
C#C—C$^+$	3.630	3.378	1.815	–	8.822
C=C=C—C$^+$	1.172	2.786	0.067	0.836	4.860
C—C—C$^-$	0	0	0	–	0
C=C—C$^-$	1.479	1.481	0.608	–	3.567
C#C—C$^-$	4.168	4.170	1.135	–	9.473
C=C=C—C$^-$	1.400	3.526	0.812	0.201	5.939
C—C—N	0	0.023	0.475	–	0.498

TABLE I *(Continued)*

Compound	Atom 1	Atom 2	Atom 3	Atom 4	Total
C=C—N	1.759	1.764	0.536	–	4.059
C#C—N	4.536	4.378	1.146	–	10.062
C=C=C—N	0.721	3.404	0.694	0.532	5.351
C—C—N$^+$	0	0.058	1.140	–	1.197
C=C—N$^+$	2.133	1.952	0.802	–	4.887
C#C—N$^+$	4.317	4.095	1.551	–	9.962
C=C=C—N$^+$	1.166	3.483	0.806	0.746	6.201
C—C—N$^-$	0	0	0.025	–	0.025
C=C—N$^-$	1.483	1.476	0.791	–	3.750
C#C—N$^-$	4.126	4.118	1.379	–	9.623
C=C=C—N$^-$	1.396	3.521	0.800	0.401	6.117
C—C—O	0	0.108	0.796	–	0.904
C=C—O	1.712	1.798	0.498	–	4.008
C#C—O	4.453	4.375	1.158	–	9.987
C=C=C—O	0.720	3.353	0.725	0.498	5.296
C—C—O$^+$	0	0.230	1.408	–	1.638
C=C—O$^+$	3.290	3.252	2.256	–	8.798
C#C—O$^+$	5.334	5.259	1.089	–	11.682
C=C=C—O$^+$	1.149	4.567	2.058	2.146	9.919
C—C—O$^-$	0	0.040	0.337	–	0.377
C=C—O$^-$	1.901	1.922	0.425	–	4.247
C#C—O$^-$	4.462	4.479	1.111	–	10.052
C=C=C—O$^-$	1.371	3.899	1.206	0.675	7.152
C—C—P	0	0	0	–	0
C=C—P	1.785	1.783	0.304	–	3.873
C#C—P	4.615	4.455	0.847	–	9.917
C=C=C—P	0.722	3.433	0.714	0.297	5.165
C—C—P$^+$	0	0	0	–	0
C=C—P$^+$	1.190	0.977	1.098	–	3.266
C#C—P$^+$	3.445	3.199	2.093	–	8.737
C=C=C—P$^+$	1.183	2.578	0.142	1.143	5.046
C—C—P$^-$	0	0.030	0	–	0.030
C=C—P$^-$	1.839	1.853	0.233	–	3.925
C#C—P$^-$	4.572	4.587	0.715	–	9.875
C=C=C—P$^-$	1.404	3.895	1.194	0.184	6.677
C—C—S	0	0	0	–	0
C=C—S	1.771	1.766	0.306	–	3.842
C#C—S	4.553	4.391	0.898	–	9.842
C=C=C—S	0.724	3.415	0.694	0.302	5.134
C—C—S$^+$	0	0	0.238	–	0.238
C=C—S$^+$	1.775	1.567	0.606	–	3.949
C#C—S$^+$	3.983	3.735	1.659	–	8.377
C=C=C—S$^+$	1.168	3.138	0.429	0.673	5.409
C—C—S$^-$	0	0	0	–	0
C=C—S$^-$	1.842	1.837	0.248	–	3.927
C#C—S$^-$	4.505	4.499	0.803	–	9.807
C=C=C—S$^-$	1.398	3.887	1.165	0.155	6.604

[a] Energy values are in arbitrary units.

TABLE I bis.
Electronic energy differences of simple organic compounds[a]

Compound	Atom 1	Atom 2	Atom 3	Atom 4	Atom 5	Total
C—C—C—C	0	0	0	0	–	0
C=C—C—C	1.773	1.768	0.307	0	–	3.848
C#C—C—C	4.568	4.405	0.888	0	–	9.861
C=C=C—C—C	0.722	3.418	0.698	0.303	0	5.141
C—C—C⁺—C	0	0	0.424	0.020	–	0.444
C=C—C⁺—C	1.416	1.202	0.796	0.062	–	3.477
C#C—C⁺—C	3.626	3.376	1.841	0.061	–	8.903
C=C=C—C⁺—C	1.171	2.781	0.064	0.861	0.062	4.940
C—C—C⁻—C	0	0	0	0	–	0
C=C—C⁻—C	1.471	1.472	0.611	0.090	–	3.643
C#C—C⁻—C	4.160	4.160	1.140	0.088	–	9.548
C=C=C—C⁻—C	1.399	3.518	0.805	0.205	0.091	6.017
C—C—N—C_2	0.021	0.028	0.062	0.034	–	0.179
C=C—N—C_2	1.757	1.769	0.275	0.036	–	3.873
C#C—N—C_2	4.539	4.383	0.878	0.038	–	9.875
C=C=C—N—C_2	0.725	3.401	0.695	0.270	0.037	5.164
C—C—N⁺—C_2	0	0.067	0.245	0.077	–	0.465
C=C—N⁺—C_2	2.109	1.943	0.045	0.169	–	4.436
C#C—N⁺—C_2	4.292	4.083	1.033	0.154	-	9.715
C=C=C—N⁺—C_2	1.170	3.464	0.798	0.025	0.167	5.790
C—C—N⁻—C_2	0	0	0	0	–	0
C=C—N⁻—C_2	1.422	1.416	0.670	0.272	–	4.053
C#C—N⁻—C_2	4.066	4.060	1.244	0.270	–	9.911
C=C=C—N⁻—C_2	1.399	3.467	0.746	0.264	0.272	6.420
C—C—O—C	0	0.120	0.265	0.131	–	0.515
C=C—O—C	1.709	1.806	1.222	0.145	–	3.782
C#C—O—C	4.446	4.377	0.769	0.163	–	9.754
C=C=C—O—C	0.720	3.350	0.732	0.122	0.146	5.070
C—C—O⁺—C	0.020	0.238	0.573	0.254	–	1.085
C=C—O⁺—C	3.201	3.195	1.456	0.739	–	8.590
C#C—O⁺—C	5.286	5.227	0.274	0.616	–	11.403
C=C=C—O⁺—C	1.162	4.549	2.039	1.377	0.663	9.790
C—C—O⁻—C	0	0.046	0.078	0.053	–	0.177
C=C—O⁻—C	1.814	1.841	0.202	1.032	–	4.889
C#C—O⁻—C	4.403	4.427	0.833	0.945	–	10.608
C=C=C—O⁻—C	1.395	3.852	1.163	0.196	0.971	7.576
C—C—P—C_2	0	0	0	0	–	0
C=C—P—C_2	1.784	1.781	0.302	0	–	3.867
C#C—P—C_2	4.612	4.452	0.848	0	–	9.912
C=C=C—P—C_2	0.721	3.430	0.714	0.295	0	5.160
C—C—P⁺—C_2	0	0	0	0	–	0
C=C—P⁺—C_2	1.186	0.975	1.112	0.088	–	3.449
C#C—P⁺—C_2	3.424	3.177	2.128	0.085	–	8.898
C=C=C—P⁺—C_2	1.170	2.558	0.163	1.174	0.088	5.240
C—C—P⁻—C_2	0	0.025	0.044	0.027	–	0.124
C=C—P⁻—C_2	1.815	1.824	0.232	0.113	–	4.096
C#C—P⁻—C_2	4.537	4.548	0.726	0.111	–	10.032
C=C=C—P⁻—C_2	0.140	3.862	1.157	0.176	0.112	6.819

TABLE I bis. *(Continued)*

Compound	Atom 1	Atom 2	Atom 3	Atom 4	Atom 5	Total
C—C—S—C	0	0	0	0	–	0
C=C—S—C	1.771	1.766	0.319	0	–	3.856
C#C—S—C	4.551	4.389	0.914	0	–	9.853
C=C=C—S—C	0.721	3.414	0.694	0.316	0	5.146
C—C—S⁺—C	0	0.023	0.025	0.030	–	0.078
C=C—S⁺—C	1.768	1.567	0.505	0.349	–	4.189
C#C—S⁺—C	3.979	3.734	1.556	0.346	–	9.615
C=C=C—S⁺—C	1.171	3.137	0.426	0.571	0.347	5.652
C—C—S⁻—C	0	0	0	0	–	0
C=C—S⁻—C	1.829	1.825	0.261	0.440	–	4.355
C#C—S⁻—C	4.492	4.486	0.817	0.433	–	10.228
C=C=C—S⁻—C	1.397	3.874	1.154	0.142	0.440	7.008

[a] Energy values are in arbitrary units.

three kinds of possible variations: atom type (C, N, O, S, and P), bond order (saturated, ethylenic, allenic, acetilenic), presence of H or C on the heteroatom. Besides there are neutral and positively or negatively charged molecules.

In this case we will only analyse some numerical results.

Some groups of molecules based on the calculated electronic energy sums are reported in Table II. Some interesting data can be found in these values. First, the molecules group according to the bond order with few exceptions; among these, the positions of the oxygenated ethylenic compounds positively charged are interesting because they are in the group of the acetylenic derivatives. The presence of a O^+ apparently perturbs the electronic energy of the double bond so much so to make it very different from the standard double bond.

The anomalous behaviour of the oxygen atom is also present if we take into account the order of the atom energies according to the atom type and the bond order; these are always the same inside an homologous series excluding the oxygen atom (in all cases but in the negatively charged molecules without H on the heteroatom) and for the positively charged nitrogen atom (see Table III).

Looking at the total energy values we can also find the following similarities:

1) single bonds; the group C—O—C is similar to the group C—N; the atoms C, S, and P, are similar in neutral molecules; the atoms C and S are also similar in negatively charged molecules; the order $P < S < C$ holds in positively charged molecules.
2) Double bonds; the group C—N—C is similar to the group C—S—C; the atoms C, S, and P, are similar in neutral molecules; the atoms C and P are similar in positively charged molecules.
3) Allenic bonds; the group C—P—C is similar to the group C—N—C; all

TABLE II
Groupings of compounds of Tables I and I bis according to their total electronic energy differences[a]

EED range	Neutral	Positively charged	Negatively charged
0	C—C—C C—C—P C—C—S C—C—C—C C—C—P—C C—C—S—C	C—C—P C—C—P—C	C—C—C C—C—S C—C—C—C C—C—S—C C—C—N—C
0–0.250	C—C—N—C	C—C—S C—C—S—C	C—C—N C—C—P C—C—P—C C—C—O—C
0.250–0.600	C—C—O—C C—C—N	C—C—C C—C—C—C C—C—N—C	C—C—O
0.600–1.650	C—C—O	C—C—N C—C—O C—C—O—C	
3.000–3.900	C=C—O—C C=C—N—C C=C—C C=C—C—C C=C—S C=C—S—C C=C—P—C C=C—P	C=C—C C=C—C—C C=C—S C=C—P C=C—P—C	C=C—C C=C—C—C C=C—S C=C—N
3.900–4.500	C=C—O C=C—N	C=C—N—C C=C—S—C	C=C—O= C=C—N—C C=C—S—C C=C—P C=C—P—C
4.500 –5.000		C=C—N C=C=C—C C=C=C—C—C	C=C—O—C
5.000–6.000	C=C=C—O C=C=C—O—C C=C=C—N—C C=C=C—N C=C=C—S—C C=C=C—S C=C=C—P—C C=C=C—P C=C=C—C C=C=C—C—C	C=C=C—N—C C=C=C—S—C C=C=C—S C=C=C—P—C C=C=C—P	C=C=C—C

TABLE II *(Continued)*

EED range	Neutral	Positively charged	Negatively charged
6.000–7.000		C=C=C—N	C=C=C—N—C C=C=C—N C=C=C—P—C C=C=C—P C=C=C—C—C C=C=C—S
7.000–8.000			C=C=C—O C=C=C—O—C C=C=C—S—C
8.000–9.000		C#C—C C#C—C—C C#C—P—C C#C—P C#C—S C=C—O C=C—O—C	
9.000–9.900	C#C—C C#C—C—C C#C—N—C C#C—S C#C—S—C C#C—O—C	C#C—N—C C=C=C—O—C	C#C—C C#C—C—C C#C—N C#C—P C#C—S C#C—S—C
9.900–12.000	C#C—N C#C—P—C C#C—P C#C—O	C#C—N C#C—O C#C—O—C C=C=C—O	C#C—N—C C#C—P—C C#C—S—C C#C—O C#C—O—C

[a] Energy values are in arbitrary units.

TABLE III
Order of electronic energy difference for the heteroatom[a]

Atom	Neutral	Unsubstituted		Substituted		
		Positive	Negative	Neutral	Positive	Negative
C	1 c 2 3	1 2 c 3	1 c 2 3	1 c 2 3	1 2 c 3	1 c 2 3
N	1 c 2 3	c 2 1 3	1 c 2 3	1 c 2 3	c 2 1 3	1 c 2 3
O	1 c 2 3	3 1 c 2	1 2 c 3	c 1 3 2	3 1 c 2	1 c 2 3
P	1 c 2 3	1 2 c 3	1 c 2 3	1 c 2 3	1 2 c 3	1 c 2 3
S	1 c 2 3	1 2 c 3	1 c 2 3	1 c 2 3	1 2 c 3	1 c 2 3

[a] The charges are positioned on the heteroatom; 1, 2, 3, c, represent the corresponding bond orders (1 = single, 2 = double, 3 = triple, c = allene).

the atom types are similar in neutral molecules; the atoms P and S are also similar in positively charged molecules.

4) Triple bonds; the atoms C and S are similar in neutral and negatively charged molecules; the atoms C and P are similar in positively charged molecules.

In conclusion we can observe that the method gives results in agreement with the usual classifications even if, in particular cases, it can suggest new orders.

3.2. Set 2. Heterocycles

Inside the group of heterocycles it is immediate the separation of the molecules into two subsets: 5 and 6 member cycles. In this case we will discuss some numerical results and some behaviour characteristics of the energy.

The classes of studied molecules are reported in Figure 1; these include all the homologues of pyrrol, furan, and thiophene, carrying a growing amount of ring carbons substituted by nitrogen atoms. N has been chosen because it is a neutral substitute of C. The variations of the electronic energy differences are also reported in Figure 1; it is interesting to note that the influence of the C—N substitutions on the graph appearance is very similar at similar substitution level independently from the heterocycle type (the numerical values are obviously different). On the other hand the presence of the main atom X determines a different position for the calculated values with a greater influence on the positions near the heteroatom.

It is also quite clear the different effect of the nitrogen atoms according to their position in the molecule, allowing an immediate recognition of the different subsets (1 azoles; 1,2 azoles; 1,3 azoles; etc.).

From the point of numbers we can emphasise the following behaviours:

1,3 Azoles + 2,4 azoles + 3,4 azoles. An equal decrease in the electronic energy differences (EED) is present for all the compounds in all the positions excluding the one occupied by the X heteroatom where the decrease follows the order $S < N < O$.

1,2 Azoles. A decrease in the EEDs is present also for these compounds excluding the 5 position where an increment is always present. The order of the X heteroatoms is the same as in the preceding group.

2,3,4 Azoles + 2,3,5 azoles. A different variation of the EEDs is present in this group; in 2,3,5 azoles the remaining free position shows a constant decrease of the values, whereas in 2,3,4 azoles the free position shows an increase of the values for O and N and a decrease for S: the specific behaviour of S therefore becomes evident.

2,3,4,5 Azoles. The heteroatom order is again shown; it must be emphasised

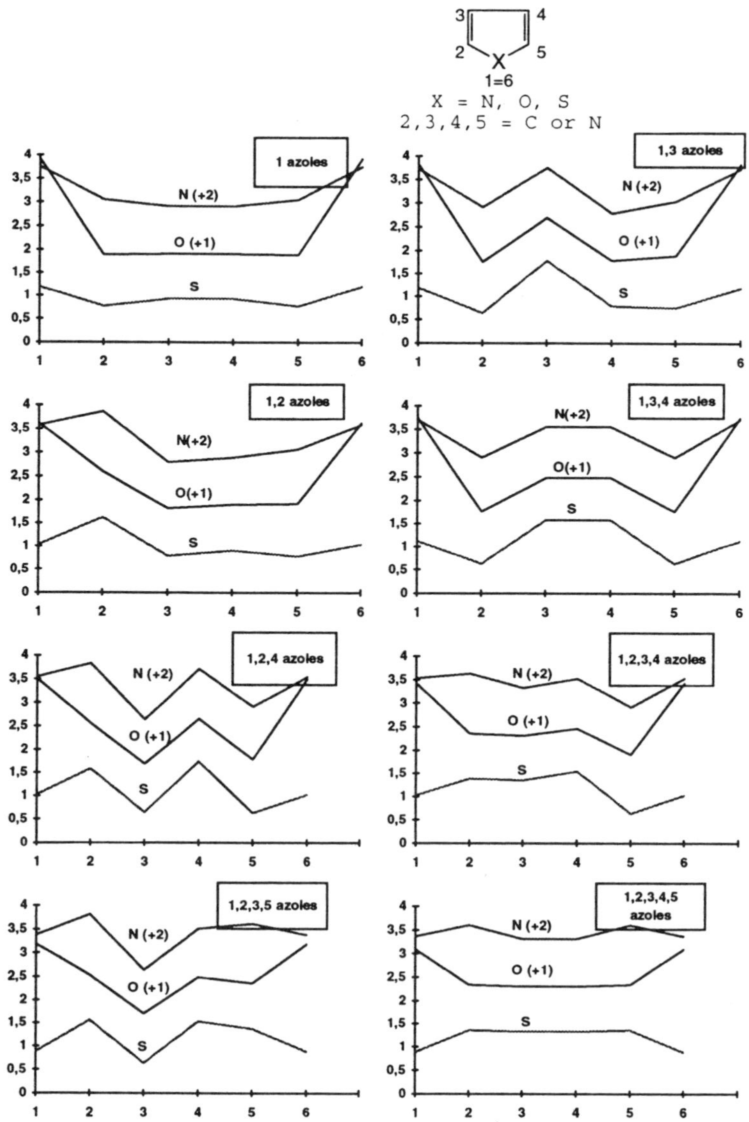

Figure 1. Azole derivatives and their EED graphs. Lines for N and O containing compounds are respectively shifted by +2 and +1 units. Abscissa numbers refer to molecular structure.

the increase of the EEDs in accordance with the increase of the number of nitrogen atoms in the heterocycle.

The sets of molecules containing 6 member cycles are reported in Figure 2; they include derivatives of pyridine, pyran, and thiopyran, carrying a growing amount of ring carbons substituted by nitrogen atoms. In this case the compounds containing O and S are positively charged.

TABLE IV
Electronic energy difference on each ring atom[a]

Compound[b]	1	2	3	4	5	6
1N	1.7567	1.0585	0.9153	0.9153	1.0585	1.7567
1S	1.1939	0.7712	0.9251	0.9251	0.7712	1.1939
1O	2.9188	0.8919	0.911	0.911	0.8919	2.9188
1,2N	1.5724	1.8717	0.7865	0.8915	1.0625	1.5724
1,2S	1.034	1.6204	0.787	0.9011	0.7774	1.034
1,2O	2.6104	1.5924	0.8165	0.8997	0.9196	2.6104
1,3N	1.726	0.9137	1.7541	0.7772	1.0344	1.726
1,3S	1.1811	0.6309	1.783	0.7872	0.7522	1.1811
1,3O	2.8274	0.7601	1.6902	0.7859	0.8856	2.8274
1,3,4N	1.6999	0.9077	1.5489	1.5489	0.9077	1.6999
1,3,4S	1.11685	0.6266	1.5763	1.5763	0.6266	1.11685
1,3,4O	2.732	0.7664	1.4856	1.4856	0.7664	2.732
1,2,4N	1.5464	1.8344	0.642	1.717	0.9163	1.5464
1,2,4S	1.0241	1.59	0.6434	1.7477	0.6372	1.0241
1,2,4O	2.5103	1.5611	0.6942	1.656	0.7935	2.5103
1,2,3,4N	1.5363	1.6295	1.3328	1.5221	0.9176	1.5363
1,2,3,4S	1.0258	1.3853	1.3533	1.5476	0.6337	1.0258
1,2,3,4O	2.4256	1.3611	1.3171	1.4677	0.907	2.4256
1,2,3,5N	1.3837	1.8113	0.642	1.5177	1.6224	1.3837
1,2,3,5S	0.8802	1.5627	0.6366	1.5376	1.3795	0.8802
1,2,3,5O	2.1848	1.534	0.7075	1.4888	1.3543	2.1848
5N	1.3683	1.6029	1.3135	1.3135	1.6029	1.3683
5S	0.8863	1.3614	1.3318	1.3318	1.3614	0.8863
5O	2.092	1.3365	1.3039	1.3039	1.3365	2.092

[a] Energy values are in arbitrary units; atom numbers refer to Figure 1.
[b] Compound names refer to Figure 1 indicating positions of added N.

The corresponding graphs are reported in Figures 2 and 3. In this case too the graph appearance strongly depends on the number of nitrogen atoms present in the cycle; it is particularly remarkable that, at the same substitution level, it is even possible to compare 5 and 6 member heterocycles. The presence of charges on the heteroatom *does not* influence the graph appearance, whereas the calculated EED value changes.

The numerical EED changes are mainly negative in this case too, even if they are less evident. We will not discuss each class because we prefer to comment on other aspects more interesting for our definition of similarity.

The analysis of these compounds permits the check of the similarity level between heterocyclic classes. We have already commented on the differences and the similarities of equisubstituted cycles; it is also clear that N, O, and S, can be considered interchangeable. But looking more deeply inside the calculated values we note that N and O are more similar, leaving S apart; the belonging to different periods and, as a consequence, the differences of the electronic behaviour can fully justify this result.

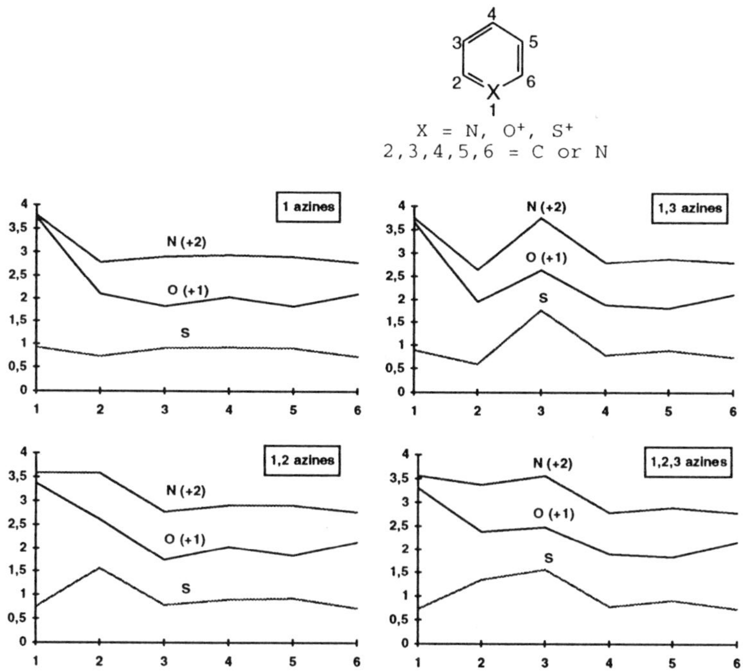

Figure 2. Azine derivatives and their EED graphs. Lines for N and O containing compounds are respectively shifted by +2 and +1 units. Abscissa numbers refer to molecular structure.

3.3. Set 3. Mono and disubstituted benzene derivatives

The sets of molecules containing mono and disubstituted (ortho, meta, and para, positions) benzene derivatives are reported in Figure 4; the substituents have been chosen between those reported as having different electronic effects on the aromatic ring. We will only discuss some aspects of the results with special attention to the graph appearances.

The graphs corresponding to the EED variations on the ring atoms are reported in Figures 4 and 5, the values have been properly scaled in order to emphasise their relevance.

In the monosubstituted compounds it is possible to note the differences in the graph appearance according to the substituent class; in particular substituents with a marked +M effect (Graph 1); substituents with a medium +M effect (Graph 2); substituents with both +M and –M effect (Graph 3); substituents with –M effect (Graph 4). Inside these groups there are obviously some anomalous points; in group 2 we find the methyl group whose effect is still partly unclear; in group 3 we find the chlorine atom (+M, –I) whose graph is as a matter of fact intermediate between groups 2 and 3 (in this case we can also take into account that the –I effect could be sufficient to flatten the energy curve). Between groups 1, 2, and 3, the difference of

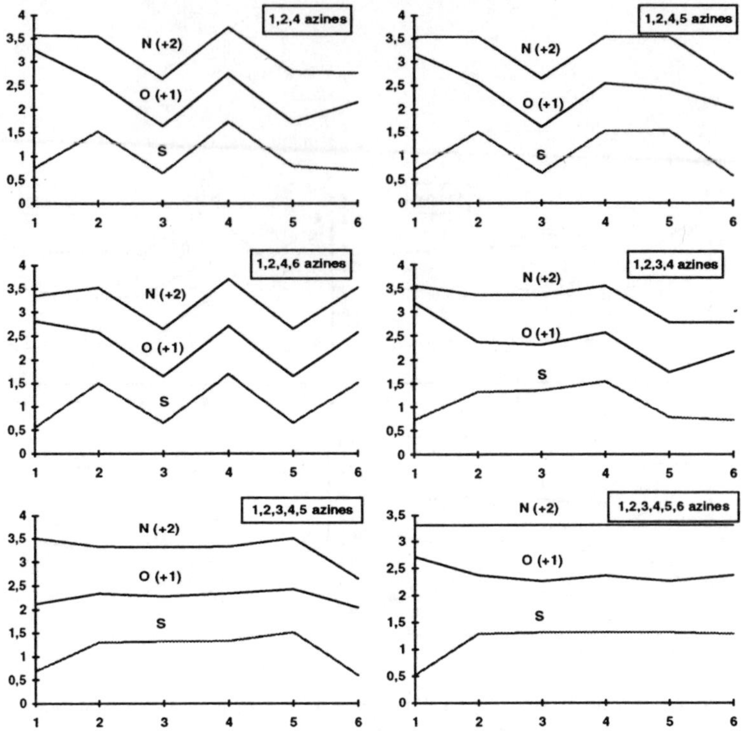

Figure 3. EED graphs of azine derivatives. Lines for N and O containing compounds are respectively shifted by +2 and +1 units. Abscissa numbers refer to the molecular structure of Figure 2.

the appearance is clearly determined by the different effects of the substituents onto the ipso and ortho atoms, whereas the meta and para positions are much less differentiated. A particular attention is required by group 4 where the conjugate effect is very evident (also look at the numeric values) and shows a clear ortho-meta-para waving. We are not intentioned to discuss the correlation between similarity measures and reactivity, but we would like to emphasise that the class of the deactivating substituents is very compact, whereas that of the activating substituents is scattered over many groups therefore in agreement with the experimental results (greater level of interchange ability of deactivating substituents).

Not all the possible combinations of two substituents have been considered, but a limited number of sets of molecules have been chosen in order to comment on the similarity behaviour.

We will not discuss all the graphs which, however, show a marked homogeneity in their appearance according to the substitution (see, for example, Graph 7 which shows the appearances of the para substituted compounds); we will on the contrary discuss some numerical results.

TABLE V
Electronic energy difference on each ring atom[a]

Compound[b]	1	2	3	4	5	6
N	1.7912	0.7891	0.9039	0.9376	0.9039	0.7891
S	0.9249	0.7356	0.9102	0.9297	0.9102	0.7356
O	2.7446	1.1023	0.83	1.0275	0.83	1.1023
1,3N	1.7561	0.646	1.7561	0.8002	0.8815	0.8002
1,3S	0.8938	0.5933	1.7664	0.7902	0.8897	0.7448
1,3O	2.6683	0.9491	1.641	0.8768	0.8077	1.1111
1,2N	1.588	1.588	0.7798	0.9159	0.9159	0.7798
1,2S	0.7467	1.5595	0.7845	0.9041	0.924	0.7243
1,2O	2.3704	1.6219	0.749	1.0301	0.8422	1.139
1,2,3N	1.5582	1.3711	1.5582	0.7902	0.8996	0.7902
1,2,3S	0.7216	1.344	1.5657	0.7814	0.9061	0.7364
1,2,3O	2.2994	1.3847	1.4862	0.9028	0.8319	1.1571
1,2,4N	1.57	1.5516	0.6393	1.739	0.7804	0.7647
1,2,4S	0.7397	1.5249	0.6389	1.7383	0.7864	0.7073
1,2,4O	2.2565	1.5879	0.6389	1.7574	0.7281	1.1513
1,2,4,5N	1.5338	1.5338	0.6361	1.5338	1.5338	0.6361
1,2,4,5S	0.7097	1.508	0.6361	1.5348	1.546	0.5844
1,2,4,5O	2.1807	1.5658	0.6292	1.5424	1.4419	1.022
1,2,4,6N	1.3464	1.5267	0.6501	1.7063	0.6501	1.5267
1,2,4,6S	0.5542	1.5074	0.6495	1.7068	0.6495	1.5074
1,2,4,6O	1.8162	1.5772	0.6501	1.7234	0.6501	1.5772
1,2,3,4N	1.5441	1.3483	1.3483	1.5441	0.7777	0.7777
1,2,3,4S	0.7174	1.3232	1.3514	1.5451	0.7829	0.7229
1,2,3,4O	2.1923	1.3673	1.3095	1.5635	0.7384	1.1703
5N	1.5103	1.3297	1.3247	1.3297	1.5103	0.6489
5S	0.6898	1.3057	1.3279	1.3315	1.5217	0.5989
5O	2.1188	1.3441	1.2791	1.3395	1.4314	1.043
6N	1.3119	1.3119	1.3119	1.3119	1.3119	1.3119
6S	0.5231	1.2937	1.3129	1.3125	1.3129	1.2937
6O	1.7159	1.3692	1.2616	1.3649	1.2616	1.3692

[a] Energy values are in arbitrary units; atom numbers refer to Figure 2.
[b] Compound names refer to Figures 2 and 3 indicating positions of added N.

The first class of compounds comprehends phenol derivatives with different substituents. Looking at the values calculated for the carbon atom carrying the OH group we can note: O, Cl, and C, in ortho position diminish the EED value, this decrease becomes less important moving toward the meta and para positions, up to changing the direction for C. The formyl group again behaves autonomously, showing a marked increase of the value in ortho-para position and a decrease in meta position.

Analysing the formyl derivatives ortho-meta-para substituted by O, Cl, and C, we can observe (looking at the atom carrying the substituent) an appearance very similar in all the cases with ortho-para and meta waving. The values, however, are very sensitive to the substituent, making a decreasing perturbation scale Cl < C < O, where we have again to note the behaviour of the

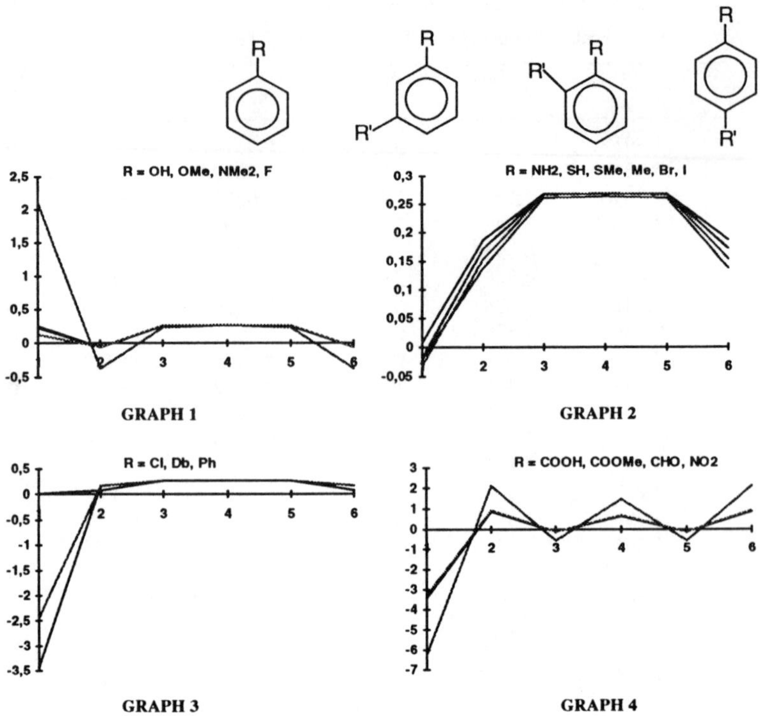

Figure 4. Mono and disubstituted benzene derivatives. EED graphs of monosubstituted derivatives. +M substituents are shown in Graph 1 and 2, ±M substituents are shown in Graph 3, −M substituents are shown in Graph 4.

methyl group which has an effect greater than Cl in negative direction in meta position and in positive direction in ortho-para position.

3.4. Set 4. Nucleic bases

The last set of compounds we considered is composed by the bases of the nucleic acids. These molecules are interesting for two reasons: first, because they represent a logical extension of the work on the heteroaromatic compounds; second, because they have a great applicative interest. In order to emphasise this second aspect we think necessary to briefly introduce the matter [11].

In living organisms there are five different kinds of nitrogenous bases: two purines (adenine (A) and guanine (G)) and three pyrimidines (cytosine (C), thymine (T), and uracil (U)). The order followed by the bases in an RNA or DNA polymer determines the genetic information of the cell. This information should be maintained constant passing from one generation to the other and should be transcribed on each complementary strain in order

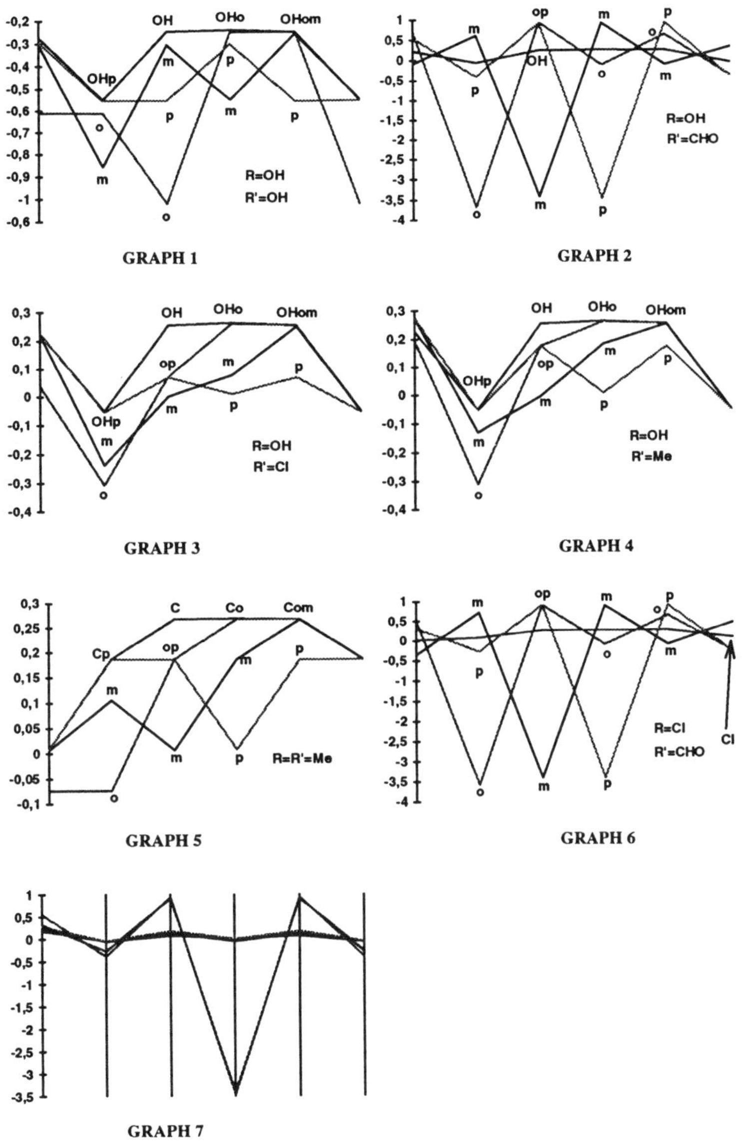

Figure 5. Mono and disubstituted benzene derivatives. EED graphs of disubstituted derivatives. Substituted phenols are shown in Graphs 1–4; Me-substituted toluenes are shown in Graph 5; CHO-substituted chlorobenzenes are shown in Graph 6; para substituted compounds are summarized in Graph 7. Lines marked by OH, C, and Cl correspond to phenol, toluene, and chlorobenzene; lines marked by o, m, and p, correspond to ortho, meta, and para, derivatives.

TABLE VI
Electronic energy difference on each ring atom[a]

R	1	2	3	4	5	6
Me	0.008	0.187	0.268	0.269	0.268	0.187
O	0.227	−0.051	0.255	0.264	0.255	−0.051
N	−0.02	0.138	0.261	0.264	0.261	0.138
S	−0.018	0.172	0.267	0.269	0.267	0.172
F	2.111	−0.372	0.236	0.271	0.236	−0.372
I	−0.018	0.172	0.267	0.269	0.267	0.172
CHO	−3.414	0.901	−0.097	0.648	−0.097	0.902
Db	−3.428	0.172	0.268	0.269	0.268	0.172
Ph	−2.447	0.171	0.267	0.269	0.267	0.171
NO_2	−6.208	2.11	−0.542	1.464	−0.542	2.11
MeO	0.249	−0.06	0.256	0.267	0.256	−0.06
Me_2N	0.129	−0.051	0.267	0.269	0.267	−0.051
MeS	−0.019	0.172	0.268	0.269	0.267	0.172
Cl	0.013	0.082	0.265	0.269	0.265	0.082
Br	−0.032	0.154	0.267	0.269	0.267	0.154
COOH	−3.275	0.868	−0.104	0.671	−0.104	0.868
COOR	−3.27	0.933	−0.105	0.672	−0.105	0.933

[a] Energy values are in arbitrary units; atom numbering starts at the C carrying the substituent in Figure 4.

TABLE VII
Electronic energy difference on each ring atom[a]

R, R′	1	2	3	4	5	6
oOH, OH	−0.111	−0.111	−0.52	0.256	0.256	−0.52
mOH, OH	0.195	−0.358	0.195	−0.046	0.247	−0.046
pOH, OH	0.203	−0.054	−0.054	0.203	−0.054	−0.054
oOH, CHO	0.701	−3.685	0.935	−0.105	0.637	−0.377
mOH, CHO	−0.087	0.614	−3.416	0.928	−0.108	0.332
pOH, CHO	0.545	−0.395	0.929	−3.464	0.929	−0.395
oOH, Cl	0.038	−0.307	0.073	0.261	0.255	−0.054
mOH, Cl	0.223	−0.237	0.001	0.078	0.251	−0.051
pOH, Cl	0.227	−0.054	0.07	0.01	0.07	−0.054
oOH, Me	0.194	−0.31	0.177	0.263	0.255	−0.052
mOH, Me	0.274	−0.131	−0.003	0.184	0.254	−0.051
pOH, Me	0.276	−0.052	0.175	0.007	0.175	−0.052
oCl, CHO	0.498	−3.58	0.908	−0.098	0.648	−0.265
mCl, CHO	−0.353	0.716	−3.415	0.901	−0.098	0.465
pCl, CHO	0.314	−0.273	0.904	−3.413	0.904	−0.273
pOH, CHO	0.545	−0.395	0.929	−3.464	0.929	−0.395
pOH, Cl	0.227	−0.054	0.07	0.01	0.07	−0.054
pOH, Me	0.276	−0.052	0.175	0.007	0.175	−0.052
pCl, CHO	0.314	−0.273	0.904	−3.413	0.904	−0.273

[a] Energy values are in arbitrary units; atom numbering starts at the C carrying the substituent cited first in the "Substituent" column.

to be used. These two main functions of the genetic material can be done only because of the high specificity of the base pairing which allows the formation of the DNA double helix and the correct transcription and translation of the information by the RNA.

In the DNA one molecule of A can pair only with a T base and a C base only with a G base owing to steric factors and to hydrogen bond formation. The steric limit is forced by the regularity of the helix structure of the sugar-phosphate skeleton of the DNA. The pairing is also strictly connected to the formation of the H bonds which are at their best according to the orientation and to the distance so to give the strongest interaction between the bases.

The four nitrogenous bases are very seldom present in their tautomeric form and therefore they can rarely give pairings different from A—T and C—G (e.g. the imine tautomer of A could pair with C). Other variations in the correct pairings can be obtained by base analogous (as 5-Br uracil or 2-aminopurine) or by mutagens. For example the nitrous acid can induce oxidative deaminations transforming A into Hypoxanthine (H) (which pairs with C) and G into xanthine (X) (which pairs with C).

The situation is different for the RNA (single helix) where there are often, in loop folded parts, faulty pairings so that some of the facing bases can be not complementary as, for example, G—U, therefore originating weak bonds. And this is exactly the characteristic needed in particular situations where the usual A—U, C—G pairings would be too strong in order to be easily broken, therefore becoming a strong limitation for the reaction rate.

For this purpose it is often present the inosine nucleoside (which contains H as the base) which can form bonds with U, C, and A.

The nitrogenous bases we studied are reported in Figure 6; these comprehend the 5 main bases (A, C, T, G, U) and some of their modifications (X, M, H, A*, T*). The corresponding EED graphs are reported in Figure 7.

In order to discuss some aspects of the behaviour of the nucleic bases we made two series of calculations: one on the neutral bases and the other on the charged bases so to simulate the interactions due to the H bonds. The model puts a charge of absolute value equal to 0.5 and with positive or negative sign according to the donor or acceptor character of the atom; we made the choice of the atoms where to put the charge taking into account their participation to the base-base pairing. Obviously the charge initially positioned on one atom is then automatically distributed over the molecule according to the usual electronic effects.

The appearance of the obtained graphs are interesting both because they permit a comparison between similar bases (also overcoming the division between purines and pyrimidines) and because they show the coordinative effect. This is in any case a high decrease in terms of distance (EED scale) between coordinating atoms. Concerning the calculated values the similarity of like groups is evident with variations ranging between 0 and 6 and showing very clear separations.

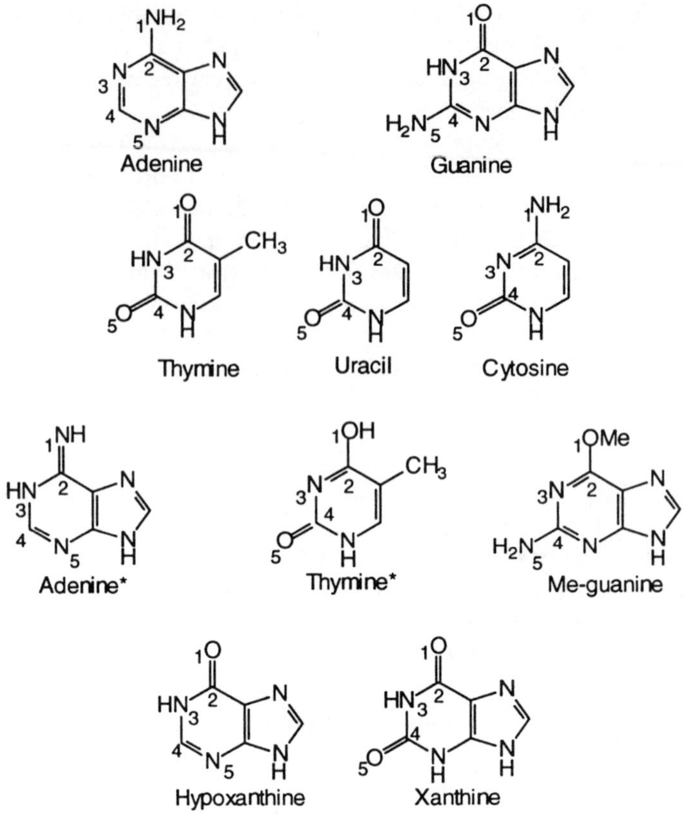

Figure 6. Nitrogenous bases of nucleic acids used in the analysis. A, G, T, U, and C, are main bases; A*, T*, M, H, and X, are main base derivatives either naturally occurring or obtained by mutation.

Values calculated after a further treatment of the usual ones are reported in Table IX. In fact using the Equation 1:

$$Z = |(N_1 - N_2)_{charged} - (N_1 - N_2)_{neutral}|$$

where N_i is the EED calculated for atom (i), we obtain values which are representative of the stabilisation level given by the base pairing.

At a first glance it is evident (underlined values) that in the sub matrix comprehending the main bases the favourite pairings are A—T, A—U, G—C (these last two bases require a three point pairing). The resulting meaning is that the Z values are expression of the preference of the natural pairings; from the similarity viewpoint we can affirm, with some cautions, that the pairing is favoured between dissimilar bases which have the maximum advantage in similarity after pairing.

Enlarging the analysis to the modified bases it becomes more complicated. In fact some experimentally found pairings are clearly favoured (X—C,

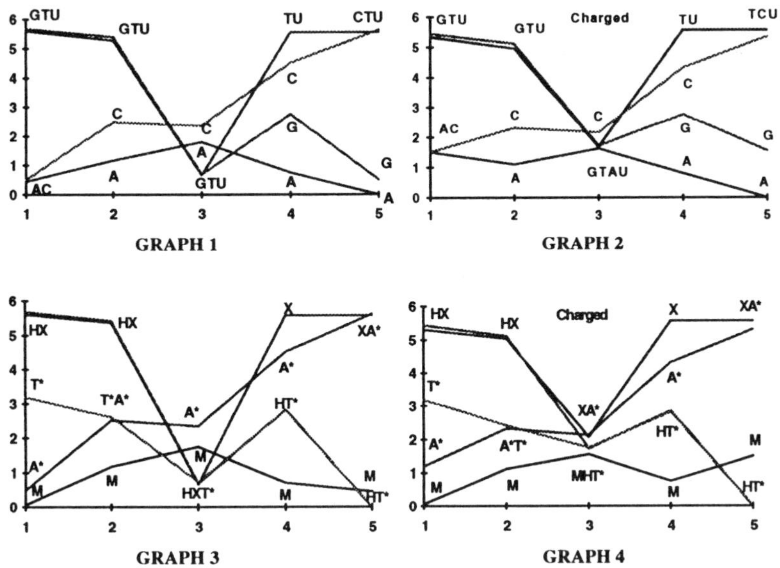

Figure 7. EED graphs of neutral and charged nitrogenous bases. Main bases are shown in Graphs 1 and 2; base derivatives are shown in Graphs 3 and 4. The bases are identified by the first letter of their names. Abscissa numbers refer to Figure 6.

H—A, A*—C, T*—G); others are doubtful (H—C); eventually some others require coordination shifts (M—T, H—U), however experimentally found. Besides there are potential pairings between modified bases (e.g. X—T*, A*—T*, M—X) which, being not much probable, are difficult to be experimentally found. Eventually those pairings requiring an inversion in the base facings have been excluded.

In conclusion we can affirm that the obtained values are interesting in order to explain some of the experimental data.

From the similarity viewpoint we can observe by comparing the results: 1) full similarity for X, T, and U, and for C and T*; 2) similarity in positions 1, 2, 3, for G, T, U, H, and X; for A* and G after the scaling of the values; 3) in positions 1, 2, 3, 4, for A and C after scaling and inversion 2–3; 4) for A and M comparing positions 1, 2, 3, of A with positions 5, 4, 3, of M (with a smaller similarity level).

4. Conclusion

The use of the electronic energy differences calculated using our model allows the comparison between atoms, atom groups, and molecules; using this comparison it is possible to determine different similarity measures both of numeric and graphic nature, therefore getting interesting information on their potential behaviours.

TABLE VIII
Electronic energy difference on each atom[a]

BASE	1	2	3	4	5
Adcnine	0.4558	1.1727	1.8013	0.7572	–
Guanine	5.6788	5.4126	0.6987	2.767	0.5154
Thymine	5.6123	5.287	0.6727	5.5595	5.5548
Cytosine	0.5116	2.4939	2.3834	4.5239	5.6349
Uracil	5.6219	5.2816	0.6733	5.5619	5.5599
Adenine*	3.1904	2.5985	0.7364	2.7972	–
Thymine*	0.5085	2.5094	2.3339	4.5213	5.6206
Xanthine	5.6082	5.366	0.6743	5.5769	5.5779
Hypoxanthine	5.684	5.4138	0.7067	2.8155	–
Me-guanine charged	0.0675	1.1893	1.7516	0.7119	0.4611
Adenine	1.5061	1.1166	1.6257	0.8069	–
Guanine	5.4444	5.1244	1.7154	2.7737	1.5635
Thymine	5.3336	4.9633	1.6868	5.5573	5.5587
Cytosine	1.531	2.3136	2.1682	4.3298	5.3383
Uracil	5.3427	4.9656	1.6822	5.567	5.571
Adenine*	3.1825	2.409	1.7575	2.8041	–
Thymine*	1.2122	2.3211	2.1263	4.326	5.3248
Xanthine	5.3142	5.0475	2.0577	5.5746	5.573
Hypoxanthine	5.4553	5.1229	1.7251	2.8399	-
Me-guanine	0.0669	1.1272	1.5608	0.7583	1.5051

[a] Energy values are in arbitrary units; atom numbering refers to Figure 6.

TABLE IX
Values of Z (Equation 1) obtained using electronic energy difference on each atom pair[a]

	A	T	G	C	U	X	H	M	A*	T*
A	0.0	0.139	0.092	0.070_i	0.145	0.215	0.085	0.022_s	0.139	0.379_i
T	0.139	0.0	0.015_s	0.069	0.006_i	0.352_i	0.054_i	0.117_i	0.278_i	0.239
G	0.092	0.015_s	0.0	1.323[b]	0.238_s	0.617_s	0.009_i	0.071_i	0.231_i	1.630
C	0.070_i	0.069	1.323[b]	0.0	0.075	0.285	0.015	0.049	0.209	0.308_i
U	0.145	0.006_i	0.238_s	0.075	0.0	0.360_i	0.060_s	0.124_i	0.284_i	0.234
X	0.215	0.352_i	0.617_s	0.285	0.360_i	0.0	0.300_i	0.236	0.089_i	0.593
H	0.085	0.054_i	0.009_i	0.015	0.06s	0.300_i	0.0	1.064_i	0.224_i	0.294
M	0.022_s	0.117_i	0.071_i	0.049	0.124_i	0.236	1.064_i	0.0	0.16_i	0.357
A*	0.139	0.278_i	0.231_i	0.209	0.284_i	0.089_i	0.224_i	0.16_i	0.0	0.517
T*	0.379_i	0.239	1.630	0.308_i	0.234	0.593	0.294	0.357	0.517	0.0

[a] Energy values are in arbitrary units; the bases are identified by the first letter of their corresponding names. Letter (i) or (s) near numbers respectively correspond to the need of *i*nversion or of *s*hift of the base to have coordination.Æ[b] The value calculated on two points (0.882) is still the best one.

The introduction of one or more similarity indexes is not, at present, particularly interesting for our work, even if their creation has been shown very easy in the case of the nucleic bases.

Eventually the use of our system of similarity measuring has not been thoroughly examined and on this matter our work is still in progress.

Acknowledgments

Partial financial support by the Consiglio Nazionale delle Ricerche, and by the Ministero dell'Universita' e della Ricerca Scientifica e Tecnologica, is gratefully acknowledged.

References

1. a) Johnson, M. A. and Maggiora, G. W., eds., *Concepts and Application of Molecular Similarity*, (Wiley, New York, 1990). b) Rouvray, D. H., *J. Chem. Info. Comput. Sci.* **32**, 580 (1992). c) For an up to date source of references see all the papers published on vol. **32**, no. 6, of *J. Chem. Info. Comput. Sci.* (1992).
2. Mezey, P. G., *J. Chem. Info. Comput. Sci.* **32**, 650 (1992).
3. Gasteiger, J., Ihlenfeldt, W. D., Fick, R., and Rose J. R., *J. Chem. Info. Comput. Sci.* **32**, 700 (1992).
4. a) Carbó, R. and Calabuig, B., *Int. J. Quantum Chem.* **42**, 1695 (1992). b) Carbó, R., Calabuig, B., Besalu, E., and Martinez, A., *Mol. Eng.* **2**, 43 (1992). c) Ponec, R. and Strnad, M., *J. Chem. Info. Comput. Sci.* **32**, 693 (1992). d) Good, A. C. and Richards, W. G., *J. Chem. Info. Comput. Sci.* **33**, 112 (1993).
5. The meanings of similarity have been thoroughly discussed. See for example ref. 1b.
6. Sello, G., *J. Chem. Info. Comput. Sci.* **32**, 713 (1992).
7. Baumer, L., Sala, G., and Sello, G., *Anal. Chim. Acta* **235**, 209 (1990).
8. Sello, G., *J. Am. Chem. Soc.* **114**, 3306 (1992).
9. Bader, R. F. W. and Legare, D. A., *Can. J. Chem.* **70**, 657 and refs. cited (1992).
10. Baumer, L. and Sello, G., *J. Chem. Info. Comput. Sci.* **32**, 125 (1992).
11. Lewin, B., *Genes IV* (Oxford University Press and Cell Press, Cambridge MA, 1990).

Study of the Root Mean Square Deviations of Bond Lengths, Bond Angles and Torsional Angles as a Measure of Molecular Similarities

NADINE J. SMEYERS* and YVES G. SMEYERS

Instituto de Estructura de la Materia, C.S.I.C., Calle Serrano 123, E-28006, Madrid, Spain;
** Permanent address: Departamento de Química Orgánica, Universidad de Alcalá de Henares, E-28716 Alcalá de Henares, Spain*

Abstract. Root mean square (*rms*) deviations of bond lengths, bond angles and torsion angles with respect to a reference molecule are used as similarty parameters in QSAR studies. In the same way, *rms* are used as similarity measures in organic parent compounds, versus the conformational changes and the structural variations with the substitution. Similarity matrices are defined. Finally, deviations through space distances, distance angles and torsion angles are also considered. In general, it is seen that the *rms* deviations can be used in order to characterize similar (or dissimilar) molecular structures, as well as to retain some intramolecular interactions.

1. Introduction

Two molecules look similar when they possess some significant common parts. If we want to deep into the similarity concept, the two molecules have to be compared.

One way to achieve this comparison is to consider the small geometrical differences existing between the molecules, and to calibrate globally these differences by performing root mean square deviation (*rms*) calculations:

$$(\Delta X^2)^{\frac{1}{2}} = \sqrt{\frac{\sum_i^N (X_i^a - X_i^b)^2}{N-1}} \tag{1}$$

where X_i^a and X_i^b are the comparable parameters in the reference molecule a and the comparable one b, and N the number of comparable parameters.

These calculations may be carried out in Cartesian coordinates, after maximizing the overlap between the molecules. However, when two molecules exhibit a significant commoun parts, modified by some structural change (such as conformational change or substitution), bond length, bond angle and torsion angle deviations seem to be a better starting point for this comparison. Such type of similar molecules will be called parent molecules.

Other geometrical parameters can be also considered, such as through space distances between some critical points, as well as their corresponding distance angles and torsion angles.

Let us mention here that there is a subjective criterium in this methology. Which are indeed the comparable parameters? These will depend of course

R. Carbó (ed.), Molecular Similarity and Reactivity: From Quantum Chemical to Phenomenological Approaches, 291–301.

on the properties to be looked for. Normally, the similarity between the common parts of the two molecules has to be close enough to avoid any ambiguity.

Another important point is the number N of parameters to be compared. The larger it is, the more significant are the root mean square deviations. Since the Z-matrix is usually used. N is the number of the comparable bond lengths, N-1 is the comparable bond angles, and N-2 is the comparable torsional angles. In the following, we shall use the *rms* deviations as similarity indexes.

2. RMS deviations of bond lengths, bond angles and torsion angles as structural parameters in QSAR studies

2.1. Substituted histamine agonists in position 4

As well known, histamine (β-(4-imidazolyl)-ethylamine (see Figure 1) produces at least two different biological effects in the organism through the interaction with different receptors, namely H_1 and H_2, which can be characterized by selective agonists and specific antagonists [1–2]. At physiological pH, histamine exists as monocation and in two different tautomeric forms, namely, N_3H and N_1H β-(4-imidazolyl) ethylammonium.

Substituted histamines in position 4 (according to the Ganellin notation), (histamine, 4-methyl-, 4-chloro-, and 4-nitrohistamine), show a clear structure-activity relationship (QSAR) for histamine H_2-receptors as a function of the mean net charge on the imidazol ring calculated theoretically. The larger is this charge, the larger is the biological activity [3]. This feature is not related

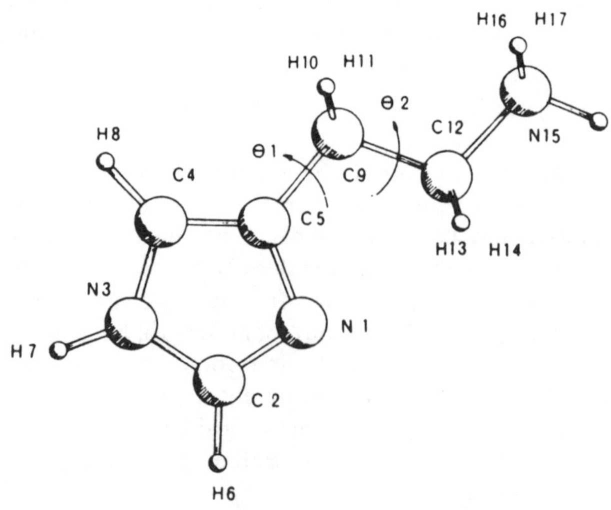

Figure 1. Structure of Histamine and atom labelling according to Ganellin [2].

with a particular atom, but with the all imidazol ring. It seemed thus to be interesting to compare the small geometrical changes appearing in the histamine frame with the substitution. The stuctures of these compounds have been determined theoretically, at RHF level using the STO-4G basis and full optimization of the geometry up to reach mean gradient values lower than 10^{-3} mdyn/A or radian [4].

Using these geometries, the *rms* deviations of bond lengths and bond angles are calculated for three substituted N_3H histamine tautomers, with respect to the N_3H histamine tautomer itself, and gathered in Table I. The molecule is assumed to be planar. In the same table, the corresponding *rms* obtained in the same way are given for the three substituted N_1H histamine tautomers, with respect to the N_1H histamine tautomer. Finally, the *rms* deviations between the N_3H and N_1H histamine tautomers is also achieved [4].

It is seen that the *rms* deviations of bond lengths and bond angles are relatively small. In particular, the bond length deviations fall into the margins of errors. Some bond angle deviation-biological activity relationship, however, is detected. This relationship could be related with that of the charges. Anyway, substitution in position 4, as well as tautomerization, do not affect seriously the geometry of the rest of the histamine molecule. As a result, it may be put forwards that the frame of these derivatives of histamine is relatively similar, so that they may be docked on the receptor just as histamine itself does, without taking into account the possible hindrance of the substituent.

TABLE I

Bond length and bond angle root mean square deviations (*rms*), in Å or degrees, as a function of the substitution in histamine and histamine derivatives and their tautomers versus the H_2 receptor biological activity

	Bonds	Angles	Activity
N_3H Tautomers			
Histimine	0	0	100
4-m-Hist.	0.16295	0.34224	43
4-Cl-Hist.	0.20094	0.50538	11
4-NO_2-Hist.	0.1572	0.64756	0.6
N_1H Tautomers			
Histimine	0	0	100
4-m-Hist.	0.16295	0.37131	43
4-Cl-Hist.	0.20323	0.54447	11
4-NO_2-Hist.	0.16172	0.64205	0.6
N_3H and N_1H Tautomers			
N_3H (π)	0	0	
N_1H (τ)	0.37142	1.19885	

$N = 16$.

2.2. Alkaloids of Vinca Minor

Vincamine and Vincamone are two indol alkaloids of clinical use against brain sclerosis, as well as post-operative states of the central nervious systems [5]. Vincamone has been also proposed for the treatment of drepanocytosis (sicklemia) [6]. The biological activity at receptor level is completely unknown. Vincamine and vincamone seem to act as an oxygen vector in living cell. Vincamone has a longer duration of activity with more pronounced effects.

Vincamine derivatives (vincamine, vincamone, apovincamine, vincaminol, and desoxyvincaminol) present identical hexacyclic frame, but small different side chain, (see Figure 2.). All these compounds, however, exhibit different biological activities. It seems thus interesting to compare the small structural changes in the frame versus the modifications of the side chain. In addition, the optical enantiomer of vincamone, d-Eburnamonine, which presents also some biological activity will be considered too. The structures of these compounds have been established at AM1 level with full optimization of the geometries up to reach mean gradient values lower than 0.01 kcal/A or degree [7]. It is seen that these molecules exhibit non-planar but similar structures for the polycyclic frame.

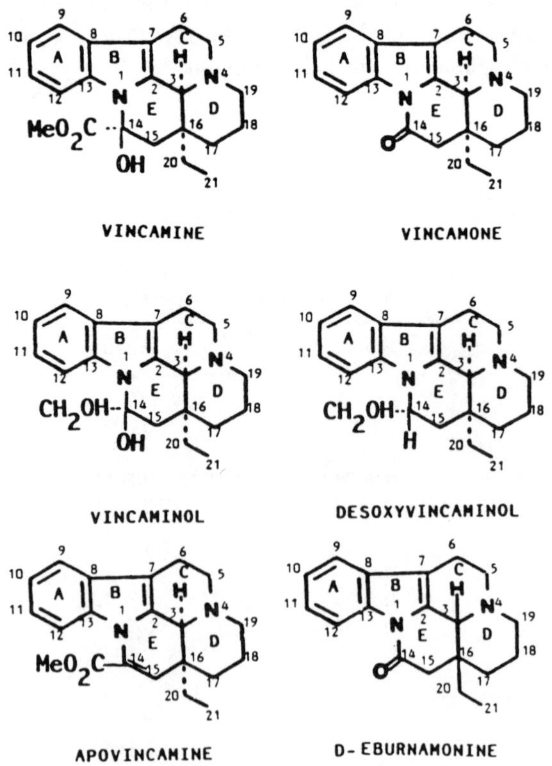

Figure 2. Structures of six derivatives of Vincamine.

Using these geometries, the *rms* deviations of bond lenths, bond angles, and torsion angles are determined with respect to the most active compound, Vincamone, for the above mentioned molecules, and gathered in Table II versus the biological activities. It is seen that the *rms* of the bond lengths are very small, and fall into error margins of the theoretical calculations. The bond angle *rms* are larger, but still small enough to be not very significant. In contrast, the torsion angle *rms* are very large, especially that of d-Eburnamonine. If these *rms* parameters are compared with the biological activities, a simple relationship cannot be established.

It may be concluded that whereas the frame of these molecules look very similar, the torsion angle *rms* are the most descriptive parameters, especially when Vincamone (l-Eburnamonine) and d-eburnamonine are compared. No simple correlation with the activities may be put forwards.

TABLE II
Root mean square deviations (*rms*) of bond lengths in Å × 10^3, bond angles in degrees × 10^2, and torsion angles in degrees, versus biological activities of some derivatives of vincamine. Vincamone is taken as reference molecule

Molecules	B. Lengths	B. Angles	Torsion An.	Biol. Activities
Vincamone	0	0	0	2.0
Vincamine	0.261	0.154	10.01	1.5
Aprovincamine	0.238	0.259	30.19	0.1
Vincaminol	0.540	0.626	11.30	–0.1
Desoxyvincam.	0.339	0.574	27.36	–0.1
d-Eburnamonine	0.000	0.000	85.41	0.2

$N = 44$.

3. RMS deviations of bond lengths, bond angles and torsion angles as structural parameters in parent molecules

3.1. GABA$_B$ ANTAGONIST PRECURSORS

The 3-azabicyclo[3.3.1]nonan-9-one derivatives are precursors of some rigid antagonists of the GABA$_B$ receptors. In these molecules, the GABA skeleton is simulated by the 3-azabicyclo[3.3.1]nonane system. These bicyclic molecules exhibit a nonplanar structure. Each of the six membered ring may have two possible conformations: chair (**C**) and boat (**B**). In addition, the N-hydrogen atom can adopt two spatial orientations: α or β. As a result, the molecule can adopt eight different conformations (see Figure 3). The organic chemists have introduced a series of substitutions in this molecule in order to favor one of the possible conformations, or even to prohibit a series of them. Here, similarity indexes will be used to analyze how these conformations differ from each other.

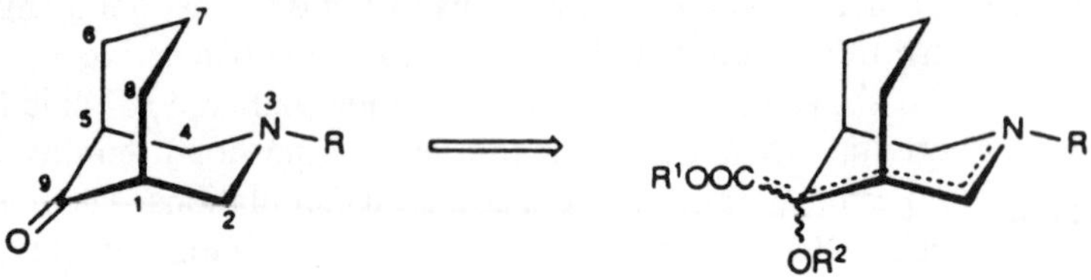

Figure 3. Structures of 3-azabicyclo[3.3.1]nonan-9-one and its N—R derivatives and labelling of the heavy atoms. The dashed lines show the molecular similarity with the $GABA_B$ skeleton.

The structures of the eight conformers of 3-azabicyclo[3.3.1]nonan-9-one and its N-methyl derivative have been determined at AM1 level with full geometry optimization up to reach mean gradient values lower than 0.01 kcal/A or degree [8]. It is seen that these conformers exhibit a non-planar, but comparable structures. In particular that the N-methyl substitution changes the conformational preference from α to β. In same way, the structure of the four allowed conformers of 2,4-diphenyl-3-azabicyclo[3.3.1] nonan-9-one, and their para-fluor derivatives, have been also established [9], being the **BC** and **BB** conformations sterically forbidden. Finally, the structures of preferred **CC** – β conformer of the 2,4-diphenyl-3-azabicyclo[3.3,1]nonan-9-hydroxy-acids, i.e., the possible antagonists of $GABA_B$, are determined [10].

At this stage of the calculations, the *rms* deviations of bond lengths, bond angles, and torsion angles are studied versus the conformational variations. For this purpose, the *rms* deviations are determined for the eight conformers of the 3-azabicyclo [3.3.1]nonan-9-one with respect to each others. In such a way, *rms* deviation 8×8 matrices, namely, simmilarity matrices, are obtained. They are given in Tables III, IV, and V. The relative formation energy values for the eigth conformers are given also in Table V. It is seen that the **CC** – α conformer is by far the most stable.

At first glance, the bond length deviation matrix shows that the bond length variations are very small with the conformational changes, except for the **BC** – β conformation because perhaps of some steric hindrance. The bond angle deviation matrix shows that the bond angle variations are also small, except for the **BB** conformations which exhibit a somewhat more flattered structure [8]. Finally, torsion angle matrix shows large torsion angle variations, as expected because of the conformational changes. Notice that some additivities can be observed in the *rms* values as a function of the conformational changes.

Next, the *rms* deviations of bond lengths, bond angles, and torsion angles are considered versus the substitution. For this aim, the *rms* deviations are determined for the two more stable conformers, **CC** – α and **CC** – β, of some 3-azabicyclo[3.3.1]nonan-9-one derivatives, i.e., the N-methyl, 2,4-diphenyl, 2,4-diphenyl-N-methyl, and their para-fluor derivatives, with respect

TABLE III

rms of bond lengths in Å × 10^2 of the eigth conformers of 3-azabicyclo[3.3.1]nonan-9-one in chair-chair (CC), chair-boat (CB), boat-chair (BC) or boat-boat (BB) conformations, and with the N—H atom in α or β orientation

	CCα	CCβ	CBα	CBβ	BCα	BCβ	BBα	BBβ
CCα	–	0.0053	0.0021	0.0034	0.0239	0.0034	0.0369	0.0147
CCβ	0.0053	–	0.0075	0.0019	0.0292	0.0087	0.0422	0.0200
CBα	0.0021	0.0075	–	0.0055	0.0217	0.0013	0.0348	0.0126
CBβ	0.0034	0.0019	0.0055	–	0.0273	0.6822	0.0402	0.0181
BCα	0.0239	0.0292	0.0217	0.0273	–	0.0205	0.0130	0.0092
BCβ	0.0034	0.0087	0.0013	0.6822	0.0205	–	0.0335	0.0113
BBα	0.0369	0.0422	0.0348	0.0402	0.0130	0.0335	–	0.0222
BBβ	0.0147	0.0200	0.0126	0.0181	0.0092	0.0113	0.0222	–

$N = 23$.

TABLE IV

rms of bond angles in degrees of the eigth conformers of 3-azabicyclo[3.3.1]nonan-9-one

	CCα	CCβ	CBα	CBβ	BCα	BCβ	BBα	BBβ
CCα	–	0.0028	0.0232	0.0057	0.0210	0.0036	0.1362	0.0889
CCβ	0.0028	–	0.0261	0.0085	0.0182	0.0008	0.1333	0.0861
CBα	0.0232	0.0261	–	0.0175	0.0443	0.0268	0.1593	0.1121
CBβ	0.0057	0.0085	0.0175	–	0.0267	0.0093	0.1418	0.0946
BCα	0.0210	0.0182	0.0443	0.0267	–	0.0174	0.1151	0.0679
BCβ	0.0036	0.0008	0.0268	0.0093	0.0174	–	0.1325	0.0853
BBα	0.1362	0.1333	0.1593	0.1418	0.1151	0.1325	–	0.0472
BBβ	0.0889	0.0861	0.1121	0.0946	0.0679	0.0853	0.0472	–

$N\text{-}1 = 22$.

TABLE V

rms of dihedral angles in degrees, and formation in kcal/mol. of the eigth conformers of 3-azabicyclo[3.3.1]nonan-9-one

	CCα	CCβ	CBα	CBβ	BCα	BCβ	BBα	BBβ
CCα	–	24.53	35.59	42.98	27.94	42.81	44.01	53.21
CCβ	24.53	–	45.55	37.91	18.97	28.96	39.54	46.37
CBα	35.59	45.55	–	24.93	46.61	57.65	32.88	42.04
CBβ	42.98	37.91	24.93	–	45.07	50.30	23.02	33.44
BCα	27.94	18.97	49.61	45.07	–	25.32	36.90	46.01
BCβ	42.81	28.96	57.65	50.30	25.32	–	43.36	36.87
BBα	44.02	39.54	32.88	23.02	36.90	43.36	–	25.22
BBβ	53.21	46.37	42.04	33.44	46.01	36.87	25.22	–
	–52.61270	–49.53223	–51.92329	–48.3228	–47.7997	–50.55501	–44.59796	–47.281948

$N\text{-}2 = 21$.

to each first compound of the series. These values are gathered in Tables VI and VII, respectively.

It is found again that the bond length and bond angle variations are very small with the substitution. In contrast, the torsion angle variations are significantly large. Especially, the NCH_3 substitution introduces in the **CC** – α series a 6.6° variation, which is additive with the 2,4-diphenyl variation of 3.3°, as it can be seen in Table VII. This significant variation of the torsion angle value with the N—CH_3 substitution in the α series can be related with an energetic destabilization observed with the N-methylation in this series with respect to the compounds of the β series [8, 9].

Finally, the *rms* deviations are studied versus the functionalization of the

TABLE VI
Mean Root Square Deviations of bond lengths, in Å, bond and torsion angles, in degree, in 3-azabicyclo[3.3.1]nonan-9-one, in CCα conformation, as a function of substitution

Molecules	B. lengths	B. angles	Torsion angles
NH	0	0	0
NCH_3	0.2132 –04	0.3095 –01	6.689
NH-2,4-diphenyl	0.2985 –04	0.1604 –01	3.458
NCH_3-2,4-diphenyl	0.4904 –04	0.2287 –01	9.899
NH-2,4-diaryl	0.4264 –04	0.2572 –01	3.531
NCH_3-2,4-diaryl	0.2132 –04	0.2350 –01	9.944

The 3-azabicyclo[3.3.1]nonan-9-one is taken as reference.
$N = 23$.

TABLE VII
Mean Root Square Deviations of bond lengths, in Å, bond and torsion angles, in degrees, in 3 azabicyclo[3.3.1]nonan 9-one, in CCβ conformation, as a function of substitution

Molecules	B. lengths	B. angles	Torsion angles
NH	0	0	0
NCH_3	0.1066 –04	0.1882 –01	1.093
NH-2,4-diphenyl	0.1706 –04	0.2480 –01	5.891
NCH_3-2,4-diphenyl	0.0639 –04	0.2466 –01	5.284
NH-2,4-diaryl	0.0512 –04	0.2281 –01	5.787
NCH_3-2,4-diaryl	0.7460 –04	0.2633 –01	5.192
NH-9-hydroxyacid	0.4256 –01	2.7505	14.231
NCH_3-9-hydroxyacid	0.6316 –01	2.7965	13.645
NH-2,4-diphenyl-9-hydroxyacid	0.4104 –01	2.7866	15.027
NCH_3-2-4-diphenyl-9-hydroxyacid	0.6333 –01	2.6582	14.421

The 3-azabicyclo[3.3.1]nona-9-one is taken as reference.
$N = 23$.

keto group. For this purpose, the *rms* of the above antagonist precursors of the β series are determined with respect of the first compound of the series versus the functionalization in hydroxyacids. The numerical values are given also in Table V. It is found that the funtionalization introduces appreciable variations in the structure. Variations of 5% are observed in the bond lengths, of 2.7° in the bond angles, and of 15° in the torsion angles. These large variations could affect the biological properties of these molecules.

4. RMS deviations of through space distances, distance angles and torsion angles as structural parameters in QSAR studies

4.1. Fosfomycin and phosphoenolpyruvic acids

Fosfomycin, (-)cis(1,2-epoxypropyl)-phosphonic acid, is a broad spectrum antibiotic (see Figure 4) [11]. This antibiotic interfers in the bacterial wall biosynthesis, by inhibiting the transfer of phosphoenolpyruvate to UDP-N-

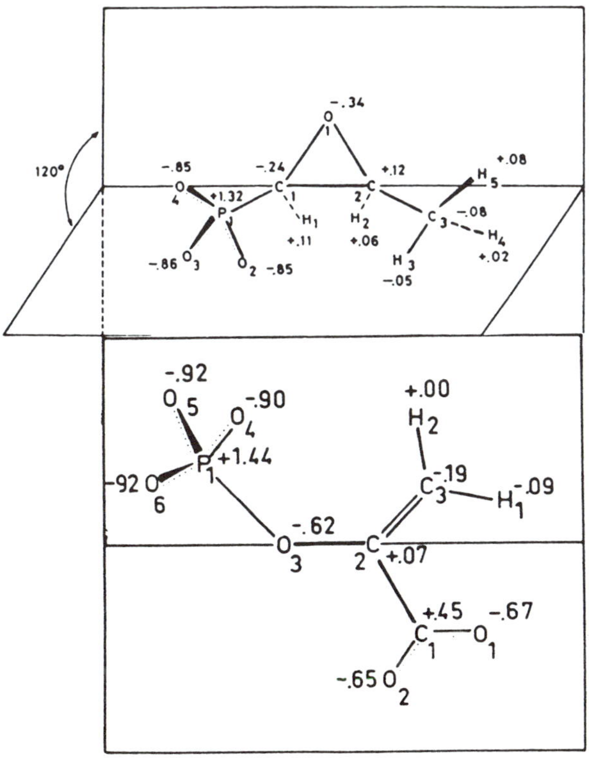

Figure 4. Structures of fosfomycin, charge distributions and atom labellings.

acetylglucosamine in the synthesis of muramic acid, which is one of the major components of the peptidoglucan network. This inhibition occurs in the following way: fosfomycin mimics phosphoenolpyruvate in the adduct formation with the enzime. In this reaction, fosfomycin was shown to interact in a irreversible way with a negative cysteinyl sulphur rest, and a proton donor, with the subsequent breaking of the epoxy ring [12].

Fosfomycin and phosphoenolpyruvate are expected to be completely ionized at physiological pH. In spite of their negative charges, these compounds were seen to react with a negative sulphur rest through the C_2 carbon atom, whereas the O_1 oxygen or the C_3 carbon atom accepts the proton. The negative phosphonic or phosphoric center is expected to play also an important role. For this reason, charge distributions were determined on both molecules, by using different approaches [13,14]. These charge distributions are given in the same Figure 4.

These two molecules do not look very similar. Therefore, a through space isosterism, between the critical centroides of charges, was put forwards in [12]. Their structures were determined by using the AM1 approximation. Next, *rms* deviations of through space distance, between the phosphorus atom, the C_2 carbon atom and the proton acceptor center, as well as the corresponding angles, were performed for fosfomycin with respect to phosphoenolpyruvate. In the same way, standard *rms* of bond length, bond angle and torional angle deviations were determined. The corresponding values are gathered in Table VIII.

TABLE VIII
Through space and bond *rms* deviations, in Å or degree, for fosfomycine and phosphoenolpyruvate anions

	N	Lengths	Angles	Torsion An.
Bond	6	0.1428	0.5086	90.54
Through space	4	0.3059	21.2669	0.00

It is seen that these to molecules are indeed dissimilar. Whereas the through space *rms* deviations of distances and angles are very large, the standard ones of bond lengths and bond angles are smaller. The torsion angle deviations are, however, very large in this last case. If we admit that the torsion angles are more deformable, the *rms* deviations bond lengths and bond angle should be, in the present case, a more reliable desciptor for the bioreactivity. In the same way, the local isosterism proposed in [12] appears not clear enough to be retained. Anyway the numbers of parameters considered in the present *rms* calculations are very low.

5. Conclusions

In the present work, *rms* deviations of bond lengths, bond angles, and torsion angles are determined as similarity measures for some series of comparable molecules, as a function of their conformational changes or substitution. In general, the bond length deviations are found to be too small for being related with the biological activities. The bond angle deviations are seen to be small too, but some bond angle deviation-biological activity relationship is detected in the case of histamine derivatives. In contrast, the torsion angle deviations are relatively larger. In the case of vincamine derivatives, it can be put forwards that the structural changes are not responsible for the bioactivity variations observed.

In the cases of the $GABA_B$ antagonist precursors, the *rms* deviations and *rms* deviation matrices are seen to be good descriptors for the structural changes, as well as for the intramolecular interactions.

Finally, in the cases of Fosfomycin, the through space distance variations seems to show that local isosterism proposed in [12] is not clear enough to be retained.

It may be concluded that the *rms* deviations of bond lengths, bond angles and torsion angles may be advantageously used as similarity (or dissimilarity) measures for a series of parent or related molecules.

References

1. Ganellin, C. R. and Durant, G. J., *Burger Medicinal Chemistry*, Part II, 4th ed. (J. Wiley, New York, 1981).
2. Ganellin, C. R., in *Pharmacology of Histamine Receptors*, C. R. Ganellin *et al.*, eds. (Wright PSG, Bristol, 1982).
3. Smeyers, Y. G., Romero-Sánchez, F. J., and Hernández-Laguna, A., *J. Mol. Struct. (Theochem)* **123**, 431 (1985).
4. Smeyers, Y. G., Hernández-Laguna, A., Rández, J. J., and Rández, F. J., *J. Mol. Struct. (Theochem)* **207**, 157 (1990).
5. Tamayo, M., *Vademecum International* (DAIMON, Madrid, 1990).
6. Yougui-Massok, M., *Médecine Praticienne* **571**, 118 (1972).
7. Smeyers, Y. G., Smeyers, N. J., Rández, J. J., Hernández-Laguna, A., and Gálvez-Ruano, E., *Mol. Engineering* **1**, 153 (1991).
8. Diez, M., Izquierdo, M. L., Arias-Pérez, M. S., Gálvez-Ruano, E., Matesanz, E., and Martin-Ripoll, J. M., *J. Mol. Struc.* **249**, 203 (1991).
9. Smeyers, Y. G., Hernández-Laguna, A., Smeyers, N. J., Arias-Perez, M. S., Fernández, M. J., *J. Mol. Struct. (Theochem)*, **282**, 105 (1992).
10. Arias-Pérez, M. S., Smeyers, Y. G., Fernández, M. J., Smeyers, N. J., Galvez, E., Fonseca I., and Sanz-Aparicio, J., *J. Org. Chem. Sc.*, submitted.
11. Hendin, D. *et al.*, *Science* **166**, 122 (1969).
12. Kaham, F. M., Kahan, J. S., Casidy, P. J., and Kropp, H., *Ann. N.Y. Acad. Sci.* **235**, 364 (1974).
13. Smeyers, Y. G., Hernández-Laguna, A., and Von Carsten-Lichterfelde, C., *J. Pharm. Sc.* **72**, 1011 (1983).
14. Smeyers, Y. G., Hernández-Laguna, A., Romero-Sánchez, F. J., Fernández-Ibáñez, M., Gálvez-Ruano, E., and Arias-Pérez, M. S., *J. Pharm. Sc.* **76**, 753 (1987).

Molecular Similarity and LFER

ROBERT PONEC
Institute of Chemical Process Fundamentals, Academy of Sciences of Czech Republic, Prague 6, Suchdol 2, 165 02 Czech Republic

Abstract. The similarity model based on the use of similarity indices as a quantitative measure of molecular alikeness was applied to the description of substituent effect on the dissociation of substituted benzoic acids and the ionization of substituted kumylchlorides. It has been shown that within a class of substituents of constant donor/acceptor character, the proposed model correctly describes the substituent effect for both types of processes. The question of the role of empirical principle of analogy as a basis for the existence of LFER is also discussed.

1. Introduction

One of the basic aims of any natural science is the formulation of simple models on the basis of which it is possible to understand the observed phenomena and, subsequently, to provide their quantitative or at least semi-quantitative description. The sophistication of these models depends, of course, on the degree of development of a given science and, also, on the complexity of the studied phenomena. One of the most complex and to the rigorous theoretical description also the most resisting fields is the field of the relations between the structure and reactivity and/or properties of organic molecules. The important role in the formulation and development of these relation thus still belongs to empirical rules and relationships.

Among the most frequently used such relationships are the so-called linear free energy relationships (LFER), the typical examples of which are, e.g., the well-known Hammett and Taft equations [1–3]. Because of immense importance of these relationships, the considerable attention was devoted to the elucidation of the factors responsible for the existence and the validity of these, basically empirical relations. In spite of remarkable progress these studies brought into the solution of many particular problems as, e.g., the sensitivity of a given skeleton to the transmission of the substituent effect [4–6], our understanding of the factors underlying the existence of these relations is still rather restricted [7]. This is due to the fact that these relations cannot be derived from theoretically well founded thermodynamic models (extrathermodynamic relationships) but that they are based only on the empirical principle of analogy the basic idea of which is that similar structural changes, represented, e.g., by the systematic substitution in a series of different but structurally related skeletons, will have similar effect also on the properties or reactivity of corresponding molecules.

Because of immense versatility of this principle in practically any area of human activity [8], the increasing effort starts to be still more frequently

R. Carbó (ed.), Molecular Similarity and Reactivity: From Quantum Chemical to Phenomenological Approaches, 303–309.

devoted to the attempts at the formulation of some more precise measures of similarity [9, 10]. Although it is clear that in dependence on the intended application such measures can be introduced in a number of various ways,[1] the main attention, especially of theoretical chemistry concentrates on the attempts to characterize the molecular similarity in terms of electron structure of corresponding molecules. Among these attempts the greatest use have found the so-called similarity indices [12, 13], whose systematic exploitation is the subject of increasing interest of various authors [13–22]. Although it is true that this increasing interest is to a considerable extent motivated by the potential exploitation of these indices in the design of new biologically active compounds [15], the nonnegligible parallel effort was also devoted to the application of similarity ideas in the chemical theory. An example in this respect is, e.g., the recent studies [22–25] dealing with the use of similarity indices in the theory of pericyclic reactivity.

Our aim in this study is to profit the possibility offered by similarity indices to characterize quantitatively the molecular alikeness in a series of molecules and to provide thus a straightforward theoretical support for the principle of analogy as an empirical basis underlying the existence of LFER.

2. Theoretical

The basis of our approach is the so-called topological similarity index [22], recently introduced by incorporating the original similarity index by Polansky and Carbó [12, 13] into the framework of overlap determinant method [26]. Within the framework of this method, the original Carbó's Definition (1) can be rewritten in the form (2), where P_A and P_B are usual charge density bond order.

$$r_{AB} = \frac{\int \rho_A(1)\rho_B(1)d\tau_1}{\left(\int \rho_A^2(1)d\tau_1\right)^{\frac{1}{2}}\left(\int \rho_B^2(1)d\tau_1\right)^{\frac{1}{2}}}, \tag{1}$$

$$r_{AB} = \frac{1}{2N} \operatorname{Tr} P_A \overline{P}_B. \tag{2}$$

Matrices of the molecules A and B, the similarity of which we are interested in and where the prime over the matrix P_B denotes the similarity transformation (3) with the matrix T express

$$\overline{P}_B = T^{-1} P_B T. \tag{3}$$

The mutual relation of AO bases $\{\chi_A\}$ and $\{\chi'_B\}$ on the molecules A and B respectively (4).

$$\chi_B = T\,\chi_A. \tag{4}$$

In spite of many interesting applications [22–25], this index had one important conceptual limitation. This index was, namely, designed only for comparison of density matrices derived from simple HMO method. This, of course, restricts the possibility of its practical exploitation to only simplest model systems. In order to overcome this important limitation we proposed in previous study [27] the generalization of the index to the level of semi-empirical *MO* methods. The result of this generalization is a strictly positionally invariant index which, because of taking into account all valence electrons, can be applied to real chemically interesting systems. The basis of this generalization is the so-called mapping analysis [28], the principal idea of which is an iterative optimization of the transformation matrix T ensuring the maximization of the similarity index.

The basic idea of this optimization is the following. The matrix T is expected to be block diagonal with 4×4 blocks corresponding to s, p_x, p_y and p_z basis orbitals on the heavy atoms and 1×1 blocks corresponding to s orbitals on hydrogens. The 4×4 blocks are then constructed analogously as in the Jabobi diagonalization procedure by subsequent unitary transformations describing the rotation around x, y and z axes on individual centers. The corresponding the angles φ, ϑ and ω are chosen to ensure the maximization of the index. The iteration process is terminated after reaching the selfconsistence when the change of the index is lower than preselected limit.

After this presentation of the methodological background it is possible, in the following part, to apply the similarity indices to the problem of LFER. As an example of these relationships let us consider first the well-known Hammett equation correlating the rate or equilibrium constants for nonsubstituted (k_0) and substituted (k_x) reaction partners in a given process by the empirical relation (5).

$$\log\left(\frac{k_x}{k_0}\right) = \rho\sigma_x. \tag{5}$$

The so-called reaction constant ρ which characterizes the sensitivity of a given skeleton to the transmission of the substituent effect depends only on the type of the reaction and σ_x is the parameter characterizing a given sustituent. The scale of substituent constants was derived from the reference process of the dissociation of substituted benzoic acids (Scheme I) for which the ρ was arbitrarily set to 1 so that σ constants are given simple by the differences of pK values of substituted with respect to parent nonsubstituted acid (6).

$$pK_0 - pK_x = \sigma_x. \tag{6}$$

$$X\text{—}C_6H_4\text{—}COOH \rightarrow X\text{—}C_6H_4\text{—}COO^- + H^+$$

Scheme 1.

Because of importance of this process as a standard for the Hammett equation we shall start our study just by the analysis of this process. In this connection it is necessary to mention the previous studies [29–32] in which the same process was analysed by usual quantum chemical approach approximating the substituent induced trends in the dissociation constants *pK* by the values of corresponding reaction energies ΔE. Although such an approach is usually able to reproduce the substituent effect fairly well, the basic problem to what extent the existence of linear relation (6) is conditioned by the empirical principle of analogy and underlying similarity ideas has remained untouched by these studies. The analysis of this topic is the main concern of this study.

For this purpose it is first necessary to propose a model in terms of which the primary problem of substituent effect on the dissociation process (I) can be characterized by similarity indices as the quantitative measures of molecular alikeness. In proposing such a model let us consider first the dissociation of parent nonsubstituted benzoic acid. Since the only structural change it in this case the conversion of the acid into its conjugated anion it seems quite natural to characterize the extent of this structural change by the value of the similarity index $r^0_{AH,\,A-}$ between the acid and the anion. If now the same approach is repeated for the whole series of substituted acids, the result is a set of similarity indices $r^x_{AH,\,A-}$, whose values reflect the changes induced in the dissociation process by the presence of substituent *X*. This opens the possibility to test the legitimacy of the similarity model since if this model is correct, the correlation or at least a parallel should exist between the values of similarity indices $r^x_{AH,\,A-}$ and the substituent constant σ_x.

3. Results and discussion

The similarity indices characterizing the dissociation equilibrium (I) for a series of 7 substituted benzoic acids X = H, *p*-Cl, *p*-$COCH_3$, *p*-NO_2, *p*-CH_3, *p*-CH_3O and *p*-NMe_2) were calculated from the density matrices generated by semiempirical CNDO calculations [33]. All the molecules were considered in optimized geometries. The results of calculations are summarized in Table I and for better clarity also graphically in the form of dependence $r^x_{AH,\,A-}$ vs σ_x (Figure 1). Let us attempt now to discuss the main conclusions suggested by this dependence. First of all it is possible to see that the figure has not the form of simple linear correlation as it is in the case of Hammett correlation but rather

TABLE I
Calculated values of similarity indices $r^x_{AH,\,A-}$ for a series of substituted benzoic acids X—C_6H_4—COOH

X	H	*p*-Cl	*p*-$COCH_3$	*p*-NO_2	*p*-Me	*p*-MeO	*p*-NMe_2
r	0.9770	0.9796	0.9827	0.9830	0.9795	0.9816	0.9832

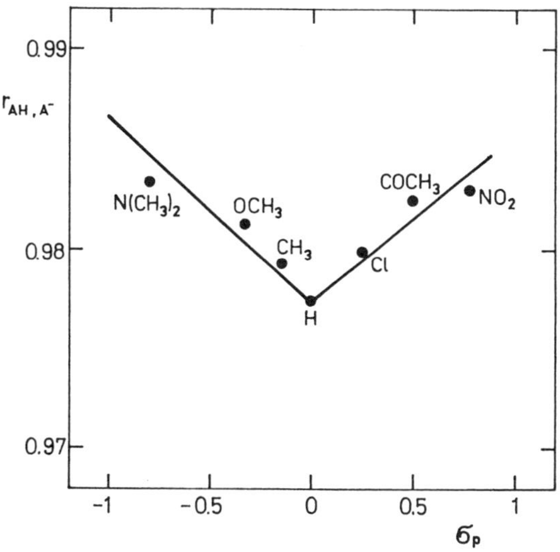

Figure 1. Dependence of calculated similarity indices $r^x_{AH, A-}$ for the dissociation of substituted benzoic acids on the substitutent constants σ_x.

it is splitted into two partial correlations of which one corresponds to donor ($\sigma_x < 0$) and the other to the acceptor ($\sigma_x > 0$) substituents. This implies that the expected parallel between the similarity index and the substituent constants does not hold in this case universally but only within the series of substituents of constant donor/acceptor character. From this then follows that the validity of the proposed model does not reach the universality of the Hammett equation but within the range of the substituents of the same type the use of similarity index as a measure of the substituent efect is quite plausible. The origin of this certain restriction is not yet completely clear to us but it is probable that it is due to the inherent limitation of the similarity model. That this could be the case seems to be supported by the fact that the above picture of similarity index vs σ_x correlations is not restricted only to the dissociation of benzoic acids but that the completely analogous behaviour is observed for other reactions as well. An example is the dissociation of substituted kumylchlorides (Scheme II), which is a standard for the introduction of another set of substituent constants, namely σ_x^+ [34].

$$X—C_6H_4—C(CH_3)_2Cl \rightarrow X—C_6H_4—C(CH_3)_2^+ + Cl^-$$

Scheme II.

This process was analysed by a similarity model in which the structural change associated with the ionisation was measured by the similarity index $r^x_{RCl, R+}$ between the substituted kumylchlorides and corresponding carbocations.

Similarly as in the previous case the calculations were based on density matrices derived from CNDO calculations. The molecules were again considered in fully optimized geometries. The results of the calculations are displayed graphically in Figure 2. As can be seen, the form of the dependence is completely analogous to what was observed for benzoic acids. This suggest that subject to the similar limitations, the ability of the model to describe the substituent efect is probably quite general. This result is very interesting since the generality of the model opens the possibility of its use also for other structure/(re)activity studies and provides thus the desired theoretical support for the intuitive similarity ideas underlying the powerfull but still basically empirical principle of analogy.

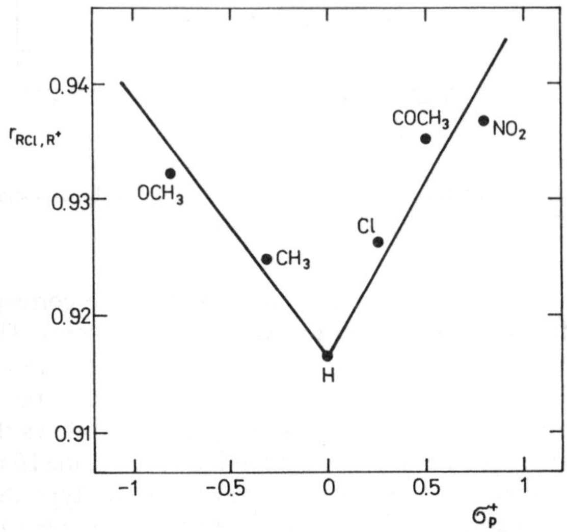

Figure 2. Dependence of calculated similarity indices $r^x_{RCl,R+}$ for the ionization of substituted kumylchlorides on the substitutent constants σ_p^+.

Note

1. Numerous examples of the possible use of molecular similarity can be found in the special issue of *J. Chem. Inf. Comp. Sci.* [11], collecting the contributions presented on the 3rd Beistein Workshop on Similarity in Organic Chemistry.

References

1. Hammett, L. P., *Trans. Faraday Soc.* **34**, 96 (1938).
2. Taft, R. W., *J. Am. Chem. Soc.* **75**, 4231 (1953).
3. Shorter, J., *Chem. Britain* **5**, 269 (1969).
4. Ponec, R., *Coll. Czech. Chem. Commun* **45**, 1646 (1980).

5. Pietrzycki, W. and Tomasik, P., *J. Chim. Phys.* **90**, 27 (1993).
6. Krygowski, T. M. and Perjessy, A., *Bull. Acad. Sci. Polon.* **22**, 437 (1974).
7. Wold, S., *Chem. Scripta* **5**, 97 (1974).
8. Rouvray, D. H., *J. Chem. Inf. Comp. Sci.* **32**, 580 (1992).
9. Similarity Models in Organic Chemistry, Biochemistry and Related Fields, R. I. Zalewski, T. M. Krygowski, and J. Shorter (eds.), Elsevier Science Publishers, Amsterdam 1991.
10. Concepts and Applications of Molecular Similarity, M. A. Johnson, G. M. Maggiora (eds.) J. Wiley & Sons 1990.
11. Proceedings of 3rd Beilstein Workshop on Molecular Similarity, May 1992 Bolzano, Italy, *J. Chem. Inf. Comp. Sci.* **32**, 577–748 (1992).
12. Polansky, O. E. and Derflinger, G., *Int. J. Quant. Chem.* **1**, 379 (1967)
13. Carbó, R., Leyda, L., and Arnau, M., *Int. J. Quant. Chem.* **17**, 1185 (1980).
14. Bowen-Jenkins, P. E., Cooper, D. L., and Richards, D. G., *J. Phys. Chem.* **89**, 2195 (1985).
15. Richards, W.G., *Pure & Appl. Chem.* **62**, 231 (1992), 1342 (1986).
16. Hodgkin, E. E. and Richards, W. G., *Int. J. Quant. Chem., Quant. Biol. Symp.* **14**, 105 (1987).
17. Cooper, D. L. and Allan, N. L., *J. Comp. Aided. Mol. Design* **3**, 253 (1989).
18. Cooper, D. L. and Allan, N. L., *J. Am. Chem. Soc.* **114**, 4773 (1992).
19. Carbó, R. and Calabuig, B., *J. Chem. Inf. Comp. Sci.* **32**, 600 (1992).
20. Amovilli, C. and McWeeny, R., *J. Mol. Struct. (Theochem)* **227**, 1 (1991).
21. Cioslowski, J. and Fleischmann, J., *Am. Chem. Soc.* **112**, 64 (1991).
22. Ponec, R., *Coll. Czech. Chem. Commun.* **52**, 555 (1987).
23. Ponec, R. and Strnad, M., *Int. J. Quant. Chem.* **42**, 501 (1992).
24. Ponec, R. and Strnad, M., *J. Phys. Org. Chem.* **4**, 701 (1991).
25. Ponec, R. and Strnad, M., *J. Chem. Inf. Comp. Sci.* **32**, 693 (1992).
26. Ponec, R., *Coll. Czech. Chem. Commun.* **49**, 455 (1984).
27. Ponec, R. and Strnad, M., *Croat. Chim. Acta* **66**, 123 (1993).
28. Trindle, C., *J. Am. Chem. Soc.* **92**, 3251 (1970).
29. Böhm, S. and Kuthan, J., *Int. J. Quant. Chem.* **26**, 21 (1984).
30. Danihel, I. and Kuthan, J., *Coll. Czech. Chem. Commun.* **45**, 2201 (1980).
31. Kang, S. and Beveridge, D. L., *Theor. Chim. Acta.* **22**, 312 (1971).
32. Botrel, A. and Guerillot, C. R., *J. Chim. Phys.* **71**, 974 (1974).
33. Pople, J. A. and Beveridge, D. L., *Approximate MO Theory* (McGraw-Hill, New York 1970).
34. Okamoto, Y. and Brown, H. C., *J. Org. Chem.* **22**, 485 (1957).

Evaluation of Molecular Similarity Using Topological Fragment Spectra

YOSHIMASA TAKAHASHI and YUICHI ISHIYAMA
Department of Knowledge-Based Information Engineering, Toyohashi University of Technology, Tempaku, Toyohashi 441, Japan

Abstract. This paper describes an approach for the evaluation of structural similarity between molecules. This approach is made of three major steps: (1) Enumeration of all possible substructures from each chemical structure. (2) Description of the molecular profile with substructure characteristics. (3) Calculation of the similarity between them. The present method has been tested using several preliminary set of chemical structures. For the description of the molecular profile the basic approach uses fragmental molecular weight which defined by the sum of atomic weights for the fragment. This approach can be easily extended to describe it with the other type of substructure characteristics.

1. Introduction

The authors have been investigating on the problem of automatic identification of common structural features (possessing explicitly defined characteristics) among sets of chemical structures. We have proposed and computerized algorithms for the identification of common substructures not only from the topological point of view but also considering the three dimensional aspect of the molecules [1–3]. Moreover, we are conducting research on methods to identify structural similarity based on the utilization of a knowledge base [4].

All these methods focus the similarity problem from the point of view of local structural characteristics. However, we realize that an approach is required to process structural information in a more flexible way in order to allow somehow the automatic evaluation of the more ambiguous global structural similarity; in other words, a method to examine the similarity of structures when they are regarded as whole entities.

In this study we present a method to evaluate automatically the global similarity of structures considering only the topological characteristics of the chemical structure which is regarded as one graph. We also explain the attempts to develop the pertinent software.

2. Methods

In what follows we analyze a method for the quantitative evaluation of structural similarity based only on the topological information of chemical

R. Carbó (eds.), Molecular Similarity and Reactivity: From Quantum Chemical to Phenomenological Approaches, 311–316.

structures excluding from any consideration of three dimensional geometry of the molecules.

Representation of the structural information: In order to distinguish types of atoms and bonds of a chemical structure, this is represented by means of a weighted graph whose vertices are the atoms and the edges are the bonds of the structure.

Enumeration and characterization of the substructures: For a given structure represented as a chemical graph (hydrogen suppressed graph), all the possible substructures embedded in it are enumerated. Here, all the substructures of the parent structure (the original chemical graph) possessing different elements but represented by the same subgraph are taken into account. The algorithm used to enumerate all the substructures for a given structure is described elsewhere [5]. Subsequently, all the individual substructures are characterized quantitatively. To perform this characterization we have used two methods in the present study as follows: (i) The overall sum of degrees of the nodes composing each subgraph. (ii) The overall sum of the mass numbers of the atoms (atomic groups) represented by the nodes of the subgraph.

With the first method the chemical structure is represented by a simple graph where the hydrogen atoms are omitted, thus, the characterization of the structure depends only on the topology of the structural skeleton. In the second method attached hydrogen atoms are taken into account as augmented atoms and are represented by weighting correspondingly their respective graph nodes.

Fragment spectrum: We define as "topological fragment spectrum" (fragment spectrum in what follows) the histogram resulting from displaying the frequency distribution of a set of individually characterized substructures (structural fragments) according to the value of their characterization index. The fragment spectrum generated according to this definition is a representation of the topological profile of the structure of the chemical compound in question. Figure 1 is an example of the fragment spectrum for 2-Methylbutane when the characterization index is the sum of degrees of the graph nodes.

Quantitative evaluation of molecular similarity: The fragment spectrum obtained in the above can be described as a kind of multidimensional pattern vector. Consequently, using this pattern representation of a spectrum it is possible to apply diverse quantitative methods for the evaluation of simi-

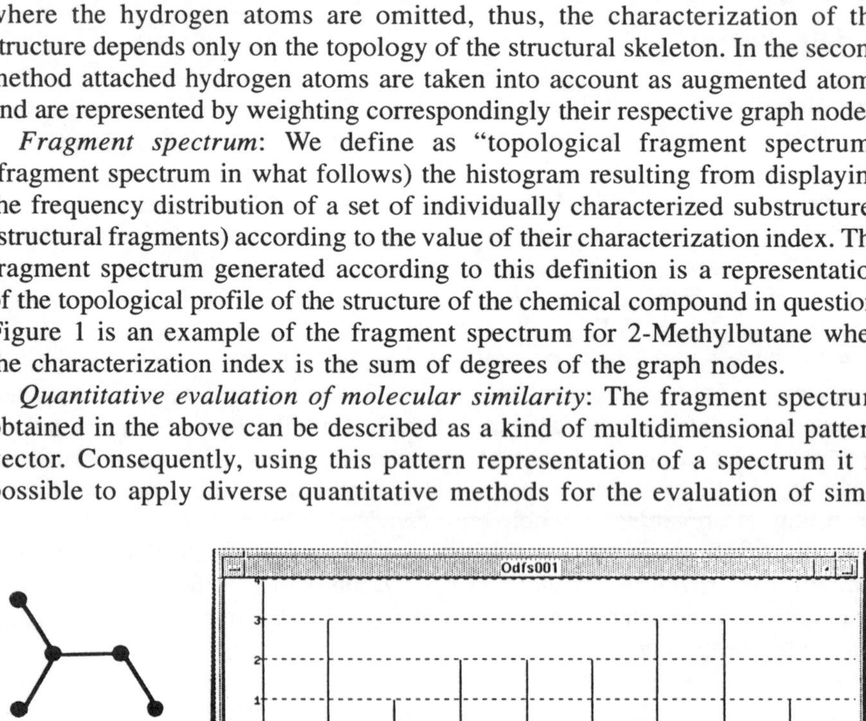

Figure 1. A fragment spectrum of 2-Methylbutane weighted by the sum of degree of the fragmental graph nodes.

larity. Here, we have examined the following two functions as quantitative measures for the evaluation of similarity.

(i) Similarity (S) based on the normalized distance:

$$S(x_i, x_j) = 1 - D(x_i, x_j)/D_{max} \tag{1}$$

x_i and x_j are the fragment spectra for the ith and jth molecule respectively. $D(x_i, x_j)$ is the Euclidian distance between the patterns x_i and x_j. D_{max} is the largest distance among the set of molecules under analysis.

(ii) Correlation coefficient (R):

$$R(x_i, x_j) = \frac{\sum_{k=1}^{p} (x_{ik} - \bar{x}_i)(x_{jk} - \bar{x}_j)}{\sqrt{\sum_{k=1}^{p} (x_{ik} - \bar{x}_i)^2 \sum_{k=1}^{p} (x_{jk} - \bar{x}_j)^2}} \tag{2}$$

x_{ik} and x_{jk} are the kth quantitative characteristic index of the patterns x_i and x_j respectively. $\bar{x}_i$ and $\bar{x}_j$ are the averages of the quantitative characteristic indices and p is the number expressing the dimension of the pattern vector.

3. Results and discussion

Applying the method described above to compounds represented by their carbon skeleton graphs, we have attempted to perform an evaluation of the structural similarity of a set of molecules comprising thirteen of hydrocarbon isomers constituted by five carbon atoms (three chains, five monocyclics, and five bicyclics). Figure 2 illustrates the spanning tree obtained by com-

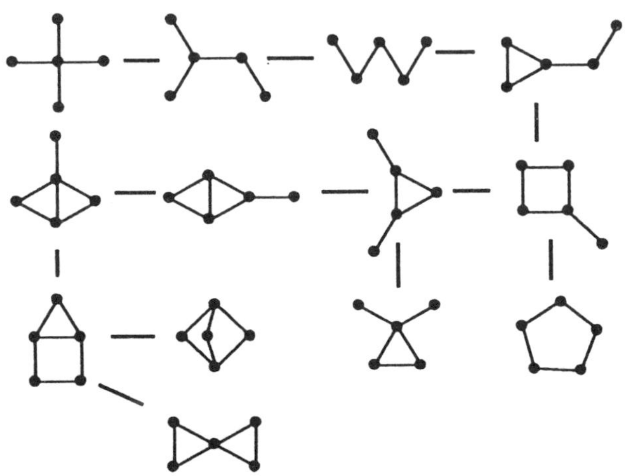

Figure 2. The spanning tree of thirteen hydrocarbons obtained by the similarity analysis of the fragment spectra which were weighted by the sum of degree of the fragmental graph nodes. The similarity measure defined in equation (1) was used for the analysis.

paring the fragment spectrum for the individual compounds. Here, the comparisons were carried out by means of the normalized distance; and the fragment spectra were obtained using the sum of the degree of the atoms at each vertex of the subgraphs. As is evident from the figure, the set of molecules is properly divided in chain, monocyclic and bicyclic types. This fact shows that the fragment spectrum reflects properly the topological constitution of the molecular skeleton and furthermore the applicability of the present method for the evaluation of structural similarity.

Nonetheless, to evaluate structural similarity of compounds involving unsaturated bonds and heteroatoms it is necessary to properly express these differences at the stage of characterizing the structural fragment.

Figure 3 shows the fragment spectra for five psychotropic agents of which chemical structures are similar, when the fragments are characterized by the fragment weight. The resolution of the spectra is substantially increased by this fragment characterization. The results of the evaluation of structural similarity for the spectra in Figure 3 using the normalized distance are illustrated in Figure 4. Again, from this figure, the usefulness of the fragment spectrum method for the evaluation of structural similarity is evident even for relatively similar structures.

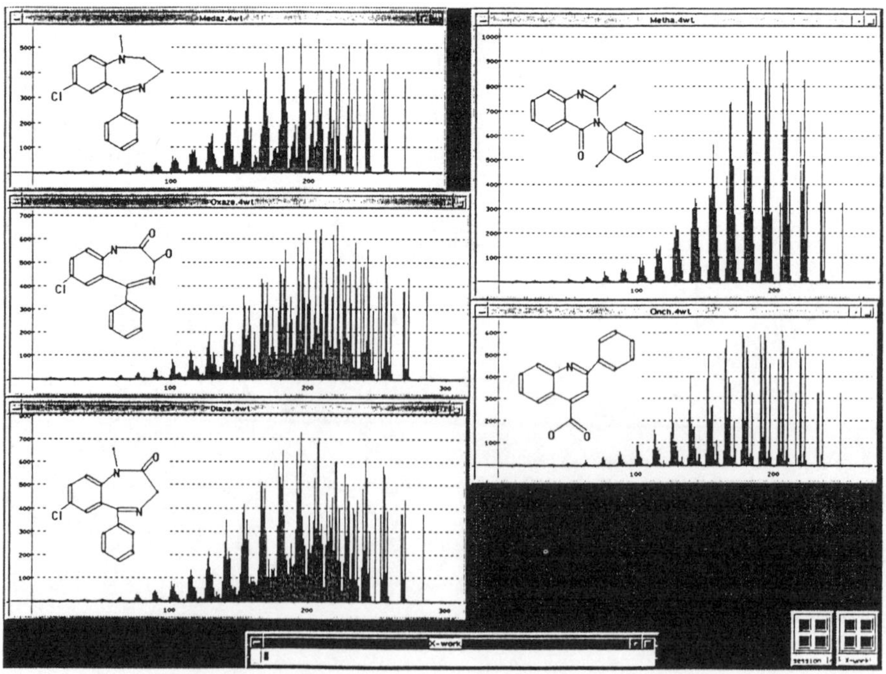

Figure 3. Chemical structures of five psychotropic agents and their fragment spectra weighted by fragment mass.

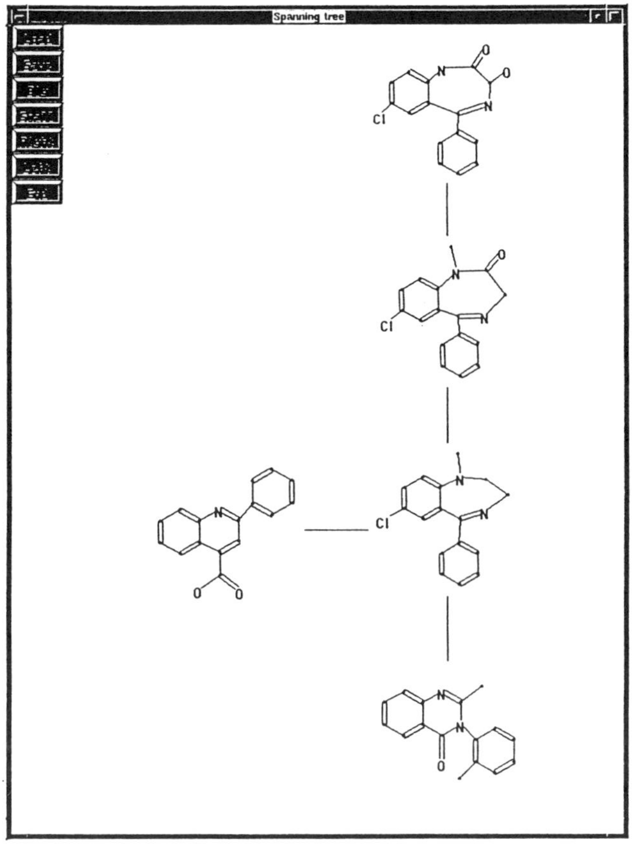

Figure 4. The spanning tree of five psychotropic agents obtained by the similarity analysis of fragment spectra which were weighted by fragment mass. The similarity measure defined in equation (1) was used for the analysis.

4. Conclusions

The goal of the present study is the evaluation in a rather objective and quantitative way the problem of structural similarity of chemical compounds which until now has been performed by the chemist in an essentially intuitive way based on his visual perception. The method introduced here is completely different to the maximal common substructuresearch or the conventional one based on the search of explicitly defined common substructural units. On the other hand, as stated before, the problem of computer handling of chemical structures is not only of the utmost importance in attempts to use computers in chemistry and related areas but it is also specific to each of those areas.

Within this framework, we believe that methods like the one developed here to evaluate structural similarities of chemical compounds are indispensable regarding higher applications of computers in chemistry and at the same time they constitute new techniques for chemical information processing which are so strongly required in this field.

Acknowledgments

This work was performed through Special Coordination Funds of the Science and Technology Agency of the Japanese Government. We also thank The Hori Information Science Promotion Fundation for partial support of this work.

References

1. Takahashi, Y., Satoh, Y., Suzuki, H., and Sasaki, S., *Anal. Sci.* **3**, 23 (1987).
2. Takahashi, Y., Maeda, S., and Sasaki, S., *Anal. Chim. Acta.* **200**, 363 (1987).
3. Takahashi, Y. and Sasaki, S., *J. Syn. Org. Chem. JAPAN* **45**, 1087 (1987).
4. Takahashi, Y., Sukekawa, M., and Sasaki, S., *J. Chem. Inf. Comput. Sci.* **32**, 639 (1992).
5. Takahashi, Y. and Ishiyama, Y., in preparation.

Index

Understanding Chemical Reactivity

1. Z. Slanina: *Contemporary Theory of Chemical Isomerism.* 1986 ISBN 90-277-1707-9
2. G. Náray-Szabó, P. R. Surján, J. G. Angyán: *Applied Quantum Chemistry.* 1987 ISBN 90-277-1901-2
3. V. I. Minkin, L. P. Olekhnovich and Yu. A. Zhdanov: *Molecular Design of Tautomeric Compounds.* 1988 ISBN 90-277-2478-4
4. E. S. Kryachko and E. V. Ludeña: *Energy Density Functional Theory of Many-Electron Systems.* 1990 ISBN 0-7923-0641-4
5. P. G. Mezey (ed.): *New Developments in Molecular Chirality.* 1991 ISBN 0-7923-1021-7
6. F. Ruette (ed.): *Quantum Chemistry Approaches to Chemisorption and Heterogeneous Catalysis.* 1992 ISBN 0-7923-1543-X
7. J. D. Simon (ed.): *Ultrafast Dynamics of Chemical Systems.* 1994 ISBN 0-7923-2489-7
8. R. Tycko (ed.): *Nuclear Magnetic Resonance Probes of Molecular Dynamics.* 1994 ISBN 0-7923-2795-0
9. D. Bonchev and O. Mekenyan (eds.): *Graph Theoretical Approaches to Chemical Reactivity.* 1994 ISBN 0-7923-2837-X
10. R. Kapral and K. Showalter (eds.): *Chemical Waves and Patterns.* 1995 ISBN 0-7923-2899-X
11. P. Talkner and P. Hänggi (eds.): *New Trends in Kramers' Reaction Rate Theory.* 1995 ISBN 0-7923-2940-6
12. D. Ellis (ed.): *Density Functional Theory of Molecules, Clusters, and Solids.* 1995 ISBN 0-7923-3083-8
13. S.R. Langhoff (ed.): *Quantum Mechanical Electronic Structure Calculations with Chemical Accuracy.* 1995 ISBN 0-7923-3264-4
14. R. Carbó (ed.): *Molecular Similarity and Reactivity: From Quantum Chemical to Phenomenological Approaches.* 1995 ISBN 0-7923-3309-8

Kluwer Academic Publishers – Dordrecht / Boston / London